Volume II: Thermal Behaviour, Energy Efficiency in Buildings and Sustainable Construction

Volume II: Thermal Behaviour, Energy Efficiency in Buildings and Sustainable Construction

Editor

Paulo Santos

MDPI • Basel • Beijing • Wuhan • Barcelona • Belgrade • Manchester • Tokyo • Cluj • Tianjin

Editor
Paulo Santos
University of Coimbra
Portugal

Editorial Office
MDPI
St. Alban-Anlage 66
4052 Basel, Switzerland

This is a reprint of articles from the Topic published online in the open access journals *Energies* (ISSN 1996-1073), *Administrative Sciences* (ISSN 2076-3387), *Behavioral Sciences* (ISSN 2076-328X), and *Clean Technologies* (ISSN 2571-8797) (available at: https://www.mdpi.com/journal/energies/special_issues/Buildings_and_Sustainable_Construction_2022).

For citation purposes, cite each article independently as indicated on the article page online and as indicated below:

LastName, A.A.; LastName, B.B.; LastName, C.C. Article Title. *Journal Name* **Year**, *Volume Number*, Page Range.

ISBN 978-3-0365-5237-8 (Hbk)
ISBN 978-3-0365-5238-5 (PDF)

© 2022 by the authors. Articles in this book are Open Access and distributed under the Creative Commons Attribution (CC BY) license, which allows users to download, copy and build upon published articles, as long as the author and publisher are properly credited, which ensures maximum dissemination and a wider impact of our publications.

The book as a whole is distributed by MDPI under the terms and conditions of the Creative Commons license CC BY-NC-ND.

Contents

About the Editor vii

Preface to "Volume II: Thermal Behaviour, Energy Efficiency in Buildings and Sustainable Construction" ix

Antonino D'Amico, Giuseppina Ciulla, Alessandro Buscemi, Domenico Panno, Michele Zinzi and Marco Beccali
Road Thermal Collector for Building Heating in South Europe: Numerical Modeling and Design of an Experimental Set-Up
Reprinted from: *Energies* **2022**, *15*, 430, doi:10.3390/en15020430 1

Gianmarco Fajilla, Emiliano Borri, Marilena De Simone, Luisa F. Cabeza and Luís Bragança
Effect of Climate Change and Occupant Behaviour on the Environmental Impact of the Heating and Cooling Systems of a Real Apartment. A Parametric Study through Life Cycle Assessment
Reprinted from: *Energies* **2021**, *14*, 8356, doi:10.3390/en14248356 25

Rui Oliveira, Ricardo M.S.F. Almeida, António Figueiredo and Romeu Vicente
A Case Study on a Stochastic-Based Optimisation Approach towards the Integration of Photovoltaic Panels in Multi-Residential Social Housing
Reprinted from: *Energies* **2021**, *14*, 7615, doi:10.3390/en14227615 47

Rosa Francesca De Masi, Antonio Gigante, Valentino Festa, Silvia Ruggiero and Giuseppe Peter Vanoli
Effect of HVAC's Management on Indoor Thermo-Hygrometric Comfort and Energy Balance: In Situ Assessments on a Real nZEB
Reprinted from: *Energies* **2021**, *14*, 7187, doi:10.3390/en14217187 63

Jaqueline Litardo, Ruben Hidalgo-Leon and Guillermo Soriano
Energy Performance and Benchmarking for University Classrooms in Hot and Humid Climates
Reprinted from: *Energies* **2021**, *14*, 7013, doi:10.3390/en14217013 93

Georgios E. Arnaoutakis and Dimitris A. Katsaprakakis
Energy Performance of Buildings with Thermochromic Windows in Mediterranean Climates
Reprinted from: *Energies* **2021**, *14*, 6977, doi:10.3390/en14216977 111

Paulo Santos and Telmo Ribeiro
Thermal Performance Improvement of Double-Pane Lightweight Steel Framed Walls Using Thermal Break Strips and Reflective Foils
Reprinted from: *Energies* **2021**, *14*, 6927, doi:10.3390/en14216927 125

Kristian Fabbri and Jacopo Gaspari
A Replicable Methodology to Evaluate Passive Façade Performance with SMA during the Architectural Design Process: A Case Study Application
Reprinted from: *Energies* **2021**, *14*, 6231, doi:10.3390/en14196231 141

Piotr Michalak
Experimental and Theoretical Study on the Internal Convective and Radiative Heat Transfer Coefficients for a Vertical Wall in a Residential Building
Reprinted from: *Energies* **2021**, *14*, 5953, doi:10.3390/en14185953 157

Sultan Kobeyev, Serik Tokbolat and Serdar Durdyev
Design and Energy Performance Analysis of a Hotel Building in a Hot and Dry Climate: A Case Study
Reprinted from: *Energies* **2021**, *14*, 5502, doi:10.3390/en14175502 179

Raluca Buzatu, Viorel Ungureanu, Adrian Ciutina, Mihăiță Gireadă, Daniel Vitan and Ioan Petran
Experimental Evaluation of Energy-Efficiency in a Holistically Designed Building
Reprinted from: *Energies* **2021**, *14*, 5061, doi:10.3390/en14165061 197

Raphaele Malheiro, Adriana Ansolin, Christiane Guarnier, Jorge Fernandes, Maria Teresa Amorim, Sandra Monteiro Silva and Ricardo Mateus
The Potential of the Reed as a Regenerative Building Material—Characterisation of Its Durability, Physical, and Thermal Performances
Reprinted from: *Energies* **2021**, *14*, 4276, doi:10.3390/en14144276 219

Alessandro Franco, Lorenzo Miserocchi and Daniele Testi
Energy Intensity Reduction in Large-Scale Non-Residential Buildings by Dynamic Control of HVAC with Heat Pumps
Reprinted from: *Energies* **2021**, *14*, 3878, doi:10.3390/en14133878 239

About the Editor

Paulo Santos

Paulo Santos [ORCID 0000-0002-0134-6762] is Assist. Prof. in Dep. of Civil Eng. of University of Coimbra, PT. Member of ISISE (Institute for Sustainability and Innovation in Structural Engineering) research centre. The main actual scientific research fields are Thermal Behaviour, Energy Efficiency in Buildings and Sustainable Construction. He is author of around 160 scientific publications and was supervisor of around 50 Doctoral and master's theses. He participated in around 12 funded European and Portuguese research projects, being nowadays the Principal Investigator of the Tyre4BuildIns: "Recycled tyre rubber resin-bonded for building insulation systems towards energy efficiency" research project. Furthermore, he is member of iiSBE: International Initiative for a Sustainable Built Environment (2011,-) and Technical Committee TC14: Sustainability and Eco-efficiency of Steel Construction of the ECCS: European Convention for Constructional Steelwork (2008,-).

Preface to "Volume II: Thermal Behaviour, Energy Efficiency in Buildings and Sustainable Construction"

As is well known, energy and sustainability are two of mankind's major concerns at present. Given the share of energy consumption belonging to the buildings sector, it is very important to search for innovative design solutions and the optimal thermal performance of buildings to reduce energy bills and greenhouse gas emissions, while maintaining the occupants' comfort levels. Additionally, given the environmental burdens of the construction sector, the search for more environmentally responsible processes and a more efficient use of resources is currently attracting attention.

This second volume of this Special Issue (SI), published in the Energies journal, is dedicated to the analysis of recent advances in the following main issues: (1) improvements in the thermal behavior of a building's elements (walls, floors, roofs, windows, doors, etc.), (2) energy efficiency in buildings, and (3) sustainable construction. The main goal is to compile scientific works within these topics, making use of different possible research approaches, such as: experimental, theoretical, numerical, analytical, computational, case studies, and mixtures of these. This book compiles a set of original research works with academic excellence and scientific soundness.

The Guest Editor would like to express their sincere and deep gratitude for all the scientific contributions from authors among prestigious worldwide scientists, as well as to the reviewers who significantly improved the quality of these manuscripts. Moreover, I would like to acknowledge the research project Tyre4BuildIns: "Recycled tyre rubber resin-bonded for building insulation systems towards energy efficiency", supported by FEDER funds through the Competitivity Factors Operational Programme (COMPETE) and by national funds through Foundation for Science and Technology (FCT), within the scope of the project POCI-01-0145-FEDER-032061, which contributed three scientific articles to the first volume and one paper to this SI (Volume II). Additionally, the Guest Editor also wants to acknowledge the support provided by the following companies, who were partners of the research project Tyre4BuildIns: Pertecno, Gyptec Ibéria, Volcalis, Sotinco, Kronospan, Hulkseflux, Hilti and Metabo.

Paulo Santos
Editor

Article

Road Thermal Collector for Building Heating in South Europe: Numerical Modeling and Design of an Experimental Set-Up

Antonino D'Amico [1], Giuseppina Ciulla [2], Alessandro Buscemi [2], Domenico Panno [2], Michele Zinzi [3] and Marco Beccali [2,*]

1 Consiglio Nazionale delle Ricerche, ITAE "Nicola Giordano", 98126 Messina, Italy; damico@itae.cnr.it
2 Dipartimento di Ingegneria, Università degli Studi di Palermo, Edificio 9, Viale delle Scienze, 90128 Palermo, Italy; giuseppina.ciulla@unipa.it (G.C.); alessandro.buscemi@unipa.it (A.B.); domenico.panno@unipa.it (D.P.)
3 Enea, Agenzia Nazionale per le Nuove Tecnologie e lo Sviluppo Economico Sostenibile, 00196 Roma, Italy; michele.zinzi@enea.it
* Correspondence: marco.beccali@unipa.it

Abstract: The combination/integration of renewable energy and storage systems appears to have significant potential, achieving high-energy results with lower costs and emissions. One way to cover the thermal needs of a building is through solar energy and its seasonal storage in the ground. The SMARTEP project aims to create an experimental area that provides for the construction of a road solar thermal collector directly connected to a seasonal low-temperature geothermal storage with vertical boreholes. The storage can be connected to a ground-to-water heat pump for building acclimatization. This system will meet the requirements of visual impact and reduction of the occupied area. Nevertheless, several constraints related to the radiative properties of the surfaces and the lack of proper thermal insulation have to be addressed. The project includes the study of several configurations and suitable materials, the set-up of a dynamic simulation model and the construction of a small-scale road thermal collector. These phases allowed for an experimental area to be built. Thanks to careful investigation in the field, it will be possible to identify the characteristics and the best operation strategy to maximize the energy management of the whole system in the Mediterranean area.

Keywords: road thermal collector; borehole thermal storage; alternative energy system

1. Introduction

The exploitation of traditional energy resources has, to a large extent, led to a critical point, where it is now critical to reconsider the terms of energy and focus on sustainable energy reasonableness, so that energy satisfies the energy of the present without compromising the ability to meet the needs of future generations. Currently, the energy sector is strongly committed to a policy of replacing fossil fuels with renewable and low-impact energy carriers; the transition is needed, by increasing energy consumption around the world. Renewable energy, such as wind, solar, and geothermal energy, can replace the current primary energy supply sources, especially if improvements can be made to energy conversion, transport, and distribution systems. Thanks to the great efforts of the scientific community, these systems have been optimized in order to function as efficiently as possible in terms of cost and CO_2 emissions. In particular, there seems to be great potential, in achieving the same results with lower costs and low emissions, through the combination and integration of renewable energies with storage systems. There is great theoretical potential for the use of solar energy capable of covering the heat demands of buildings; the technical potentials at the latitudes of the Sun Belt are greatly influenced by the possibility of storing heat from the summer to the heating period. One way to cover the heating demands of a building is represented by the use of solar energy and its seasonal accumulation in the ground, where the stored energy can be reused when necessary [1].

The collection of solar energy is commonly entrusted to solar collectors installed on a roof or ground. An alternative solution is represented by road thermal collectors (RTC), i.e., installation of pipes embedded in paved surfaces such as asphalt surfaces, sidewalks, roads, and parking lots. Asphalt, being a material with high thermal capacity, represents an alternative low-cost heat storage system. These surfaces are already present, and since the road surface must be resurfaced periodically, the energy system can be installed very easily during a generic maintenance phase. From the point of view of energy production, the significantly larger areas of the asphalt collectors can compensate for their expected lower efficiencies, justifying the investment for the power produced per unit. In addition, asphalt surfaces can continue producing energy, even during the night. The solar energy captured in this way can be used for different applications: in combination with accumulation systems (wells), for road safety and maintenance, keeping the roads free from snow during the winter period, reducing the cost of regular roofing road, or for the decrease in surface temperatures by reducing the effect of the heat island. Unlike traditional manifolds, this system meets the needs of reducing visual impacts, not occupying surface areas that are not always available where necessary. Finally, this system will make it possible to exploit the absorption capacity of the road surface in areas characterized by high sunshine, energetically redeveloping areas that are generally dedicated to something else (roads and parking lots).

Based on these considerations, the University of Palermo, through the SMARTEP project "sustainable model and thinking of renewable energy parking" [2], is creating an experimental area, where, in addition to the development of an innovative intelligent energy parking system, it envisages the construction of a road thermal collector for collection of solar energy, directly connected to seasonal low-temperature geothermal storage, with vertical probes. This research project's various objectives include being able to investigate the best solution and boundary conditions for a commercial diffusion of an RTC system with geothermal storage in the Mediterranean area, maximizing the accumulation phase in periods of greater sunshine. In this research, we experimented in the field of a thermal RTC integrated with a geothermal storage system in the Mediterranean area. In particular, after careful analysis of the sector bibliography, thanks to the SMARTEP experimental area, it will be possible to investigate, in the field, the best solution, configuration, and main boundary conditions that can maximize the efficiency and effectiveness of an alternative solar system, and innovativeness, such as solar collectors embedded in paved surfaces (asphalt, sidewalks, roads, parking lots), integrated with a geothermal seasonal storage system.

Creating favorable conditions for solar absorption and thermal accumulation leads to a local increase of surface and air temperature, exacerbating hazardous phenomena, such as urban heat islands and overheating [3]. It is, thus, important to identify surface materials that are able to optimize this complex task: ensuring high enough thermal gains for the collectors without comprising the outdoor thermal quality. Crucial parameters include the thermal emissivity, the solar reflectance, and the color, which is an indicator of the visible component of the latter. Several construction materials and natural stones, such as concrete and marble, have higher solar reflectance compared to asphalts, which can enhance thermal mitigation; new technologies, such as near-infrared reflecting coatings, dynamic coatings, and phase change materials, can also be applied to the purpose [4]. For this reason, a research group from the University of Palermo, with collaboration from Enea [5], are simultaneously studying a solution that attempts to counter this phenomenon. In particular, the possibility of using road surfaces in an alternative color to the traditional black, to be used on the RTC, is being evaluated, in order to guarantee high efficiency of the collection system, and at the same time be able to mitigate the heat island effect. Simultaneously, special cool materials could be used in the remaining areas surrounding the RTC areas in order to balance the heating effects with the cooling of the surrounding areas.

Indeed, experimentation in the field will allow calibrating the RTC system with geothermal storage in a highly sunny area in the summer in order to make it competitive with other renewable technologies and, therefore, replicable in other contexts. The results of the proposed research will provide valid support for the design and implementation of sustainability and circularity actions, based on a holistic and life cycle approach that integrates the three spheres of sustainability, which will allow a systemic assessment of energy—environmental, economic, social, and circularity.

2. State-of-the-Art

The solar collectors on asphalt consist of pipes embedded in the road surface in which the pipes are crossed by a heat transfer fluid. Thanks to a process of heat transfer from the flooring to the fluid, a lowering of the temperature of the flooring occurs, which mitigates the heat island effect, simultaneously reducing the risk of permanent deformations. However, what makes asphalt solar collectors interesting is the capability of exploiting the temperature increase by the circulating fluid to collect the thermal energy. In general, RTCs are coupled to low temperature geothermal heat pumps, achieving the right balance between efficiency and running costs. As already mentioned, the energy obtained from an RTC system is generally used for snow melting systems, to decrease the temperatures of the heat islands that are generated in large metropolitan cities, to maintain thermal comfort in the buildings adjacent to the alternative energy system. These collectors can also be placed at the top of a concrete surface, such as pavement, but since the black color increases the absorption coefficient; it has been estimated that higher performance is linked to manifolds made positioned in the asphalt road pavement.

The patent of 1979, entitled "Paving and solar energy system and method" [6], is one of the first works, which describes a system with pipes embedded in the pavement of a roadway or a roof. The inventor, Ion L. Wen-del, says that the fluid circulating in the duct is heated by the flooring, cooling it at the same time, extending its useful life, and reducing the transmission of heat through the roof to the interior of the building. Furthermore, this application, if installed in a large parking lot, could also represent the preheating of the water of some systems, such as swimming pools or spas. The Swiss SERSO system is one of the pioneering applications of this system applied to melt, and consisted of tubes embedded in the deck of a bridge of the Swiss motorway network on road 8 to Därligen in Bern. This project was carried out by Polydynamics SA, Zurich (Swiss), to collect the heat from the road surface of the bridge in the summer, storing the heat produced in an underground heat sink, and using the heat in the winter to heat the road surface of the bridge by providing any ice formation. It has been shown that, in the summer, the system can store 20% of the solar radiation incident on the road surface [7].

The GAIA Snow-Melting system, installed in 1995 in Ninohe, Japan, uses a thermally insulated inner tube and reverse circulation to increase heat extraction efficiency, i.e., cold fluid flows along the ring and hotter fluid flows upward through the inner tube. This apparatus features a control center that manages the system when road conditions meet specific criteria for snowmelt or heat charge [8]. In this system, the heat absorbed by the road pavement is recovered and stored in the ground during the summer period, to then be used to melt snow in winter.

Following these evolutions, many companies have specialized in the construction and installation of systems, capable of producing energy by exploiting the solar radiation collected from road surfaces, and many of these are commercialized systems, such as Dutch Ooms International Holding with its Road Energy Systems (RES) and Winner Way [9]. One of these is the ICAX™ Ltd. Company (London, UK), which has designed and installed a system capable of melting road snow in a playground of a school, and the thermal energy collected is used to improve the indoor comfort condition of the building [10]. The latter was designed using the playground as a surface, where the heat collectors are positioned underneath, formed by reinforced plastic tubes, and the heat-carrying liquid circulates inside, which, once heated by the surface of the park, is transferred to two heat accumulations positioned under the school floor (skins of power—buildings as energy collectors—Bill Holdsworth). The alternative companies and Novotech, Inc. (USA), in collaboration with the Worcester Polytechnic Institute, have developed a system called Roadway Power System, which uses manifolds, embedded in the asphalt road pavement, connected to a turbine system to create electric energy [11]. In order to provide a more complete explanation, some of the main applications and studies of the RTC system from the 1980s to today, are listed below.

In 1981, Sedgwick and Patrick experimentally designed a swimming pool heating system in the summer using a grid of plastic pipes laid 20 mm below the surface of a tennis court in the UK. In general, a classic heating system for a swimming pool environment reaches temperatures of 20–27 °C in the summer. In particular, in the summer, due to the climatic conditions typical of the United Kingdom, the pools reach comfortable temperatures with an air temperature of 22 °C and a solar radiation of 610 W/m^2. The system designed by Sedgwick and Patrick was technically feasible and cost effective, as it was capable of providing a temperature of 22 °C [12]. Around 1986–1987, Turner developed a theoretical study—a simple 1D model in the stationary field used to analyze the performance of a road collector in both the winter and summer. In the first case, it envisaged the reaching of the maximum road temperature of 15 °C, necessary for the defrosting of roads and bridges; in the second case, it assessed the achievement of the road temperature of about 70 °C, i.e., a temperature such as to guarantee the heating of the water for heat pumps and swimming pools [13]. Nayak et al. in [14] experimented with a solar system embedded in a layer of concrete and placed on a roof. The collector is made of 10 mm polyvinyl chloride (PVC) pipes and the roof surface has been painted black to increase its solar absorption capacity. The results showed that solar collectors, placed on a concrete surface, could be used as a valid and economically feasible alternative to conventional systems when the air temperature is 35 °C and solar radiation is 1000 W/m^2 [12]. In 2009, Mallick et al. in [15] published a work in which they studied the heat-treated and experimented asphalt pavements for the collection of energy in order to reduce the heat island effect. Using a finite element model, they showed that, in a system with recessed pipes and placed about 40 mm below the pavement, the surface air temperature could be reduced by up to 10 °C. Furthermore, some laboratory tests showed that to increase the system efficiency can be painted the pavement with the black acrylic paint and/or replace limestone aggregates with aggregates containing quartz (increase in water temperature) by 50%, respectively, and 100%. Similar to the previous case, Wu et al. [16] experimented with the installation of road collectors embedded in a bituminous conglomerate and observed that the surface temperature of the pavement could be significantly reduced. Furthermore, in order to improve thermal conductivity and energy exchange efficiency, they also experimented with the use of graphite powders in asphalt pavements. The addition of graphite may slightly increase the leaving water temperature, but to increase the temperature, longer piping and a larger area are required.

Recently, the authors in [17] developed a solar collector made of ordinary and vanadium–titanium black ceramic. For ordinary ceramic, raw materials mainly refer to feldspar, quartz, and porcelain clay, etc. Black vanadium–titanium ceramic, on the other hand, was used as a solar absorbing coating material with a stable solar absorption value of 0.93–0.97. Three systems were developed in this study: a metal plate solar collector system, an all-ceramic solar system, and a glass vacuum tube solar collector system. Maximum thermal efficiency has been verified for the all-ceramic solar system. The all-ceramic solar system could find the right application in a roof of a building where the roofing must feature a structure made of concrete, a waterproof layer, and the insulating layer. Furthermore, the integral ceramic solar collectors being characterized by excellent thermal stability, long life, high thermal efficiency, and compatibility with building materials, can integrate well with buildings, to produce domestic hot water [17].

3. Pavement as Thermal Collector

The high absorption capacity of classic road pavement (temperatures up to 70 °C is reached [18]) has prompted the scientific community to investigate the possibility of using these surfaces as large star collectors. In fact, the absorbed solar energy can be used to heat a collector positioned below the road surface. The collected heat can be used for various applications, such as storage in an underground tank [19] or under the pavement structure [20], to heat adjacent buildings, to defrost streets in winter, to provide hot water [12], or to convert energy using a thermoelectric generator [21]. Furthermore, this type of system would respond simultaneously to two impact phenomena, such as the lively impact and the overheating effect. In fact, the installation of a collector under the road pavement on the one hand would avoid the occupation of large surfaces, not always available near the site of interest; on the other hand, it would guarantee a reduction of the urban heat island (UHI) effect [15,18].

3.1. Energy Balance

The balance equation of an asphalt solar collector takes into account the presence of asphalt and pipes, the air, and the fluid flowing in the pipes [22]. In general, the first heat exchange is in the pavement–atmosphere interface, where are simultaneously present the convection and thermal radiation. This heat flow causes a temperature difference between the asphalt surface and a point located in the pavement at a certain depth, transporting the heat by conduction from the pavement surface to the interior. Inside of the asphalt pavement, the heat transfer is represented by the conduction exchange between the pavement-pipe and the pipe wall. Then, due to the difference between the temperature of the inner surface of the tube and the fluid, the convection transmission causes the fluid temperature to rise.

According to Fourier's law, in an RTC, the heat transfer by conduction occurs from the pavement surface towards the interior; that is to say:

$$q_{cond} = -\lambda n \, \nabla Tn \qquad (1)$$

where along the n direction, q_n (W/m^2) is the specific heat flux; λn is the thermal conductivity (W/mK), and ∇Tn is the temperature gradient (K/m). The phenomenon of heat transmission by convection occurs in two different conditions: between the surface of the flooring and the air above it and between the fluid circulating inside the collector and the walls of the collector itself [23]. The two equations are:

Convection heat flux at the pavement surface:

$$q_{conv,pav} = h_c \, (T_{air} - T_s) \qquad (2)$$

where h_c is the average convection coefficient of the surface (W/m^2 K); T_{air} is the air temperature (K), and T_s is the surface temperature (K).

Convection heat flux inside the collector

$$q_{conv,col} = h \, (T_s - T_f) \qquad (3)$$

where h is the average convection coefficient of the surface (W/m^2 K) and T_f is the fluid temperature (K).

To reduce heat losses, the best condition is represented by a natural convection between air and pavement, but, obviously, this condition depends on atmospheric conditions and wind speed.

On the other hand, to achieve the maximum heat transfer rate in the collector, the system must be designed for slow flow. There are several empirical models, published in the sector bibliography, for the calculation of the transfer coefficient (h_c) [24]. Table 1 shows the most used empirical equations to predict the temperature profiles in pavements.

Table 1. Empirical equations to determine the pavements temperature profiles' [25].

Equations	Model	Reference
$h_c = 698.24((0.00144(T_{ave})^{0.3})(v_w)^{0.7}) + 0.00097(T_0 - T_{air})^{0.3}$	Vehrecamp	[26]
$T_{ave} = (T_0 - T_{air})/2$		[27]
$h_c = 5.6 + 4 v_w$ For $v_w \leq 5$ m/s	Jurges	[28]
$h_c = 7.2 + v_w^{0.78}$ For $v_w > 5$ m/s		[29]
$h_c = (k \times Nu)/L_c$	Horizontal flat plate approach	[30]
$Nu = 0664 \times Re^{0.5} \times Pr^{0.33}$		[31]
$Re = (v_w \times L_c)/v$		[32]

3.2. Concreate and/or Asphalt Pavements

In the literature, there are Pavement Heat Collector (PHC) applications with asphalt pavements and concrete pavements. In general, concrete is made up of a mixture of aggregates (large and fine), water, and cement; a concrete floor is therefore composed of a base and/or a substrate with a cementitious concrete slab or pavement quality concrete (PQC) on top. Concrete flooring is a valid solution if it represents a durable, weather-resistant, robust, and economical surface [33]. As indicated in [34], in tropical countries, this type of pavement is often used in airport aprons, taxiways, and on the runways and pavements of busy highways. Asphalt, on the other hand, is made up of aggregates that are linked together by a binder (e.g., bitumen) and have a viscous behavior at high temperature and elastic behavior at low temperature. Due to this "viscoelastic" behavior, asphalt pavements are very susceptible to permanent deformations linked to high temperatures and/or longer loading times. However, in particular conditions, asphalt pavement can be an alternative to the concrete pavement; in [35] the PQC mix design parameters are collected. A typical asphalt pavement, on the other hand, consists of a compacted surface, in which there is a subgrade build, a, base and a surface layer directly in contact with traffic loads. In addition, the surface layer, besides ensuring the right friction, smoothness, noise control, and drainage, protects the underlying layers.

Compared to solar collectors embedded in concrete pavements—asphalt surfaces, despite having poorer conductive and thermal storage capacity (thanks to the high absorption coefficient), are able to capture more solar energy; an aspect that can be partially resolved in concrete pavements by painting the surface with dark paints in order to increase the absorption coefficient. Generally, there are surface absorption/emissivity values of 0.9/0.91 for asphalt and 0.65/0.91 for concrete. A disadvantage of using asphalt and not concrete is that is that the high temperatures of the mixture reached during installation can damage the piping system, especially for plastic piping. For this reason, metal pipes were previously used, mainly made of steel (43–54 W/mK), iron (80.4 W/mK) or copper (372 W/mK), however subject to corrosion [25]. In the literature, it is possible to find a comparison among four types of materials that are currently used for the piping in radiant heating systems or road collectors used for melting snow:

- PEX-AL radiant (0.43 W/mK). Pexal® is the system composed of a multilayer pipe made of cross-linked polyethylene (PE xb external layer) and butt-welded aluminum for water conduction (intermediate layer), and another entire layer in PE.
- PEX radiant (0.43 W/mK).
- ONIX radiant (0.29 W/mK). Onyx is a composite of nylon and short carbon fibers that gives 3D printed parts greater strength and a matte black finished surface. Compared to traditional nylon, this new material is approximately 3.5 times stronger, has greater hardness, and an HDT of 140 °C.
- Copper.

Obviously, the greatest temperature difference between the inlet and the outlet was obtained from the copper piping, followed by that of radiant PEX-AL [23]. In this case, a difference between the two systems of only 3 °C was measured, justifying the opportunity to use PEX-AL as a cheaper material [23]. An aspect of fundamental importance for the realization of an efficient system is the identification of an adequate depth of installation of the piping system; that is, one that is able to maximize the temperature increase of the heat transfer fluid, and at the same time does not compromise the integrity of the pipe itself and the surface of the pavement. In order to avoid negatively affecting the durability and the floor covering without damaging the drowned piping, it is necessary to consider as the laying depth the one in which the "reflective cracks" no longer occur under traffic load. From an analysis of existing commercial systems and literature studies, this depth generally varies from about 40 mm to about 120 mm. If, on the one hand, the installation of the piping network at a depth of less than 50 mm, i.e., very close to the surface layer, obviously guarantees the achievement of higher temperatures of the carrier fluid, on the other hand it presents structural problems related to surface cracks, due to stresses concentrated near the pipes, and technical problems for the remaking of the road surface.

To prevent cracking, a three-dimensional reinforcement grid has been developed in the Ooms Avenhorn Holding commercial system to secure and protect during the laying and compaction of the bituminous conglomerate and to reduce the stresses of the pipes. In addition, a special polymer-modified bitumen known as Sealoflex® [36], a polymer-modified bitumen (PMB), it was developed to improve the quality of the asphalt mix between the pipes and the grid; this mix strengthens the asphalt, doubling its durability over time even if subjected to intense use [23].

Currently, Ooms Avenhorn Holding states that resurfacing is a practical problem when the pipes have been installed to a depth of 40 mm. In order to guarantee an efficient PHC system with pipes installed at a safe distance, it will therefore be necessary to modify the asphalt mixture normally used in order to maximize the temperatures in the heat extraction layer. In particular, the laying of the pipe between a lower layer with high resistance and an upper layer with high thermal conductivity, can increase the transmission of heat by increasing the temperature of the pipes and the water at the outlet.

As an example, in [23], the subsoil temperatures are analyzed using three asphalt mixes and two installation depths of the polyethylene piping at 40 mm and 120 mm, in which the main aggregate of the mixture is limestone. In detail, the reference blend has the following specifications (Table 2):

Table 2. Principal parameters of some aggregates or elements of the road pavement.

Parameters	Recommended Values
Bitumen content	4.9 ± 0.4%
Type of bitumen	100/150
Air voids	4 ± 1.5%
Aggregate classification	Table 5.2 of [23]

An alternative solution is represented by the total replacement of the limestone aggregates with quartzite aggregates in the surface layer and the replacement of the limestone aggregates with a nominal size of less than 10 mm with Lytag for the layer under the pipe. The quartz and Lytag setup has been shown to achieve the same temperatures at a depth of 120 mm as those obtained from the unmodified reference pavement at a depth of 40 mm. The presence of a layer of flooring with high thermal conductivity above a depth of 120 mm, combined with a layer of flooring with high thermal resistance, considerably increases the temperature in the installation area of the pipes. In particular, it was verified that the temperature of the fluid in the pavement at 40 mm and 120 mm depth, compared to the reference pavement, increased by approximately 8 °C and 10 °C, respectively. In fact, in [23] the analysis of the temperature trend of the pavement surface shows that the reduction of the surface temperature of the flooring is greater in the case in which the pipe has been positioned at a depth of 40 mm compared to that laid at 120 mm. In general, it is evident that increasing the conductivity reduces the temperature of the asphalt surface regardless of the heat removed from the fluid; this effect must be taken into consideration for the reduction of the phenomenon of heat islands. However, the authors of the study [23] found a lower efficiency of the experimental system compared to the simulated model, due to the presence of an imperfect connection at the interface between the pipe and the asphalt. This problem could be solved by a possible layer of concrete in correspondence with the pipe. The concrete, due to its greater fluidity during positioning compared to asphalt, shows a good bond with the copper and polyethylene pipes and no signs were observed on the interface area. Therefore, it is probable that, due to its solar absorption capacity, a concrete pavement characterized by a low heat transfer at the pipe/pavement interface and by a higher conductivity, shows better performance than asphalt. Furthermore, the addition of highly absorbent colored surface coatings, a bituminous layer, or a dark additive (fly ash) could guarantee an increase in the solar absorption phenomenon by the pavements.

On the other hand, the construction of concrete solar collection surfaces would move away from the objective of the following work, namely that of exploiting existing low-cost surfaces. In fact, the asphalt surfaces are already present and since the road surface must be resurfaced periodically, the energy system can be installed very easily during a generic maintenance phase. From the point of view of energy production, significantly larger areas of asphalt collectors can compensate for their lower expected efficiency and improve their efficiency in terms of investment for the power produced per unit. Furthermore, asphalt surfaces can continue producing energy even during the night when the solar collectors are not working. Solar energy, captured by asphalt surfaces, can be used for different applications, e.g., in combination with accumulation systems (wells), for road safety and maintenance, keeping the roads free from snow during the winter period and reducing the cost of regular road surfacing; and, finally, the heat island effect can be improved by decreasing the temperatures of the asphalted surfaces.

Additives to Improve the Conductivity of Asphalt Mixes

An improvement of the energy properties of an asphalt RTC is certainly obtained by the correct addition of the solar system. It has been shown that in addition to quartz, which has optimal properties, it is possible to use other aggregates and or additives to improve the thermal conductivity of the asphalt. In [37], the thermal properties of asphalt mixtures with additives with graphite and carbon black (a pigment, produced by the incomplete combustion of heavy petroleum products). Thermal testing indicated that graphite and carbon black improve the thermal properties of asphalt mixes and that combined conductive fillers are more effective than single fillers. Four graphite and carbon black contents of 5, 10, 15, and 20% by volume of the asphalt binder were used to produce the asphalt mix. For all mixes in the study, PG 64–22 asphalt binder was used. Graphite and carbon black are added to replace the traditional aggregate (dimensions of the sieve through 0.075 mm). From the results it is clear as the thermal conductivity increases if increases the content of graphite or carbon black. The mixing was carried out, both dry and wet, showing that a slightly increased thermal conductivity is achieved by the wet process compared to the dry process.

Graphite apparently appears to be better than carbon black from a thermal conductivity point of view, but it has been noted that using a conductive filler mixed with carbon fibers is more beneficial than using a single conductive filler. However, the rheological properties of the modified asphalt binders have indicated that both graphite and carbon black can increase tear resistance on the one hand, and on the other reduce their resistance to thermal cracking. In fact, the two substances at high temperatures increase the shear parameter, but reduce the resistance to cracks at low temperatures.

From study [38], it is evident that the variation in the conductivity of the bituminous conglomerate can be influenced by the content of the binder and the air gap, remaining in any case unchanged for air gap dimensions between 0.5 and 2 mm in diameter. Instead, the addition of graphite with different percentages, from 10 to 40% by the mass of binder, shows an increasing trend in the value of thermal conductivity, almost tripling the value from 0.39 to 0.95 W/mK.

The bibliography illustrates how many and varied experiments and analyses have been in the field for the evaluation of the best configuration to be used to maximize the efficiency of an RTC system. In summary, the analysis of [37–44] made it possible to state that:

- The addition of graphite to asphalt binders causes an increase in the thermal conductivity of the bituminous concrete, resulting in an increase of 24%.
- An increase in thermal conductivity of about 50% of dense bituminous concrete (4% air gap and 6% bitumen) is guaranteed when mixing aggregates, such as diabase and quartzite. This increase is attributed to the thermal conductivity of quartzite, which is approximately twice that of diabase.
- Asphalt mix with high void content has lower thermal conductivity and a lower specific heat capacity than with low void content, because the air has a lower thermal conductivity than the asphalt mix. The conductivity values of bituminous concrete decrease by 3% as the vacuum content increases by 1%.
- A dense asphalt mix is characterized by an high heat conductivity and specific heat capacity, and has a lower heating and cooling rates than porous asphalt mix.
- It has been shown that thermal conductivity is also influenced by the degree of humidity and frost; in particular, in conditions of humidity saturation and freezing, they respectively have a thermal conductivity of 5% and 7% higher than that of its dry state. In general, the thermal conductivity of saturated asphalt concrete is higher than that in the dry state, while the thermal conductivity in the freezing state is higher than in conditions saturated with moisture.
- The higher the volute content of an asphalted surface, the higher the operating temperature. The average temperature of the asphalt mixture during heating and cooling is almost independent of the air gap content in the mixture.
- The total amount of energy accumulated in asphalt mixes with different content of air voids, but made from the same materials, during heating and cooling depends only on the density of the mixes.

3.3. Pavement Types and Characteristics

In general, from a technological point of view, asphalt pavement from the road surface to the inner layers is structured as follows:

1. Surfacing: (a thin asphalt surface and a binder course): it is divided into surface and binder layers; it is the layer in contact with traffic loads and protects the lower layers from bad weather;
2. Base (asphalt/unbound granular material) represents the main structural layer that provides most of the strength and load distribution properties of the flooring;
3. Sub-base (unbound granular material/Bitumen or cement stabilized material): located between the base and the sub-base serves as structural support, particularly when building upper layers;
4. Subgrade: in situ compacted material/cement stabilized/ unbound granular material; usually natural ground or artificial ground;
5. The asphalt, present in the first layers of the pavement, is generally made up of aggregates that are bonded together by means of a binder (bitumen);

From a thermo-physical point of view, the typical values of the thermal properties of asphalt surfaces are [25]:

- Thermal conductivity between 0.74 and 2.89 W/mK;
- Specific heat between 800 and 1853 J/kgK;
- Thermal diffusivity between 1.2 and 16.8 10^{-7} m^2 s.

While, as regards the mixture of bituminous materials, it is possible to distinguish mainly two types:
- Hot asphalt (hot macadam asphalt: HMA);
- Cold asphalt.

HMA is a blend in which all materials are mixed at a high temperature; there is the asphalt concrete (AC) (known as Dense Bitumen Macadam (DBM)) and hot rolled asphalt (HRA). As it is known, the mixing method affects the thermal performance of the flooring itself. In Table 3, there are some specifications of the aggregates normally used.

Table 3. Thermal conductivity of most used aggregates of the road pavement.

Aggregates	Thermal Conductivity [W/mK]
Quartzite	5.5–7.5
Granite	3.0–4.0
Limestone	1.5–3.0
Basalt	1.3–2.3
Bitumen	0.15–0.17
Cement	0.29
Waterfall	0.6
Air	0.024

4. Case Study

To evaluate the feasibility of an RTC system integrated with a geothermal storage system installed in the Mediterranean area, to accumulate thermal energy in the summer for winter air conditioning—at the University of Palermo campus, a theoretical-experimental study was launched, which involved the installation of a small-scale RTC prototype, in which to investigate some alternative scenarios and a large-scale prototype that will see the installation of an RTC system directly connected to a geothermal field. In particular, the entire development of the work involved:

Phase 1: identifying some possible solutions for laying an RTC system by varying the stratigraphy and the solar reflectance;
Phase 2: designing an experimental set-up for small-scale laboratory testing;
Phase 3: implementing the experimental set-up in the small-scale laboratory;
Phase 4: designing an experimental set-up for large-scale field verification;
Phase 5: developing a dynamic model of the RTC system in TRNSYS (Transient Systems Simulation Program) environment [45];
Phase 6: implementing an experimental set-up in the field on a large scale;
Phase 7: validating the model with data monitored in the field;
Phase 8: identifying the best RTC system solution to maximize performance in the field coupled to a BTES (Borehole Thermal Energy Storage).

The research carried out so far has allowed us to develop the project from Phase 1 to Phase 5. In this work, after a careful bibliographic analysis, the authors illustrate the main results obtained so far, while in the paragraph "Future Objectives" briefly explain the continuation of the work whose results will be published in a second work.

4.1. Identification of Some Possible Stratigraphy (Phase 1)

Based on what has been analyzed above, it is obvious how the activation of the surface layers, or in which the pipes are installed, leads to an increase in the performance of the RTC system. The activation of these surfaces, on the other hand, leads to a significant increase in costs and to greater attention in the installation and management phase of the RTC system. To identify the best solution that in energy, environmental, and economic terms is the most effective and sustainable, the first analyses were based on the performance of the RTC, in which the surface layer is formed of asphalt or concrete, and the geometric configuration of the drowned pipe is made to vary. In detail, the first analyses included the following three configurations:

First configuration

Hydraulic configuration: serpentine piping with a pitch of 12.5 cm.
Stratigraphy:

- Layer 1 mat, made of asphalt about 4 cm;
- Layer 2 binder in bituminous conglomerate of about 7 cm;
- Layer 3 reinforced concrete screed of about 7 cm;
- Layer 4 of 20 cm cement mix;
- Layer 5 of the drainage system of about 25 cm, made in such a way as to convey the water collected during the rain in the central and lower part of the caisson;
- Layer 6 of 10 cm insulation.

Second configuration

Hydraulic configuration: parallel piping with 12.5 cm pitch.
Stratigraphy:

- Layer 1 mat, made of asphalt about 4 cm;
- Layer 2 binder in bituminous conglomerate of about 7 cm;
- Layer 3 reinforced concrete screed of about 7 cm;
- Layer 4 of 20 cm cement mix;
- Layer 5 of the drainage system of about 25 cm, made in such a way as to convey the water collected during the rain in the central and lower part of the caisson;
- Layer 6 of insulation of 10 cm.

Third configuration

Hydraulic configuration: parallel piping with 12.5 cm pitch.
Stratigraphy:

- Layer 1 reinforced concrete screed of about 18 cm;
- Layer 2 of 20 cm cement mix;
- Layer 3 of the drainage system of about 25 cm, made in such a way as to convey the water collected during the rain in the central and lower part of the caisson;
- Layer 4 of 10 cm insulation.

Table 4 shows the main thermophysical characteristics of the project and the thicknesses of the RTC system to be built in the experimental area within the UNIPA (University of Palermo) campus.

Table 4. Thermophysical characteristics of the RTC system.

Layer	λ [W/mK]	c_p [J/kgK]	ρ [kg/m^3]	ε	α	s [m]
Asphalt	0.85	850	2400	0.9	0.9	0.11
Binder						
Reinforced concreate	1.4	840	2100			0.07
Cement mix	1.2	840	1700			0.2
Drainage	0.4		1500			0.25
Eps	0.034	1130	30			0.2

In all three configurations, the piping will be installed in the middle of the third layer.

The radiative properties of the external layers are crucial to collect the impinging solar irradiation: the lower the solar reflectance (either the higher the solar absorptance), the higher the solar gains. Conventional materials exhibit favorable properties, the solar reflectance of the new asphalt is about 5%, increasing to 15% after ageing; concrete is about 25%. Materials with high solar absorptance and very low water permeability, equally, create favorable conditions for Urban Heat Island (UHI), taking into account that urban pavements account for 29–45% of the city footprint [18]. The UHI mitigation thus calls for the application of high reflective urban materials, for pavements as well [46]. With the objective of balancing these two opposite design requirements, some solutions were developed and optically characterized.

Spectral reflectance measurements were carried out, three 21 × 21 cm samples, colored in black, brown, and dark brown. The measurements were conducted using a Perkin Elmer Lambda 950 dual-beam spectrophotometer, equipped with a 6-inch Spectralon coated sphere. The measurement was of relative type and it was performed against a calibrated Spectralon reference; the measurement uncertainty was 1%. Spectral measurements and the calculation of the broadband parameters solar (sol), visible (vis), and near infrared (nir) ranges were carried out in accordance with the relevant standard [47]. Given the texture of the material and the geometry of the incident beam (about 15 × 5 mm), three measurements were made on each sample to evaluate the repeatability of the measurement. Since the difference between each single measurement and the average of the three was always less than 1%, the latter was assumed sufficiently accurate. The reflectance values in the relevant spectral regions, referring to the 0–100% range, are presented in Table 5. The emissivity was also measured using a broadband emissometer in the 2.5–40 micron range. Measurements were carried out at 80 °C; the three samples scored the same emissivity value: 0.88.

Table 5. Experimental reflectance measurements of road surfaces.

Reflectance (%)	Black	Brown	Dark Brown
Sol	7	11	10
NIR	10	17	16
Vis	4	6	6

The proposed solutions fall within the conventional (5–15%). In general, the samples have higher reflectance in the near infrared than in the visible: 6% for black and 10–11% for browns, thus exhibiting a modest band selectivity. In this sense, the tested materials are of potential interest for road solar collectors because of the high solar absorptance but do not exhibit suitable performances for UHI mitigation purposes. For these applications, however, the solar reflectance (absorptance) might not be the only parameter shaping the thermal performance of the finishing layer. As explained above, the whole covering package of the system should create favorable conditions to transfer the absorbed heat to the buried pipes with consequent reduction of the surface temperature. The complexity of the system, thus, calls for accurate identification of the solar properties of the external layer able to optimize the solar collector performances as well as the mitigation of the urban overheating.

4.2. Design of an Experimental Set-Up for Small-Scale (Phase 2)

The research work soon led to the development and realization of a prototype set-up for the measurement of temperatures in three different configurations. The experimental set-up involves the measurement of physical quantities within a wooden tank structure, hereinafter referred to as a box. The experimental set-up includes, in addition to the box, to be built according to the following diagrams and descriptions, a hydraulic and measuring system. In order to be able to experiment at the same time with different configurations and different boundary conditions, it was decided to design a system capable of containing three different stratigraphy (Figure 1).

The system provides for the construction of the box with the presence of internal insulating partitions, in each of which a previously waterproofed layer of insulation will be inserted. The waterproofing, which will be carried out by bagging the insulation in polyethylene sheets, must also be guaranteed in the contact surfaces between the wooden panel and the material present in the three configurations.

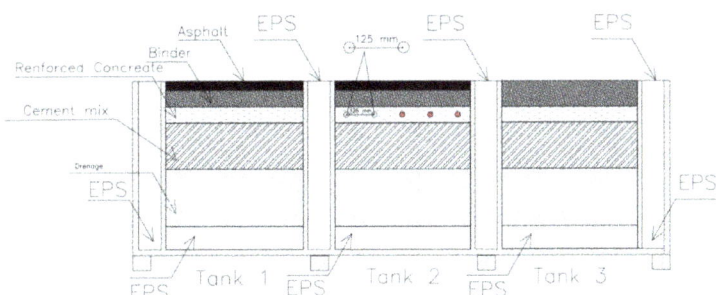

Figure 1. Experimental set-up schema configuration.

In each tank, the installation of the three configurations previously foreseen were analyzed. Figure 2 shows the three stratigraphy, i.e., tank 1 refers to the first configuration; tank 2 refers to the second, and the last tank to the third configuration.

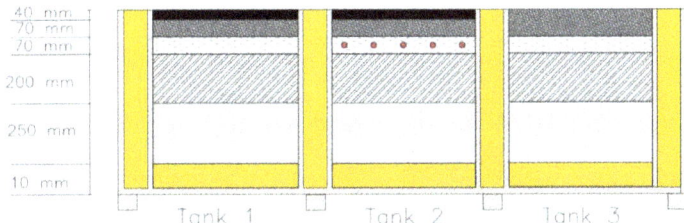

Figure 2. Experimental set-up schema geometrical configuration.

For the project to be successful and since the system will be exposed to atmospheric agents, the following specifications are expected:

- Wooden panels for the construction of formwork for reinforced concrete framed structures, Resistant to water and bad weather according to the En 13353 standard (SWP/2). In particular, a 27 mm thick wooden panel is foreseen for the external and bottom surfaces and 21 mm thick for the internal separating baffles.
- The panels were treated with a suitable resin-based paint or similar to give lasting resistance to atmospheric agents, according to the En 13353 (SWP/2) standard.
- The box is not in direct contact with the floor, but is placed on a wooden support base for outdoor use.
- The insulating panels (sections in yellow) are rigid PIR type (polyiso polyurethane foam), i.e., rigid high density insulated material with conductivity between 0.02 and 0.03 W/mK
- The use of a vapor barrier polyethylene sheets) for waterproofing the contact layer between wooden panels and the insulation layer
- At the base of each tank, there is a light draining layer of 25 cm. Furthermore, to ensure water drainage from this Section, it was necessary to prepare a through-hole (as per drawing), in each centerline of the three tanks, in which to house a rigid PVC drainage pipe included in the supply.
- The installation of the exhaust pipe guarantees the seal with the vapor barrier.

Geometry was designed for the body; the gross body dimensions are as follows:

- Length: 2.445 m;
- Width: 1.796 m;
- Height: 0.627 m;
- Total area: 4.41 m^2;
- Total volume: 2.76 m^3.

The hydraulic system that is being built will follow the following system scheme (Figure 3).

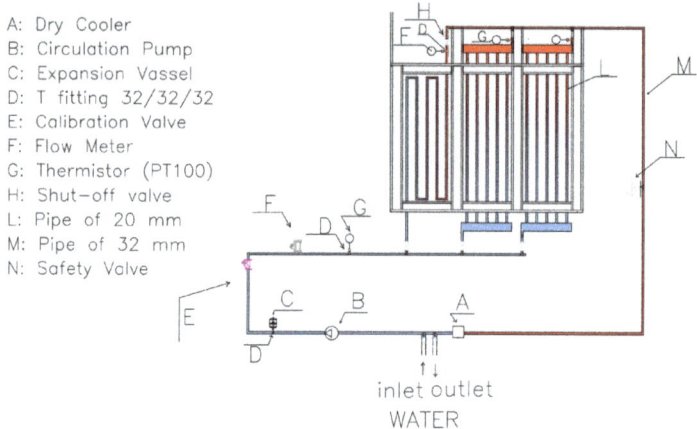

A: Dry Cooler
B: Circulation Pump
C: Expansion Vassel
D: T fitting 32/32/32
E: Calibration Valve
F: Flow Meter
G: Thermistor (PT100)
H: Shut-off valve
L: Pipe of 20 mm
M: Pipe of 32 mm
N: Safety Valve

Figure 3. Hydraulic schema.

As indicated in the picture, there is also a measurement system with temperature sensors on the cold side and hot side of the system and flow meter, which will be installed during the implementation of the experimental set-up. Furthermore, the climatic data in the field at the same time will be monitored.

4.3. Experimental Set-Up in the Small-Scale Laboratory (Phase 3)

In order to validate the results and to be able to hypothesize the best design solution that maximizes the efficiency of the RTC system installed in Palermo, it is necessary to carry out an experimental campaign. For this reason, at the same time as the theoretical study and after the design, a small-scale experimental prototype installed at the University of Palermo campus is being installed. The test bench faithfully follows the previously designed schemes and stratigraphy. Below are the photos of some phases of the installation to date (Figures 4 and 5).

Figure 4. Some phases of the installation of the RTC Box.

In particular, the insulation layer, the drainage layer, and the first layer of lean concrete were put in place. Two constantan copper thermocouples were embedded in the latter for each single tank. In fact, the entire system will be monitored along with all the layers up to the surface where it will be laid the road mat. At the same time, the air temperature, wind speed, direction, and horizontal global solar irradiation will be collected.

  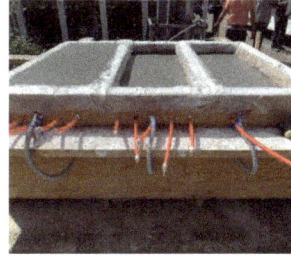

Figure 5. Some phases of the installation of the RTC Box.

The prototype will be the experimental base in which to experiment in the field and on a small scale all the alternative solutions previously hypothesized and others that will be considered appropriate, in this way, it will be identified as the best solution to be implemented in the Mediterranean area. In detail, the small-scale prototype was born from the idea of being able to experiment on a reduced model with different geometric configurations, materials and alternations of stratigraphy's that can represent an alternative and improved solution to the model envisaged on a large scale. In detail, the research group decided to develop a dynamic model that is calibrated with the data monitored in the field. A dynamic model, created with the TRNSYS software (Thermal Energy System Specialists, LLC, Madison, USA) with data in the field will then be validated, guaranteeing the reliability of the results. In this way, the effects of the application of alternative solutions, configurations, and/or materials to the basic configuration will be predicted numerically, also allowing a parametric analysis for the identification of the best solution. In order to assess whether a solution, configuration, and/or choice of alternative material packages highly reliable in the field, a small-scale field trial is necessary. This set-up, thanks to its small size, will allow one to experiment with various solutions, to select those that make the system more efficient, reducing time and, above all, costs. For example, the step of adding a small area of road pavement, which may not lead to expected results, does not impact economically as that of adding the entire experimental area. Figure 6 presents some stages of laying of the RTC system at the UNIPA experimental area; this phase is being completed.

Figure 6. Experimental area in the UNIPA Campus.

4.4. Design of an Experimental Set-Up for 1:1-Scale (Phase 4)

The study of the RTC system, the object of the research in the Mediterranean area, has, as its final aim, the subsequent integration with a BTES system to create a seasonal thermal energy storage system. In this case, the RTC plant system will in fact be aimed at collecting thermal energy from the solar source through the piping system embedded in the ground, to be directly connected to a geothermal storage system. In fact, the project involves the installation of the entire RTC + BTES system in an experimental area within the university campus of Palermo. The area, identified for the realization of the large-scale experimental set-up, is represented in the following Figure 6.

For the large-scale set-up, on the basis of what has been indicated above, and of the available area, 15 pipes 20 m long connected in parallel, with 25 cm spacings, intercepted by two collectors (intake manifold and delivery manifold). The piping supplied is of the PE-Xa type Rehau Rautherm S type characterized by an external diameter of 32 mm and a thickness of 2.9 mm (Figure 7).

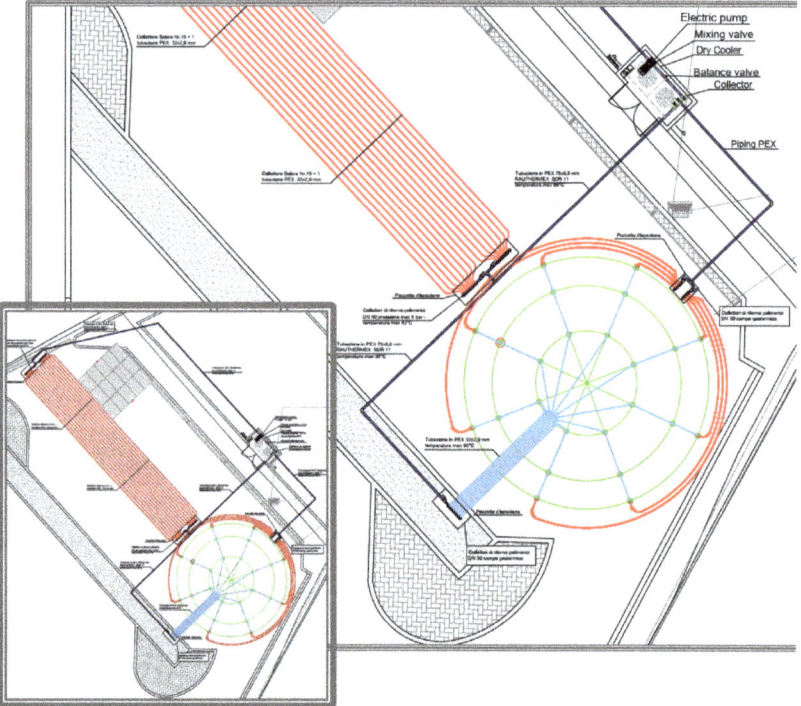

Figure 7. RTC + BTES schema of the experimental set-up.

The entire collection system is buried at a depth of about 14 cm (axis of the pipeline). For the correct sizing of the system, pure water was considered as working fluid (Table 6) and a turbulent motion with Re equal to 10,000 was imposed.

Table 6. Specific characteristics of the thermal fluid (Water).

	Thermal Fluid	Unit
ρ	1000	kg/m^3
c_p	4189.99	J/kgK
μ	0.00089	kg/ms
λ	0.5944	W/mK

The chosen pipe has a diameter of 0.0291 m, or an area of 6.65×10^{-4} m^2; based on these data and the hypothesis of turbulent motion, the value of the estimated optimal flow rate is equal to

$$m = Re \cdot \mu \cdot (A/d) = 0.204 \text{ (kg/s)} \quad (4)$$

which corresponds to a volumetric flow rate of

$$V = m/1000 = 0.000204 \text{ (m3/s)} \quad (5)$$

and with a speed of

$$v = V/A = 0.307 \text{ (m/s)} \quad (6)$$

Assuming an RTC area of 80 m^2, i.e., 4 m wide and 20 m long, it is possible to place 15 pipes with 0.25 m pitch, for a total mass flow rate of 3.058 kg/s, or approximately 11,008.84 kg/h. The piping system must be positioned and anchored to the depth, as planned, by means of a 6/8 mm electro-welded metal mesh, which will be previously placed on top of the mixed cement layer. Downstream of the RTC system, a temperature probe (PT100 type), an automatic vent valve, and a safety valve are installed on the delivery pipe to the BTES system. The BTES plant system is aimed at storing thermal energy from a solar energy source through a system of 24 "U" probes. It will consist of eight head probes connected in parallel (one per circular sector) and three probes arranged in series in each branch of the circuit. Each probe is characterized by a length of 15 m each (30 m of pipe) and a starting depth of 0.58 m. Each pipe is characterized by an external diameter of 28 mm corresponding to an internal diameter of 26.2 mm. The entire BTES storage system is intercepted by two collectors, as shown in Figure 8. The stratigraphy of the ground above the BTES system is shown below in Table 7.

Table 7. Thermophysical properties of the stratigraphy to be carried out in the experimental area above the BTEES.

	λ (W/mK)	c_p (J/kgK)	ρ (kg/m^3)	ε	α	s (m)
Wear mat	0.85	850	2400	0.9	0.9	0.11
Binder	0.85	850	2400			0.11
Mixed cement	1.2	840	1700			0.2
EPS	0.034	1130	30			0.2
Ground	0.85	800	1890			20

The total mass flow will be sent to eight geothermal probes, corresponding to a head mass flow of 0.328 kg/s per probe. For a hypothesized pipe with a diameter of 0.026^2 m, or with an area of 5.39×10^{-4} m^2, we obtain a Reynolds number equal to 20,825.3. The piping that connects the manifolds of the two systems is characterized by an external diameter of 40 mm corresponding to an internal diameter of 36.3 mm in multilayer coated with a suitable layer of insulation. In this Section, the operating conditions are characterized by a total mass flow rate of 3058 kg/s, a speed of 2955 m/s and a Re number equal to 120,247.93.

The following components will be installed on the return pipe from the BTESS to the RTC:

1. Dry cooler;
2. Circulation pump;
3. Expansion vessel;
4. PT100 temperature meter;
5. Flow rate adjustment valve;
6. Flowmeter;
7. Two ball valves with cock for filling and discharging the hydraulic circuit;
8. Shut-off valves.

The final circuit also includes a By-pass branch of the BTESS system. Below, in Figure 8, a detail of the RTC + BTES system.

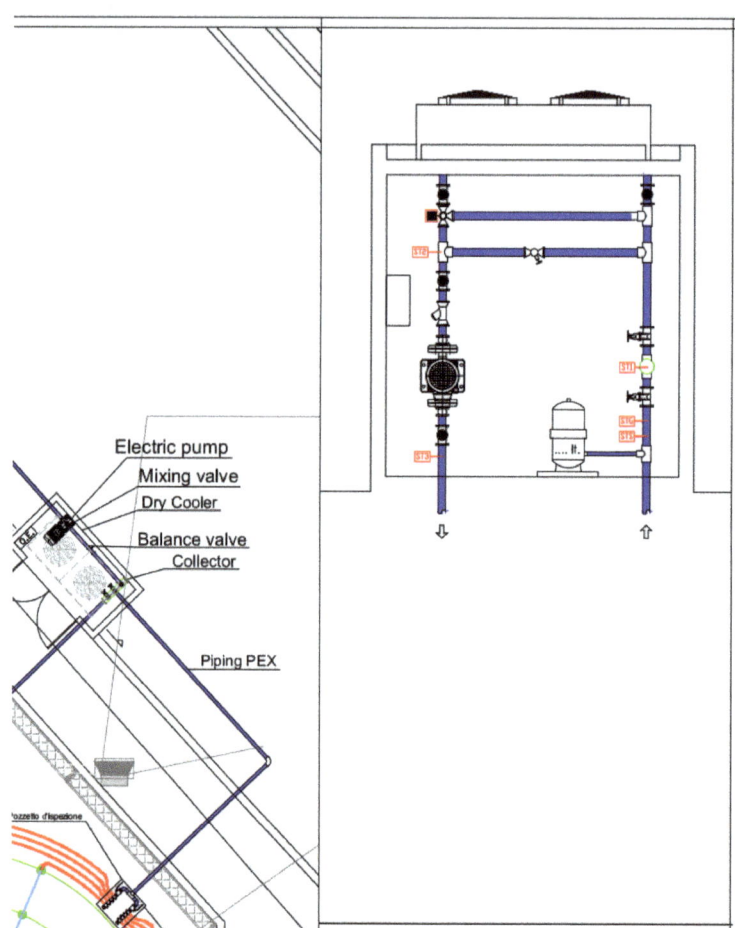

Figure 8. A detail of RTC + BTES of the experimental set-up.

4.5. Development of a Dynamic Model of the RTC System (Phase 5)

In order to evaluate the operation and management of the thermal loads of the RTC system, a model was created with TRNSYS software [47]. Figure 9 shows the scheme of the model built and implemented on the TRNSYS dynamic simulation platform.

In this first phase, the 1:1 scale model of the RTC system was created with the thermophysical characteristics of the stratigraphy indicated in Section 4.4, taking into account the stratigraphy considered in Table 7 and all the geometric and thermophysical parameters, the boundary conditions, and the management system envisaged for the 1:1 model, faithfully respecting the provisions of the experimental area.

Obviously, this model will be calibrated with the data that will be monitored in the field. Furthermore, thanks to the experimental set-up on a small scale, alternative and improvement solutions of the entire RTC system can be evaluated and then proposed in the large-scale system.

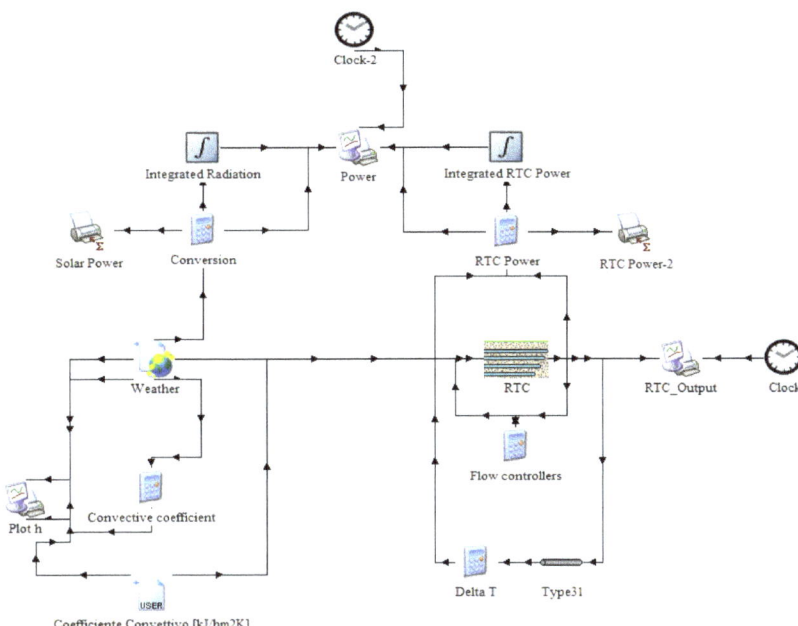

Figure 9. TRNSYS model schema.

The results of the first simulations, from April to October, are illustrated below (Figure 10); in more detail, the first graph shows the trend of temperatures inside the soil in ordinary conditions, i.e., in the absence of an RTC that removes heat.

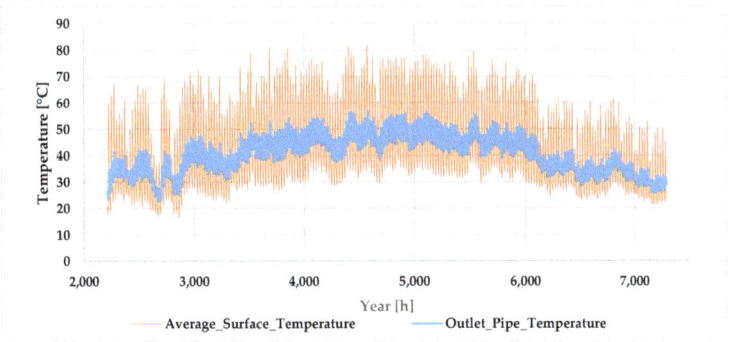

Figure 10. Temperature trend of the surface pavement and at the depth where the pipeline is located without the RTC system.

Table 8 shows the data related to the surface temperature and into the soil at the depth where the pipeline is located, indicating the maximum, minimum, and average values.

Table 8. Temperature value in the surface of the pavement and in the soil.

Temperature	Pipe Depth (°C)	Surface (°C)
Max	56.86	81.88
Average	40.91	42.54
Min	22.79	16.41

In order to respect the evaluations previously carried out in Section 4.4, the temperature trend was evaluated for Re = 10,000 (Figure 11). In this case, the RTC system influences the trend temperature, decreasing the values.

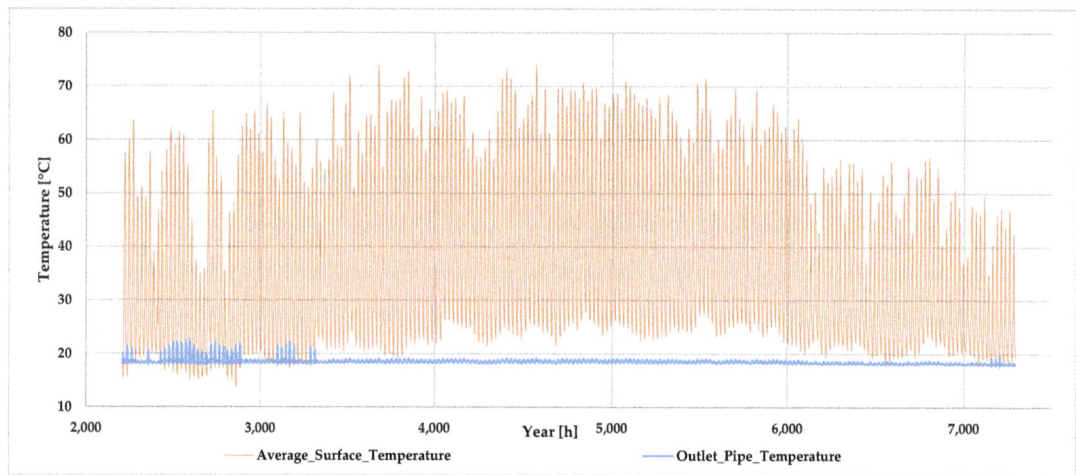

Figure 11. Temperature trend of the surface pavement and the pipeline for the RTC system with Re = 10,000.

The maximum, minimum, and average values of the water temperature at the outlet of the pipe are indicated in the Table 9.

Table 9. Temperature value in the surface of the pavement and in pipeline with Re = 10,000.

Temperature	Pipe (°C)	Surface (°C)
Max	22.53	74.08
Average	18.63	13.64
Min	18.06	36.18

The comparison between the solar and RTC power trend is illustrated in Figure 12.

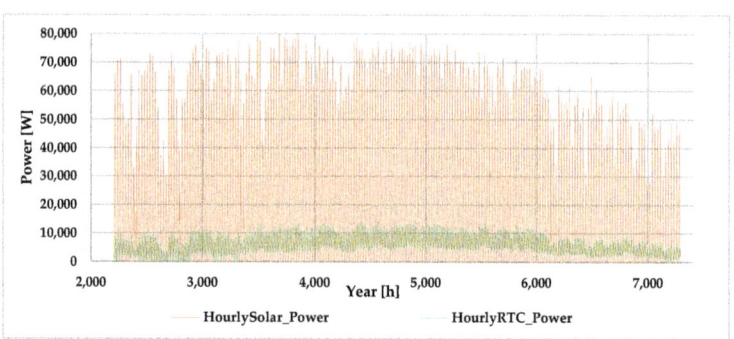

Figure 12. Solar and RTC power trend for the RTC system with Re = 10,000.

Generally, a maximum power of 14.3 kW and an average power of 6.17 kW have been evaluated, compared to an incident solar energy of 105,096 kWh, 31,529 kWh have been removed from the RTC system, for an overall efficiency of 30%.

This evaluation was carried out considering an inlet temperature to the system of 18 °C and setting a double control system that stops the circulation pump when:

- The temperature gradient between the surface and the ground, at the pipe laying depth, is less than and/or equal to zero.
- The temperature of the pipes is lower than the water inlet temperature, which is 18 °C.

In the first case (Figure 10), the trend of temperatures inside the soil in ordinary conditions is show, i.e., in the absence of an RTC that removes heat, while Figure 11 describes the trend temperature of the RTC system in ordinary conditions (Section 4.4). The comparison underlines as the RTC system reduces the temperature of the surface of about 10 °C; on the contrary, the RTC system deeply influenced the temperate in the soil, reducing all values even more than half.

This model will subsequently be calibrated with the data monitored in the field and then hypothesize alternative solutions to those identified in the first phase. In this way, it will be possible to choose the solution that will allow optimizing the system and identifying the best configuration. At the same time, this solution will be integrated with the contemporary study that will be carried out on the BTES system.

4.6. Future Objectives (From Phase 6 to Phase 8)

The possibility of being able to install a 1:1 scale system of an RTC system directly connected to a BTES at the Palermo campus, will allow to investigate this technology and evaluate its performance in the field.

The simultaneous field experimentation in a small-scale system will allow identifying the best geometric, technical, and thermophysical choices that can improve the efficiency and effectiveness of the system and, therefore, identify the best solution to be applied in the system on a real scale.

In this way, it will be possible to investigate the use of an RTC for the construction of a low temperature geothermal field, in order to explore the potential for seasonal heat storage and subsequent use in heat pump systems or for pre-heating of the water. The integration between a useful energy source and a storage medium is the real challenge of this research, especially in a highly sunny area in the summer, such as southern Italy; that is, the identification of the best solutions to be adopted to maximize the efficiency of the system in the Mediterranean area, and the identification of the best practices for the development of a renewable system that is competitive with today's market.

5. Conclusions

The collection of solar energy is commonly committed to solar collectors installed on roofs or the ground. An alternative solution is instead represented by the installation of solar collectors embedded in paved surfaces, such as asphalt surfaces, sidewalks, roads, and parking lots (road thermal collector). Solar energy, captured by asphalt surfaces, can be used for different applications, e.g., in combination with accumulation systems (wells), for road safety and maintenance, keeping the roads free from snow during the winter period and reducing the cost of regular road surfacing. Furthermore, the heat island effect can be improved by decreasing the temperatures of the asphalt surfaces. The present research proposes investigating the use of an RTC for the construction of a low-temperature geothermal field, in order to explore the potential for seasonal heat storage and subsequent use in heat pump systems or pre-heating of the water. This system makes it possible to exploit the absorption capacity of the road surface in areas characterized by high sunshine, energetically redeveloping areas that are generally dedicated to something else (roads and parking lots).

The study, design, and validation of a RTC with seasonal geothermal storage could guarantee the development of an alternative, sustainable, and low global impact renewable energy system. Preliminary results deduced from numerical analysis, carried out using a model developed in TRNSYS, show that an RTC (with a capturing surface of about 80 m^2) can produce thermal energy from the sun, with an annual average global efficiency of about 30%. On average, in correspondence of a total water flow rate of about 3.06 kg/s distributed over 15 exchanger pipes, having a length of 20 m and spaced 25 cm, it is possible to reach a water temperature rise of about 2 °C. Moreover, the same numerical results show that, with an inlet water temperature of 18 °C, it is possible to reduce the average pavement surface temperature from 42 °C down to 14 °C during the operation period of the RTC (from April to October).

The entire study, which refers to the SMARTEP project, involves the design and construction of an experimental area within the University of Palermo (Italy), unique in the whole panorama of the Mediterranean area.

Author Contributions: Conceptualization: A.D., G.C., A.B. and M.B.; methodology: A.D., G.C., D.P.; software development: A.D.; validation: A.D. and D.P.; formal analysis: A.D. and A.B.; investigation: A.D. and M.Z.; writing: G.C. and A.B.; supervision: M.B. All authors have read and agreed to the published version of the manuscript.

Funding: This work received funding from the project SmartEP (CUP: G58I18000770007) financed within the call POFESR Sicilia 2014–2020 Azione 1.1.5 "Sostegno all'avanzamento tecnologico delle imprese attraverso il finanziamento di linee pilota e azioni di validazione precoce dei prodotti e di dimostrazione su larga scala".

Institutional Review Board Statement: Not applicable.

Informed Consent Statement: Not applicable.

Conflicts of Interest: The authors declare no conflict of interest. The funders had no role in the design of the study; in the collection, analyses, or interpretation of data; in the writing of the manuscript, or in the decision to publish the results.

Nomenclature

A	area [m^2]
c_p	specific heat [J/kgK]
d	diameter [m]
h	average convection coefficient of the surface [W/m^2 K];
n	direction of vector
q	specific heat flux [W/m^2]
$q_{conv,coll}$	Convection heat flux inside the collector [W/m^2]
$q_{conv,pav}$	Convection heat flux at the pavement surface [W/m^2]
m	flow rate [kg/s]
Nu	Nusselt Number
Re	Reynolts Number
s	Thickness [m]
T_{air}	air temperature (K)
T_f	fluid temperature (K)
Ts	surface temperature (K)
v	fluid speed [m/s]
V	volumetric flow rate [m^3/s]
α	Absorption coefficient
ε	Coefficient of emissivity
λ	Thermal conductivity [W/mK]
μ	Dynamic viscosity [kg/(m s)]
ρ	Density [kg/m^3]

Acronyms

AC	Asphalt Concrete
BTES	Borehole thermal energy storage
DBM	Dense Bitumen Macadam
HMA	Hot Madacam Asphalt
HRA	Hot Rolled Asphalt
PCM	Phase Change Materials
PHC	Pavement Heat Collector
PQC	Pavement Quality Concrete
RES	Road Energy Systems
RTC	Road Thermal Collector
SMARTEP	Sustainable model and thinking of renewable energy parking
UHI	Urban Heat Island
UNIPA	University of Palermo Campus

References

1. McTigue, J.D.; Castro, J.; Mungas, G.; Kramer, N.; King, J.; Turchi, C.; Zhu, G. Hybridizing a geothermal power plant with concentrating solar power and thermal storage to increase power generation and dispatchability. *Appl. Energy* **2018**, *228*, 1837–1852. [CrossRef]
2. *SMARTEP: Sustainable Model and Thinking of Renewable Energy Parking*; PO FESR Sicilia 2014/2020–Azione 1.1.5 "SmartEP", n. 08CL4120000188-codice U-GOV 2017-NAZ-0075; Università degli Studi di Palermo: Sicilia, Italy, 2020.
3. Santamouris, M. On The Energy Impact of Urban Heat Island and Global Warming on Buildings. *Energy Build.* **2014**, *82*, 100–113. [CrossRef]
4. Wang, C.; Wang, Z.H.; Kaloush, K.E.; Shacat, J. Cool pavements for urban heat island mitigation: A synthetic review. *Renew. Sustain. Energy Rev.* **2021**, *146*, 111171. [CrossRef]
5. Enea, Agenzia Nazionale per le Nuove Tecnologie, l'Energia e lo Sviluppo Economico Sostenibile. Available online: https://www.enea.it/it (accessed on 23 October 2021).
6. Paving and Solar Energy System and Method. Available online: https://patents.google.com/patent/US4132074 (accessed on 19 October 2021).
7. Geo-Heat Center Quarterly Bulletin. Available online: https://www.osti.gov/servlets/purl/1209231 (accessed on 18 October 2021).
8. Morita, K.; Tago, M. Operational characteristics of the Gaia snow-melting system in Ninohe, Iwate, Japan. *GHC Bull.* **2000**, *21*, 5–11.
9. Eng, C.S.I.; Amice, M. Innovation in the production and commercial use of energy extracted from asphalt pavements. In Proceedings of the 6th Annual International Conference on Sustainable Aggregates, Asphalt Technology and Pavement Engineering, Liverpool, UK, 21 February 2007.
10. ICAX, Interseasonal Heat Transfer. Available online: http://www.icax.co.uk (accessed on 26 July 2021).
11. Novotech, Inc. Roadway Power System: Technical Analysis for Thermal Energy Generation. 2008. Available online: http://www.masstech.org/Project%20Deliverables/Alt%20Unlimited%20FS%20Final.pdf (accessed on 28 March 2012).
12. Sedgwick, R.H.D.; Patrick, M.A. The use of a ground solar collector for swimming pool heating. In Proceedings of the ISES Congress, Brighton, UK, 27 March 1981; Volume 1, pp. 632–636.
13. Turner, R.H. Concrete slabs as summer solar collectors. In Proceedings of the International Heat Transfer Conference; 1987; pp. 683–689.
14. Nayak, J.K.; Sukhatme, S.P.; Limaye, R.G.; Bopshetty, S.V. Performance studies on solar concrete collectors. *Sol. Energy* **1989**, *421*, 45–56. [CrossRef]
15. Mallick, R.B.; Chen, B.; Bhowmick, S. Harvesting energy from asphalt pavements and reducing the heat island effect. *Int. J. Sust. Eng.* **2009**, *23*, 214–228. [CrossRef]
16. Wu, S.; Wang, H.; Chen, M.; Zhang, Y. Numerical and experimental validation of full depth asphalt slab using capturing solar energy. In Proceedings of the 4th International Conference of Bioinformatics and Biomedical Engineering, Chengdu, China, 18–20 June 2010; pp. 1–4.
17. Sun, X.Y.; Sun, X.D.; Li, X.G.; Wang, Z.Q.; He, J.; Wang, B.S. Performance and building integration of all-ceramic solar collectors. *Energy Build.* **2014**, *75*, 176–180. [CrossRef]
18. Shaopeng, W.; Mingyu, C.; Jizhe, Z. Laboratory investigation into thermal response of asphalt pavements as solar collector by application of small-scale slabs. *Appl. Therm. Eng.* **2011**, *31*, 1582–1587. [CrossRef]
19. De Bondt, A. *Generation of Energy via Asphalt Pavement Surfaces*; Asphaltica Padova: Avenhorn, The Netherlands, 2003.
20. Carder, D.R.; Barker, K.J.; Hewitt, M.G.; Ritter, D.; Kiff, A. *Performance of an Interseasonal Heat Transfer Facility for Collection, Storage, and Re-use of Solar Heat from the Road Surface*; Published Project Report PPR 302; Transport Research Laboratory: Crowthorne, UK, 2007.

21. Hasebe, M.; Kamikawa, Y.; Meiarashi, S. Thermoelectric Generators using Solar Thermal Energy in Heated Road Pavement. In Proceedings of the 2006 25th International Conference on Thermoelectrics, ICT '06, Vienna, Austria, 6–10 August 2006; pp. 697–700.
22. Banks, D. *An Introduction to Thermogeology: Ground Source Heating and Cooling*; Blackwell Publishing Ltd.: Oxford, UK, 2008.
23. Bobes-Jesus, V.; Pascual-Muñoz, P.; Castro-Fresno, D.; Rodriguez-Hernandez, J. Asphalt solar collectors: A literature review. *Appl. Energy* **2013**, *102*, 962–970. [CrossRef]
24. Palyvos, J.A. A survey of wind convection coefficient correlations for building envelope energy systems' modeling. *Appl. Therm. Eng.* **2008**, *28*, 801–808. [CrossRef]
25. Dehdezi, P.K. Enhancing Pavements for Thermal Applications. Ph.D. Thesis, University of Nottingham, Nottingham, UK, 2012.
26. Dempsey, B.J.; Thompson, M.R. A Heat-Transfer Model for Evaluating Frost Action Temperature-Related Effects in Multilayered Pavement System. *Transp. Res. Rec. J. Transp. Res. Board* **1970**, *342*, 39–56.
27. Solaimanian, M.; Kennedy, T.W. Predicting Maximum Pavement Surface Temperature Using Maximum Air Temperature and Hourly Solar Radiation. *Transp. Res. Rec. J. Transp. Res. Board* **1993**, *1417*, 1–11.
28. Mrawira, D.M.; Luca, J. Thermal properties and transient temperature response of full-depth asphalt pavements. *Transp. Res. Rec.* **2002**, *1809*, 160–171. [CrossRef]
29. Hermansson, A. Mathematical model for paved surface summer and winter temperature: Comparison of calculated and measured temperatures. *Cold Reg. Sci. Technol.* **2004**, *401*, 1–17. [CrossRef]
30. Bentz, D.P. *A Computer Model to Predict the Surface Temperature and Time-of Wetness of Concrete Pavements and Bridge Decks*; NISTIR 6551; U.S. Department of Commerce: Washington, DC, USA, 2000.
31. CIBSE. *Guide A: Environmental Design*; Chartered Institution of Building Services Engineers: London, UK, 2006.
32. Chiasson, A.D.; Spitler, J.D.; Rees, S.J.; Smith, M.D. A Model for Simulating the Performance of a Pavement Heating System as a Supplemental Heat Rejecter with Closed-Loop Ground-Source Heat Pump Systems. *J. Sol. Energy Eng.* **2000**, *122*, 183–191. [CrossRef]
33. Delatte, N. *Concrete Pavement Design, Construction, and Performance*; Taylor & Francis: London, UK, 2008.
34. Griffiths, G.; Thom, N. *Concrete Pavement Design Guidance Notes*; Taylor & Francis: London, UK, 2007.
35. Defence Estates. *Hot Rolled Asphalt and Asphalt Concrete Macadam for Airfields*; Ministry of Defence: London, UK, 2008.
36. Available online: http://www.icopal.co.uk/Products/Liquid_Roofing_Systems/sealoflex-cold-applied-liquid-waterproofing/enviroflex-system/products/Sealoflex-Ultima-Bitumen-Primer.aspx (accessed on 3 December 2021).
37. Bai, B.C.; Park, D.W.; Vo, H.V.; Dessouky, S.; Im, J.S. Thermal properties of asphalt mixtures modified with conductive fillers. *J. Nanomater.* **2015**, *2015*, 926809. [CrossRef]
38. Mirzanamadi, R.; Johansson, P.; Grammatikos, S.A. Thermal properties of asphalt concrete: A numerical and experimental study. *Constr. Build. Mater.* **2018**, *158*, 774–785. [CrossRef]
39. Hassn, A.; Aboufoul, M.; Wu, Y.; Dawson, A.; Garcia, A. Effect of air voids content on thermal properties of asphalt mixtures. *Constr. Build. Mater.* **2016**, *115*, 327–335. [CrossRef]
40. Incropera, F.P.; Dewitt, D.P.; Bergman, T.L.; Lavine, A.S. *Principles of Heat and Mass Transfer*; Wiley: Hoboken, NJ, USA, 2013.
41. Cote, J.; Konard, J.-M. Thermal conductivity of base-course materials. *Can. Geotech. J.* **2005**, *42*, 61–78. [CrossRef]
42. Çanakci, H.; Demirboğa, R.; Karakoç, M.B.; Şirin, O. Thermal conductivity of limestone from Gaziantep Turkey. *Build. Environ.* **2007**, *42*, 1777–1782. [CrossRef]
43. Xiao, Z. Heat Transfer, Fluid Transport and Mechanical Properties of Porous Copper Manufactured by Lost Carbonate Sintering. Ph.D. Thesis, University of Liverpool, Liverpool, UK, 2013.
44. Stempihar, J.J.; Pourshams-Manzouri, T.; Kaloush, K.E.; Rodezno, M.C. Pordezno, Porous asphalt pavement temperature effects for urban heat island analysis. *Transp. Res. Rec.* **2012**, *2293*, 123–130. [CrossRef]
45. TRaNsient SYstems Simulation Program. Available online: https://sel.me.wisc.edu/trnsys/ (accessed on 19 October 2021).
46. Carnielo, E.; Zinzi, M. Optical and thermal characterisation of cool asphalts to mitigate urban temperatures and building cooling demand. *Build. Environ.* **2013**, *60*, 56–65. [CrossRef]
47. *ISO 9050:2003en*; Glass in Building—Determination of Light Transmittance, Solar Direct Transmittance, Total Solar Energy Transmittance, Ultraviolet Transmittance and Related Glazing Factors. International Organization for Standardization: Geneva, Switzerland, 2003.

Article

Effect of Climate Change and Occupant Behaviour on the Environmental Impact of the Heating and Cooling Systems of a Real Apartment. A Parametric Study through Life Cycle Assessment

Gianmarco Fajilla [1,2,*], Emiliano Borri [3], Marilena De Simone [1], Luisa F. Cabeza [3] and Luís Bragança [2,*]

1. Department of Environmental Engineering, University of Calabria, 87036 Rende, Italy; marilena.desimone@unical.it
2. Department of Civil Engineering, University of Minho, 4800-058 Guimarães, Portugal
3. GREiA Research Group, Universitat de Lleida, 25001 Lleida, Spain; emiliano.borri@udl.cat (E.B.); luisaf.cabeza@udl.cat (L.F.C.)
* Correspondence: gianmarco.fajilla@unical.it (G.F.); braganca@civil.uminho.pt (L.B.); Tel.: +39-393-093-3551 (G.F.); +351-966-042-447 (L.B.)

Abstract: Climate change has a strong influence on the energy consumption of buildings, affecting both the heating and cooling demand in the actual and future scenario. In this paper, a life cycle assessment (LCA) was performed to evaluate the influence of both the occupant behaviour and the climate change on the environmental impact of the heating and cooling systems of an apartment located in southern Italy. The analysis was conducted using IPCC GWP and ReCiPe indicators as well as the Ecoinvent database. The influence of occupant behaviour was included in the analysis considering different usage profiles during the operational phase, while the effect of climate change was considered by varying the weather file every thirty years. The adoption of the real usage profiles showed that the impact of the systems was highly influenced by the occupant behaviour. In particular, the environmental impact of the heating system appeared more influenced by the operation hours, while that of the cooling system was more affected by the natural ventilation schedules. Furthermore, the influence of climate change demonstrated that more attention has to be dedicated to the cooling demand that in the future years will play an ever-greater role in the energy consumption of buildings.

Keywords: life cycle assessment (LCA); ReCiPe indicator; global warming potential (GWP) indicator; environmental impact; heating and cooling systems; occupant behaviour; climate change

Citation: Fajilla, G.; Borri, E.; De Simone, M.; Cabeza, L.F.; Bragança, L. Effect of Climate Change and Occupant Behaviour on the Environmental Impact of the Heating and Cooling Systems of a Real Apartment. A Parametric Study through Life Cycle Assessment. *Energies* **2021**, *14*, 8356. https://doi.org/10.3390/en14248356

Academic Editors: Shi-Jie Cao, Adrian Ilinca and Chi-Ming Lai

Received: 2 November 2021
Accepted: 9 December 2021
Published: 11 December 2021

Publisher's Note: MDPI stays neutral with regard to jurisdictional claims in published maps and institutional affiliations.

Copyright: © 2021 by the authors. Licensee MDPI, Basel, Switzerland. This article is an open access article distributed under the terms and conditions of the Creative Commons Attribution (CC BY) license (https://creativecommons.org/licenses/by/4.0/).

1. Introduction

According to the Sustainable Development Goals Report 2021 of United Nations [1], global efforts made so far were insufficient and there still is much progress to be made for reaching the objectives of Agenda 2030 [2]. Additionally, the authors of the Sixth Assessment Report of the Intergovernmental Panel on Climate Change pointed out that global warming of 1.5 and 2 °C will be exceeded during the twenty-first century unless significant actions are taken to reduce CO_2 and other greenhouse gases (GHG) emissions in the coming decades [3].

At the European level, new ambitious targets were set with the European Green Deal [4] to reduce GHG emissions and becoming the first climate-neutral continent by 2050. Most of the European buildings present low energy performance and inefficient heating and cooling systems [5,6]. Thus, the building sector still constitutes an important issue to be addressed, being responsible for more than 40% of the worldwide energy consumption and 36% of global GHG [7–10].

1.1. Life Cycle Assessment Applied to Buildings

Life cycle assessment (LCA) is a quantitative method widely known to assess the environmental impacts of products or systems.

The literature on LCA applications to buildings is very extensive, and there is broad agreement in recognizing the importance of the operational stage during the entire cycle of the buildings. Ramesh et al. [11] reviewed 73 buildings across 13 countries and showed that the operational stage is responsible for around 80–90% of the life cycle energy demand of buildings. Similar results were obtained for the Italian context by Asdrubali et al. [12] that assessed to 77% the contribution of the operational stage in the case of detached houses, and up to 85% for office buildings. Sartori and Hestnes [13] analysed the results of 60 LCA of buildings found in the literature highlighting that there is a linear regression between their operational and total energy. Furthermore, a review conducted by Cabeza et al. [14] reported that most of the LCAs are conducted for buildings designed and constructed as low-energy buildings while the percentage of LCA for traditional buildings is lower.

Different authors focused the attention to the environmental impact of the heating and cooling systems because they are the most responsible for emissions during the entire life cycle of a house [15].

Vignali [16] conducted a comparison of the environmental impact of a traditional gas boiler and a condensing gas boiler in three Italian cities obtaining that the impact of the condensing boiler is 23% on average lower than that of the traditional boiler. Additionally, the operational stage of both the systems was responsible, on average, for more than 90% of the total impact. Greening and Azapagic [17] compared the life cycle environmental impacts of domestic heat pumps (air, ground, and water-source) and a gas boiler for the UK context. The impact of the heat pumps was always higher than that of the gas boiler due to the use of electricity. The results showed that the total impact of heat pumps could decrease as the percentage of renewable energy sources in the UK electricity mix increases, while still exceeding the impact of the gas boiler. Llantoy et al. [18] performed an LCA for a lifespan of 30 years to compare the environmental impact of an innovative hybrid energy storage system for heating and domestic hot water production in continental climates with the environmental impact of a traditional system. The results of the analysis, performed through the ReCiPe and the IPCC global warming potential GWP indicators, showed that the total impact of the innovative hybrid energy storage system was lower than the reference one. Zsembinszki et al. [19] compared the environmental performance of an innovative compact hybrid electrical-thermal storage system and a traditional system for cooling, heating, and domestic hot water production in the Mediterranean context. The authors obtained a relevant reduction in energy consumption during the operational stage with the innovative system, despite the environmental impact during its entire life cycle (30-years lifespan) being almost double than that of the traditional system. Shah et al. [15] likened the life cycle impact of three heating and cooling systems (warm-air furnace and air conditioning, hot water boiler and air conditioning, and air–air heat pump) in four cities of the USA considering a lifespan of 35 years. In general, the impact of the heat pump was always higher than that of other systems in a percentage strictly related to the diverse energy mix of the four considered cities. Similar results were obtained by Karkour et al. [20] that performed an LCA to assess the impact of residential air conditioners in Indonesia. The results suggested that the impacts are strongly dependent on the energy mix of the country but could be reduced with diverse solutions, such as introducing refrigerants with low global warming impact and encouraging the recycling of units.

1.2. Problems Related to Climate Change

Among all the mentioned studies, it can be noticed that the attention was mainly focused on the energy demand of buildings by considering the same climatic scenario and the impact of climate change was usually neglected. However, climate change is currently recognized as the most important and critical challenge to mankind globally [21]. Nowadays, just because of climate change, buildings are facing new climatic conditions for which they were not designed [22]. Thus, the impact of climate change on the energy performance of buildings and the environmental impact of the heating and cooling systems cannot be undervalued [23].

For example, Tootkaboni et al. [24] analysed the impact of climate change on the future energy performance of residential buildings in the most populated Italian climate zone. The authors found a decrease of around 30.9% of the heating energy demand and a significant increase in cooling demand up to 255.1%. Similar conclusions were reached by Ciancio et al. [25] that, simulating the energy performance of a building located in three European cities, assessed for 2080 a reduction of the heating energy demand from 36% to 80% and an increase of cooling energy needs from 142% to 2316%. The authors in [26] assessed the influence of climate change on the heating and cooling energy demand of different building prototypes located in Toronto. By 2070, they estimated an average decrease of 18–33% for heating energy use and an average increase of 15–126% for cooling energy use.

Regarding the effect of climate change on the environmental impact of the cooling system in residential buildings of Qatar, Andric and Al-Ghamdi [27] prevised an augmentation of the CO_2 emissions as well as more consumptions of water and fossil fuel, and an increase of the impact on the already strained local marine ecosystem due to the increase of the energy demand for cooling.

1.3. Problems Related to Occupant Behaviour

Another aspect usually overlooked in the studies on LCA is occupant behaviour, whose influence on the energy performance of buildings, and specifically on the heating and cooling systems, is well known [28,29]. Moreover, the AR5 report of the Intergovernmental Panel on Climate Change (IPCC) [30] highlighted that occupant behaviour, such as different thermal control of the indoor environment and natural ventilation usages, determines factors of differences from 3 to 10 worldwide in residential energy use, also in similar dwellings. The literature focused on understanding how people use a space and how their behaviour influences the energy performance of the buildings increased in the last decades [31]. For example, a recent study [32] investigated the impact of occupant behaviour on heating energy consumption and indoor temperature in residential buildings finding that there were outstanding differences in the resulting energy consumption and in the percentage of time in which thermal comfort conditions were met for the different user scenarios.

Fajilla et al. [33] introduced a novelty by analysing the influence of both occupant behaviour and climate change on the energy performance of buildings. The authors showed how the occupant preferences through the control of heating, cooling systems, and natural ventilation impact the energy needs, and how climate change can amplify this impact.

Now, however, a step forward is needed to understand how people influence the environmental impact of the heating and cooling systems.

In fact, despite the importance of the operational stage, it seems that not enough efforts were dedicated to analysing how occupants' behaviour influences the environmental performance of these systems, and the literature in this field is very scarce.

For example, Su et al. [34] performed an LCA to assess the environmental performance of the cooling and heating systems among different houses in China focusing on the families dimension and the age of families' members. The results showed that the larger households, and the families with elderly people or children are more likely to have a higher environmental impact due to the higher cooling and heating demand. The authors highlighted the importance of carrying out further studies to better address the influence of occupant behaviour.

Negishi et al. [35] proposed a framework for a dynamic LCA applied to building systems to discover the influence of the parameters time variation of the building systems such as occupant behaviour, technical performance degradations of the systems, and the variation in the energy mix. Results suggested the need for further investigation for a deeper understanding of the influence of occupant behaviour. Su et al. [36] came to similar conclusions and pointed out that traditional LCA methods have two drawbacks consisting in not considering the time variance of parameters over the entire life cycle of the buildings (e.g., climate), and the behaviours of occupants.

1.4. Aim and Objectives of the Study

As emerged from the literature review, the operational stage is the most influencing phase during an LCA of heating and cooling systems or of an entire building. Despite the importance of this stage, it was generally treated with a standard approach and not all influencing factors were investigated. The common procedure of conducting an LCA analysis was to examine the operational stage of the systems considering usage profiles suggested by national standards, with a consequent underestimation of the impact of occupant behaviour [37]. Moreover, the simulation of energy consumption was usually conducted by not considering climatic scenarios, overlooking the effect of climate change that will cause a decrease in the heating energy demand and an increase in the cooling energy demand [25,26]. To fill this gap in the literature, the aim of this study lies in to jointly analyse the effect of occupant behaviour and climate change on the environmental impact of the heating and cooling systems through an LCA analysis. The influence of occupant behaviour and climate change on the energy performance of buildings was already investigated in the literature, but the influence of both these two factors in an LCA analysis has been never explored. This combined investigation is the main novelty of the paper, especially in the study of the environmental impact of heating and cooling systems.

More in detail, chosen an existing building as a case study with heating and cooling systems installed, an environmental analysis is performed by comparing the results obtained conducting firstly an LCA including the influence of occupant behaviour, and secondly an LCA that considers both occupant behaviour and climate change. To achieve these objectives, the paper aims to answer these research questions:

- RQ1: How does occupants behaviour affect the environmental impact of heating and cooling systems' operation phase?
- RQ2: Can occupants behaviour mitigate or amplify the effect of climate change on the total environmental impact of heating and cooling systems? In addition, vice versa?

The approach used in the investigation wants to highlight the importance of considering new and actual variables in the energy simulation and environmental impact evaluation of building systems. In particular, the environmental impacts of heating and cooling systems are analysed by considering the occupant behaviour to bring the evaluations closer to the reality, and climate change to test the validity of technical solutions in future contexts. The findings of this proposed approach could provide new insights to scientists and practitioners for the improvement of design criteria, and also to policymakers for future regulations.

2. Methods

2.1. Case Study

The heating and cooling systems of an apartment located at Rende (Southern Italy), characterized by Mediterranean climate conditions and defined as "Csa" according to the Köppen climate classification [38], were considered for the study. The heating system consists of an autonomous 24 kW wall-mounted gas boiler that works with natural gas. The cooling system is composed of two heat pumps installed in the living room and the bedrooms with a design capacity of 4 kW and 6 kW, respectively.

The apartment has a gross floor area of 80 m^2 and is located on the second floor of a six-story building built in 2008. The rooms of the apartment are south and west facing, with a floor-to-ceiling height equal to 2.7 m and a window-to-wall ratio of 30%. The structure of the building is reinforced concrete, while the external walls (U-value of 0.6 W/m^2·K) are made of double hollow bricklayers with an internal air gap partially filled with expanded polystyrene. The windows consist of double glazing and a frame with thermal break (U-value 2.72 W/m²·K). The horizontal and vertical overhangs were modelled using standard component blocks considered by the software in shading calculation. Upstairs there is an adiabatic block where there is another heated apartment, while downstairs there is an unconditioned thermal zone. Three thermal zones (living area, bedrooms, and bathrooms) were considered, and the characteristic parameters were changed in terms of management of the heating and cooling systems, as both activation period and setpoint temperature, and ventilation hourly profiles. The internal heat loads (φ_{int}) were determined through the following relation provided by the Standard UNI/TS 11300-1 [39]:

$$\varphi_{int} = 7.987\, A_f - 0.0353\, A_f^2 \qquad (1)$$

where A_f is the usable floor area of the house [m²]. The calculated value amounts to 5.56 W/m² and groups all contributions of occupancy, miscellaneous equipment, catering process, and lighting.

The calibration of the building model was verified in previous work [40] according to the ASHRAE Guideline 14-2002 [41] through the Normalized Mean Bias Error (NMBE) and the Coefficient of Variation of the Root Mean Square Error (CVRMSE). Errors lower than the limit values were obtained for both metrics (NMBE < 5%, CVRMSE < 15%).

2.2. Energy Simulations

Typical hourly profiles of heating (h1, h2, and h3), natural ventilation in winter (v1, v2, and v3), cooling (c1, c2, and c3), and natural ventilation in summer (v4, v5, and v6) were obtained from a questionnaire survey [33]. The profiles are summarised in Table 1 and shown in Tables 2–5.

Table 1. Daily heating, cooling, and natural ventilation schedules summary.

Heating	Natural Ventilation in Winter
h1: Continuous for 24 h	v1: Continuous from 07:00 to 15:00
h2: Limited to the evening (from 19:00 to 22:00)	v2: Limited to the morning hours (from 08:00 to 13:00)
h3: Discontinuous during the day (from 07:00 to 09:00 and from 19:00 to 22:00)	v3: Intermittent but prolonged throughout the day
Cooling	**Natural Ventilation in Summer**
c1: During the hottest hours of the day (from 12:00 to 18:00) in the living zone, and in two-time ranges (from 08:00 to 11:00, and from 14:00 to 17:00) in the bedrooms	v4: Continuous from 07:00 to 19:00
c2: Limited to the afternoon (from 14:00 to 17:00) in the living and bedrooms zones, and the late evening (from 22:00 to 01:00) only in the bedrooms	v5: Concentrated in the coldest hours of the day in the living and bedrooms area, and continuous for 24 h in the bathrooms
c3: Limited to the late afternoon (from 19:00 to 22:00) in the living zone, and from 22:00 to 07:00 in the bedrooms	v6: Prolonged use in the coolest hours in the living and bedrooms area, and continuous for 24 h in the bathrooms

More details about the schedules can be found in a previous work published by the authors [33]. By combining these usage schedules, nine profiles for the heating (h1v1, h1v2, h1v3, h2v1, h2v2, h2v3, h3v1, h3v2, and h3v3) and cooling (c1v4, c1v5, c1v6, c2v4, c2v5, c2v6, c3v4, c3v5, and c3v6) season were adopted to represent different occupants' behaviour. Dynamic energy simulations were conducted for the heating season from 1 October to 30 April with a setpoint temperature of 20 °C, and for the cooling season from 1 May to 30 September with a setpoint temperature of 26 °C (Table 6). Further energy assessments were obtained by varying the heating and cooling setpoint temperatures of ±2 °C. The simulations were conducted through DesignBuilder (v. 6.1.6.008) [42] by using the METEONORM weather data [43] for the current climate (2020) and the data of future climate scenarios (2050 and 2080) obtained with the climate change world weather file generator (CCWorldWeatherGen) [44]. More in detail, CCWorldWeatherGen is a Microsoft Excel-based tool commonly used in this field [9,22,25,26,45] to obtain weather files for future scenarios. It implements the morphing procedure on the current weather file and provides weather files for future scenarios using outputs from the UK Hadley Centre Coupled Model (v.3, HadCM3) [46].

Table 2. Daily heating schedules (On = 1, Off = 0). Adapted from [33].

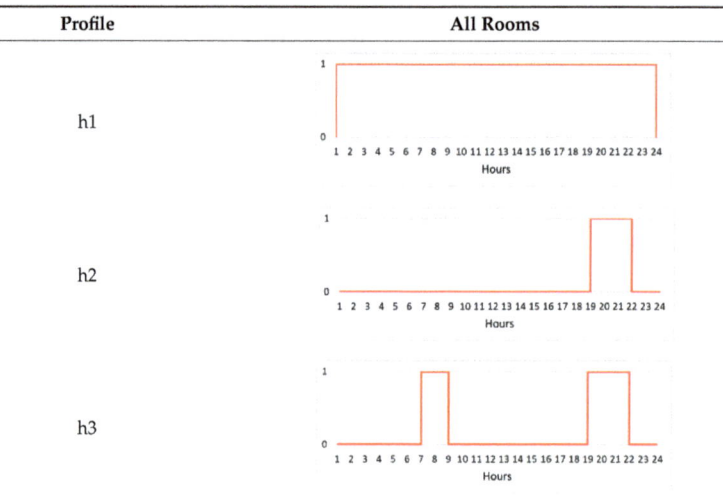

Table 3. Daily natural ventilation schedules in winter (Open = 1, Close = 0). Adapted from [33].

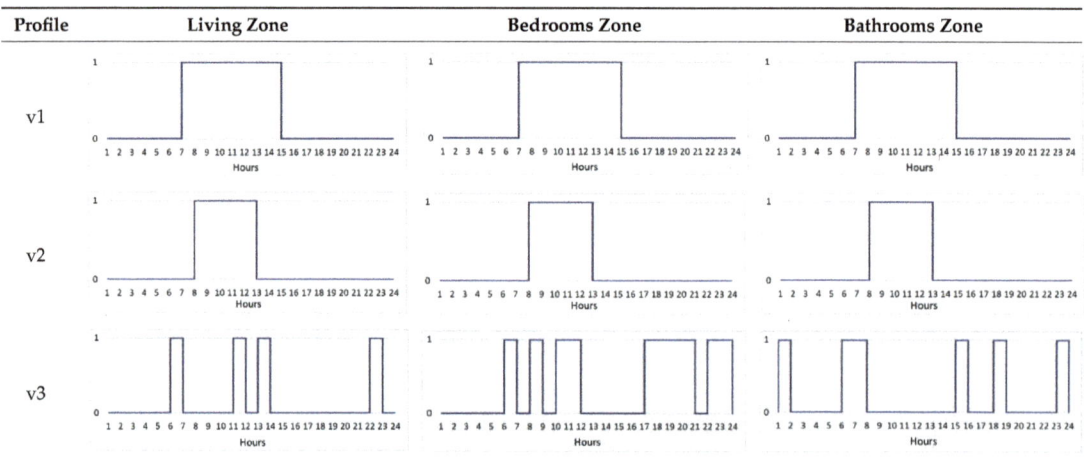

2.3. LCA Methodology

Life cycle assessment (LCA) is a tool widely common for assessing the environmental impact of a product, process, or system through its complete life cycle [47], including the manufacturing, operational, and disposal stage. The analysis was performed following the four main steps suggested by the ISO 14040 and 14044 guidelines [48,49], as shown in Figure 1.

Table 4. Daily cooling schedules (On = 1, Off = 0). Adapted from [33].

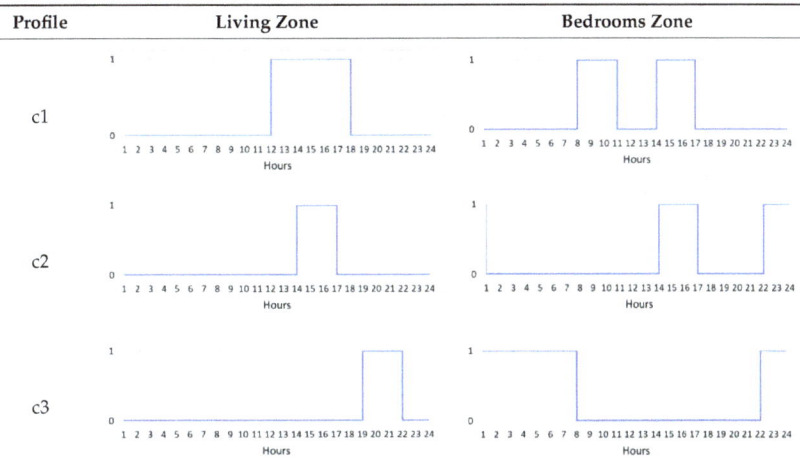

Table 5. Daily natural ventilation schedules in summer (Open = 1, Close = 0). Adapted from [33].

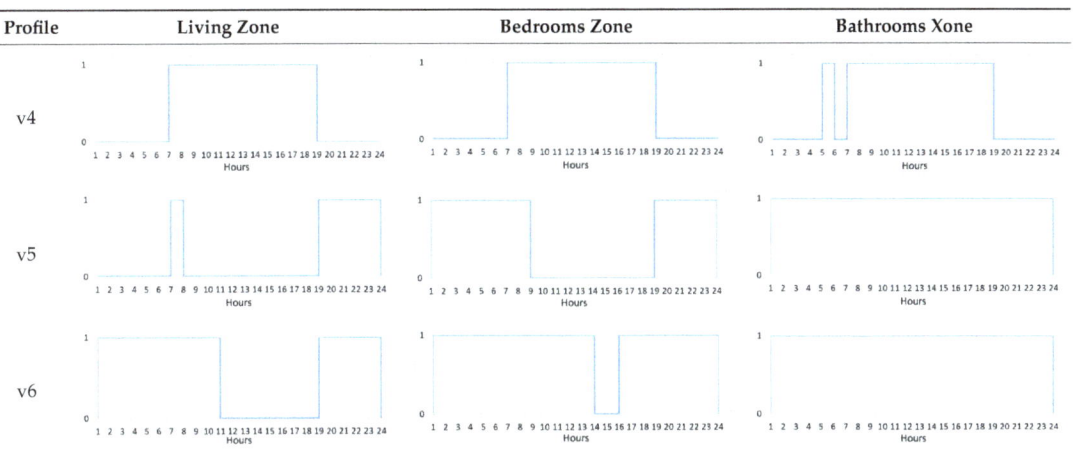

The study was carried out through the indicators IPCC 2013 Global Warming Potential for long-term effect (GWP 100 a) and short-term effect (GWP 20 a), and ReCiPe extracted from the Ecoinvent 3.7.1 database [50] by considering data related to the European context. These two indicators were selected based on the evaluation made by the Joint Research Commission of European Union that indicated the GWP indicator as the only indicator representative for all midpoint models currently used in LCA studies [51], and the ReCiPe indicator as the best method for the European context in comparison with EPS2000, Eco-indicator 99, and IMPACT2002+ [52].

2.3.1. Goal and Scope Definition

The goal of the analysis was to assess if and how the occupants' behaviour related to the use of the heating and cooling systems affects the environmental impact of these systems. Consistent with the aim of this paper, the cooling system was considered composed of a single heat pump with a design capacity of 10 kW. The details of the systems are given below in Table 7.

Table 6. Energy consumption [kWh/year] for heating and cooling with setpoint temperatures of 20 °C (heating) and 26 °C (cooling).

Season	Profiles	Consumption [kWh/year]		
		2020	2050	2080
Heating	h1v1	2299	1756	1220
	h1v2	2099	1600	1103
	h1v3	2287	1751	1220
	h2v1	720	534	350
	h2v2	682	505	328
	h2v3	771	572	380
	h3v1	1196	906	620
	h3v2	1073	809	544
	h3v3	1185	899	615
Cooling	c1v4	233	367	479
	c1v5	205	310	403
	c1v6	198	314	414
	c2v4	202	312	405
	c2v5	184	271	343
	c2v6	178	276	356
	c3v4	196	322	445
	c3v5	192	343	492
	c3v6	179	323	463

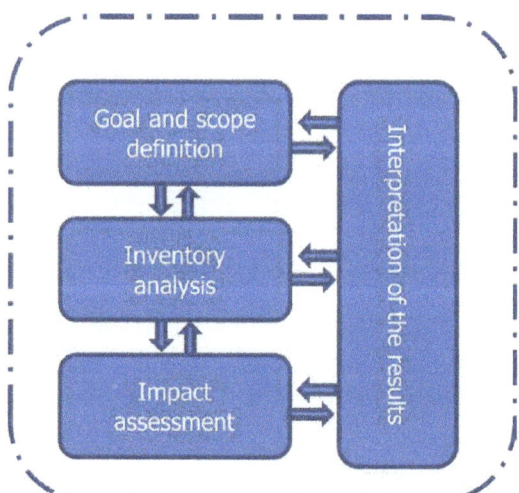

Figure 1. Life cycle assessment framework. Adapted from [48].

Table 7. Heating and cooling systems.

Component	Variable	Value	Unit
Gas boiler	Nominal heat output	24	kW
Air-source heat pump	Capacity	10	kW

Since the reference building is the same in all cases, the study was focused on the systems (operational stage) and not on the building materials; thus, only the gas boiler and the heat pump available in Ecoinvent were included in the LCA study. Failure to consider the individual materials that constitute the systems is the cause of an error. On the other hand, this approximation is reported throughout the analysis and, since it is a comparison, the error is compensated. The lifetime of the entire analysis was assumed to be 90 years, while the lifespan and the number of replacements of each system during the 90 years were calculated according to the lifetime of the various systems. Conventionally, the LCAs of products or buildings are performed with a lifetime ranging from 30 to 50 years [12,18,19,53,54], without considering any external variations (e.g., climate change) to the system boundary that could alter the environmental impact of the systems or the buildings. Following the conventional methodology, the LCA for 90 years (hereinafter referred to as LCA_b) was assessed by considering the same weather data for the entire period, as shown in Figure 2a. LCA_b was considered as a benchmark to compare the results of the LCA proposed in this analysis.

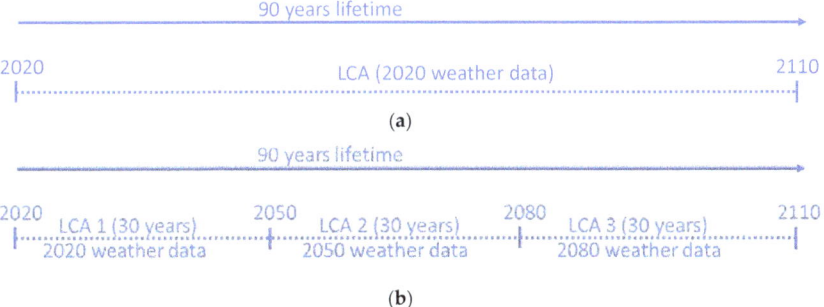

Figure 2. Life cycle assessment: (**a**) Considered as a benchmark (LCA_b); and (**b**) proposed in this study (LCA_{cc}).

This investigation was performed by considering the influence of both occupants' behaviour and climate change by varying the weather data every 30 years. In the end, the LCA for the entire lifetime of 90 years (hereinafter referred to as LCA_{cc}) was calculated as the sum of three contributions: LCA 1, LCA 2, and LCA 3, as shown in Figure 2b.

Based on previous studies carried out in this field [12,18,19,53], the functional unit of 1 m^2 of usable floor area was adopted in this analysis. The scope of the study is from 'cradle to grave'.

2.3.2. Life Cycle Inventory (LCI)

The life cycle inventory (LCI) is the phase of LCA destined to quantify all the inputs and outputs of a product system through its life cycle [48] and consists of three stages, manufacturing, operational, and end-of-life (disposal).

The inventory for the manufacturing stage of the systems is shown in Table 8. For both the gas boiler and heat pump, the disposal stage was included during the manufacturing stage.

Table 8. Inventory of the heating and cooling systems.

Element	Quantity	Unit of Measurement	Lifespan (Years)	Replacement	Total Amount
Gas boiler	1	Unit	20	4.5	4.5
Heat pump	1	Unit	20	4.5	4.5

The input data for the operational stage are the annual energy consumption for the heating and cooling systems shown in Table 6. The energy consumption of the heating and cooling systems was presented separately to consider the different impacts of electricity and natural gas.

2.3.3. Life Cycle Impact Assessment (LCIA)

The life cycle impact assessment (LCIA) is the phase of LCA in which the results of the LCI are associated with specific indicators to assess their potential environmental impacts [48]. Table 9 shows the LCIA during the manufacturing and disposal stage for both gas boiler and heat pump calculated with the GWP [$kgCO_2$-eq/m^2] and ReCiPe indicators [Impact point/m^2]. It can be noticed that the impacts for LCA_b and LCA_{cc} during the manufacturing and disposal stage are the same.

Table 9. Impact assessment during the manufacturing and disposal stage.

Element	Unit	GWP 20a [$kgCO_2$-eq/m^2]	GWP 100a [$kgCO_2$-eq/m^2]	ReCiPe [Impact Point/m^2]
Gas boiler	4.5	29.92	25.89	7.83
Heat pump	4.5	179.81	80.99	16.53

The LCIA during the operational stage of the heating and cooling systems calculated with both LCA_b and LCA_{cc} is presented in Tables 10 and 11.

Table 10. Impact assessment during the operational stage for 90-year lifetime (LCA_b).

Season	Profile	GWP 20a [$kgCO_2$-eq/m^2]	GWP 100a [$kgCO_2$-eq/m^2]	ReCiPe [Impact Point/m^2]
Heating	h1v1	856.55	694.39	67.04
	h1v2	782.13	634.06	61.21
	h1v3	852.15	690.82	66.69
	h2v1	268.33	217.53	21.00
	h2v2	254.12	206.01	19.89
	h2v3	287.11	232.75	22.47
	h3v1	445.65	361.28	34.88
	h3v2	399.82	324.13	31.29
	h3v3	441.40	357.84	34.55
Cooling	c1v4	119.20	109.38	22.33
	c1v5	104.94	96.30	19.66
	c1v6	101.35	93.00	18.98
	c2v4	103.25	94.74	19.34
	c2v5	93.92	86.18	17.59
	c2v6	90.95	83.46	17.03
	c3v4	100.24	91.98	18.77
	c3v5	97.94	89.87	18.34
	c3v6	91.67	84.12	17.17

Table 11. Impact assessment during the operational stage for 90-year lifetime (LCA_{cc}).

Season	Profile	GWP 20a [$kgCO_2$-eq/m^2]	GWP 100a [$kgCO_2$-eq/m^2]	ReCiPe [Impact Point/m^2]
Heating	h1v1	655.08	531.06	51.27
	h1v2	596.45	483.53	46.68
	h1v3	653.05	529.42	51.11
	h2v1	199.24	161.52	15.59
	h2v2	188.20	152.57	14.73
	h2v3	213.95	173.45	16.74
	h3v1	338.14	274.12	26.46
	h3v2	301.31	244.27	23.58
	h3v3	335.16	271.70	26.23

Table 11. *Cont.*

Season	Profile	GWP 20a [kgCO$_2$-eq/m^2]	GWP 100a [kgCO$_2$-eq/m^2]	ReCiPe [Impact Point/m^2]
Cooling	c1v4	183.79	168.65	34.42
	c1v5	156.54	143.64	29.32
	c1v6	157.88	144.88	29.57
	c2v4	156.24	143.37	29.26
	c2v5	135.82	124.64	25.44
	c2v6	138.03	126.66	25.85
	c3v4	164.09	150.57	30.73
	c3v5	174.92	160.52	32.76
	c3v6	164.48	150.93	30.81

3. Results and Discussion

The last step of the LCA, namely the interpretation of the results, is presented in this section.

3.1. Manufacturing and Disposal Stage

Figure 3 shows the impact of the gas boiler and the heat pump during the manufacturing and disposal stage calculated through the indicator IPPC GWP for short-term (GWP 20a) and long-term (GWP 100a) (Figure 3a), and the ReCiPe indicator (Figure 3b).

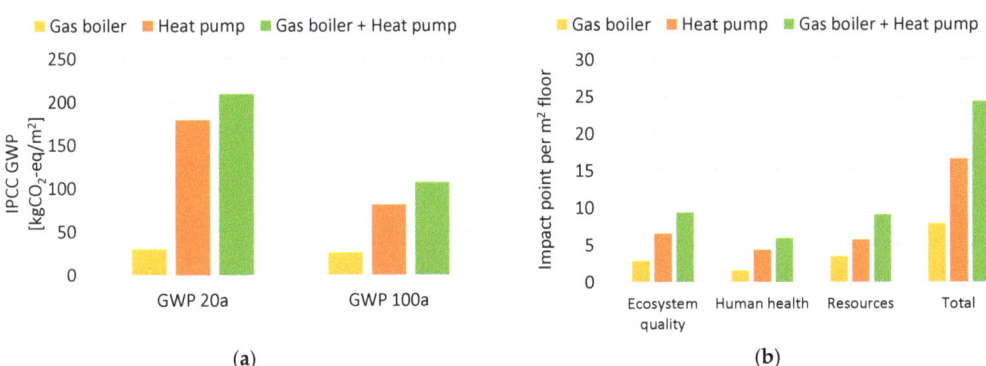

Figure 3. Results per m^2 of floor area during the manufacturing and disposal stage using the indicators: (a) IPCC GWP 20a and GWP 100a; and (b) ReCiPe.

In general, the impact of the heat pump appears to be always higher than that of the gas boiler. It is in accordance with previous studies, such as [17,19]. In fact, the environmental impacts of the heat pump ranged from 180 kgCO$_2$-eq/m^2 (GWP 20a) to 81 kgCO$_2$-eq/m^2 (GWP 100a), far higher than those of the gas boiler that reached 30 kgCO$_2$-eq/m^2 for GWP 20 and 26 kgCO$_2$-eq/m^2 for GWP 100a. Regarding the ReCiPe indicator (Figure 4b), around 16 and 8 impact point/m^2 were obtained for the heat pump and the gas boiler, respectively. More in detail, the impact of the heat pump mainly affects the ecosystem quality (around 7 impact point/m^2), while the gas boiler has a higher impact in the category of resources with an environmental impact of around 3 impact point/m^2. The higher damage of the heat pump to the ecosystem could be caused by the presence of refrigerant in the system. The heat pump found in the Ecoinvent database includes the R134a refrigerant that is one of the most damaging, by varying the typology of the refrigerant it could be possible reducing the impact of the heat pump [18]. Additionally, the damage due to the gas boiler may be reduced by decreasing the quantity of material used for manufacturing. The impacts obtained for the manufacturing and disposal stage are in line with the values of existing researches, e.g., [18,19].

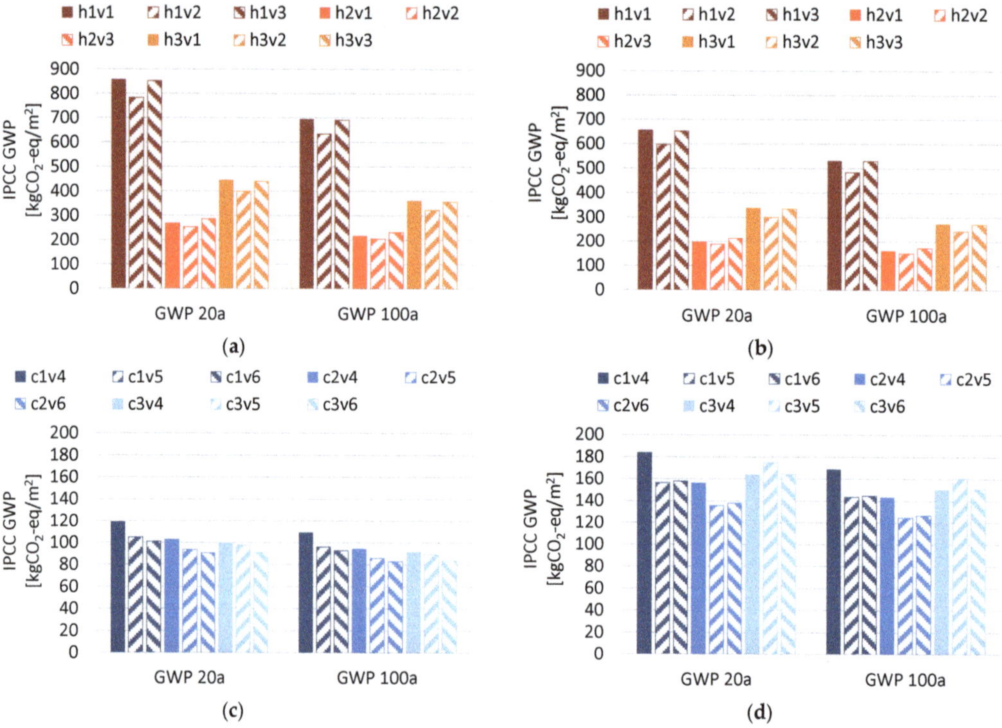

Figure 4. Comparison of the results per m² of floor area during the operational stage using the indicator IPCC GWP 20a and GWP 100a calculated: For heating with (**a**) LCA$_b$; and (**b**) LCA$_{cc}$; and for cooling with (**c**) LCA$_b$; and (**d**) LCA$_{cc}$.

3.2. Operational Stage

The results of the impacts of the heating and cooling systems during the operational stage calculated with the GWP indicator are shown in Figure 4.

LCA$_b$ and LCA$_{cc}$ provided different results: the impacts of the heating were higher for LCA$_b$ (Figure 4a) than those calculated with LCA$_{cc}$ (Figure 4b), and opposite trends were observed for the cooling (Figure 4c,d). These results can be explained by the fact that climate change, which will cause a decrease in heating consumption and an increase in cooling consumption [33,45], was included only in the LCA$_{cc}$ analysis. Moreover, it can be seen how a different occupant behaviour can influence the impact of these systems, at first glance.

Regarding the heating environmental impact, it was possible to observe variation mainly among the heating profiles. Maximum impact variations equal to −70% and −53% were obtained between the profile h1v1 and the profiles h2v2 and h3v2. For the LCA$_b$, the highest impact was obtained for the profile h1v1 (857 kgCO$_2$-eq/m² for GWP 20a and 694 kgCO$_2$-eq/m² for GWP 100a), while the lowest impact was observed for the profile h2v2 (254 kgCO$_2$-eq/m² for GWP 20a and 206 kgCO$_2$-eq/m² for GWP 100a). With LCA$_{cc}$ the maximum and minimum impacts were also found for h1v1 and h2v2, with values 24% and 26% lower than those found with LCA$_b$.

The environmental impact of the cooling system varied among the three cooling usage profiles and from one ventilation profile to another. Additionally, while the impacts calculated with LCA$_b$ decreased moving from c1 to c3 and from v4 to v6, this trend was not found with LCA$_{cc}$. The maximum impact was due to the profile c1v4 with both LCA$_b$ (Figure 4c) and LCA$_{cc}$ (Figure 4d); the minimum impact was obtained for the profile c2v6 with LCA$_b$ and the profile c2v5 with LCA$_{cc}$. The discrepancy between the results of LCA$_{cc}$ and LCA$_b$ was caused by climate change that influenced the energy consumptions of the profiles and engendered a different augmenting of the hours of operation of the cooling system during the considered 90 years. LCA$_b$ provided the maximum and minimum impact values equal to 119 and 91 kgCO$_2$-eq/m^2 for the short-term effect GWP 20a and equal to 109 and 84 kgCO$_2$-eq/m^2 for long-term effect GWP 100a, respectively. With LCA$_{cc}$, values equal to 184 kgCO$_2$-eq/m^2 (GWP 20a) and 169 kgCO$_2$-eq/m^2 (GWP 100a) for the profile c1v4 and equal to 136 kgCO$_2$-eq/m^2 (GWP 20a) and 124 kgCO$_2$-eq/m^2 (GWP 100a) for the profile c2v5 were obtained. Figure 5 shows the results of the impacts during the operational stage of the two considered systems calculated with the indicator ReCiPe. The trends observed among the usage profiles with the GWP indicator were also encountered with the indicator ReCiPe. According to the results of Figure 3b, the damage caused during the operational stage of the heating system (Figure 5a,b) mainly affects the resources category, while the impact of the cooling operation (Figure 5c,d) was particularly damaging for the ecosystem quality.

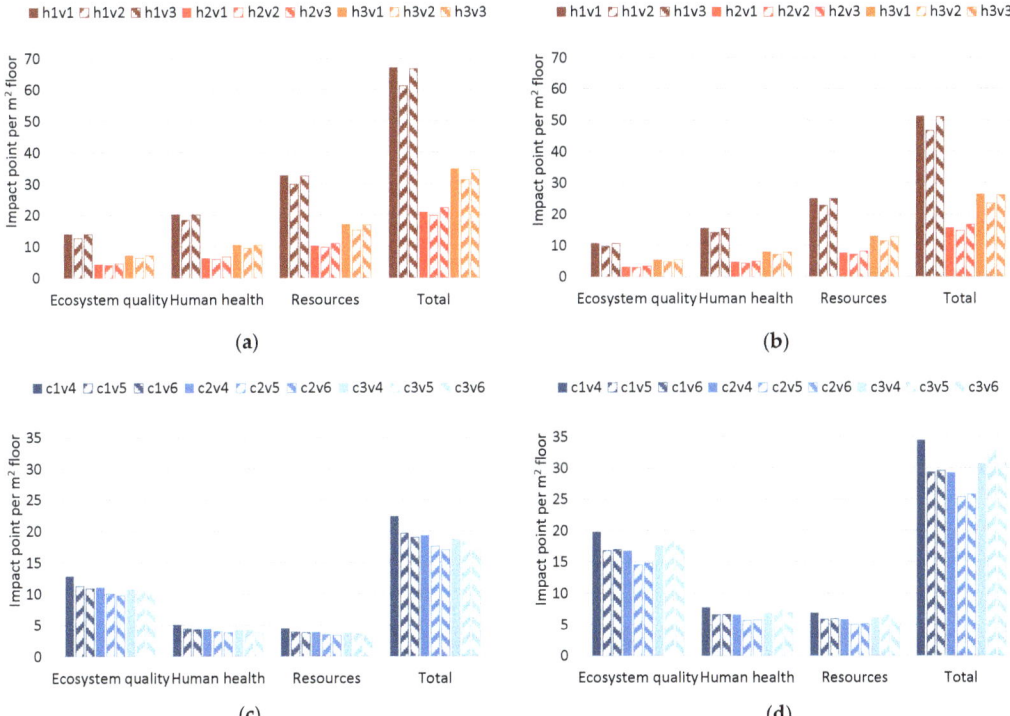

Figure 5. Comparison of the results per m^2 of floor area during the operational stage using the indicator ReCiPe calculated: For heating with (**a**) LCA$_b$; and (**b**) LCA$_{cc}$; and for cooling with (**c**) LCA$_b$; and (**d**) LCA$_{cc}$.

The impact caused by heating in the category of resources reached peaks of 33 and 25 impact point/m^2 with LCA$_b$ and LCA$_{cc}$, respectively. Regarding the cooling operation, maximum impact points per m^2 of about 13 and 20 were obtained from LCA$_b$ and LCA$_{cc}$ in the ecosystem quality category.

The effect of climate change produced a reduction of the total impacts caused by the heating system for LCA$_{cc}$, if compared with those obtained with LCA$_b$. In particular, impact values from 20 to 67 impact point/m^2 (Figure 5a) and from 15 to 51 impact point/m^2 (Figure 5b) were obtained through the indicator ReCiPe.

An opposite effect was produced to the impact of the cooling system that was higher for LCA$_{cc}$. Impact values from 17 to 22 impact point/m^2 and from 25 to 34 impact point/m^2 were obtained with LCA$_b$ and LCA$_{cc}$, respectively.

3.3. Total Impact (Manufacturing, Operational, and Disposal Stage)

The total impacts of the heating and cooling systems, namely the impact caused during the manufacturing, operational, and disposal stage, calculated through the GWP indicator are shown in Figure 6.

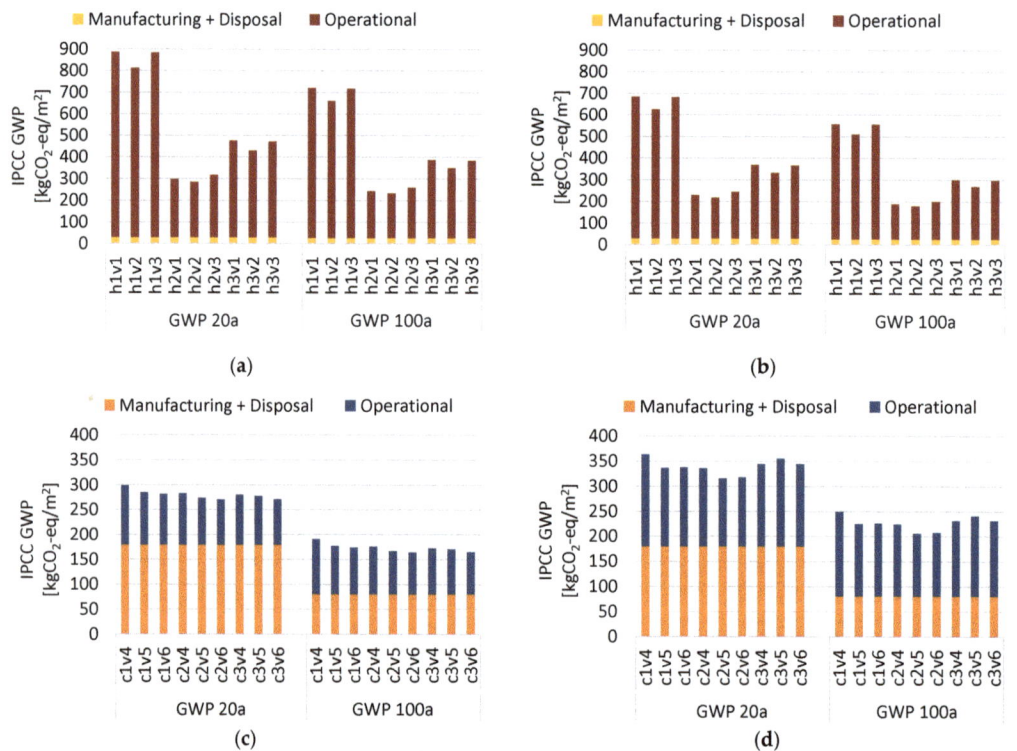

Figure 6. Comparison of the total results per m^2 of floor area (manufacturing and disposal + operational stage) using the indicator IPCC GWP 20a and GWP 100a calculated: For heating with (**a**) LCA$_b$; and (**b**) LCA$_{cc}$; and for cooling with (**c**) LCA$_b$; and (**d**) LCA$_{cc}$.

It is interesting to notice that, while for the heating system (Figure 6a,b) the total impact is almost completely due to the operational stage (more than 90%), the impact related to cooling (Figure 6d) changed between GWP 20a and GWP 100a. In fact, for all the nine cooling profiles the impact for short-term effect (GWP 20a) is mainly due to the manufacturing and disposal stage, while for the long-term effect the environmental damage related to the operational stage is predominant.

Similar results were obtained with the indicator ReCiPe (Figure 7a–d). The total impact due to heating was mainly caused by the operational stage.

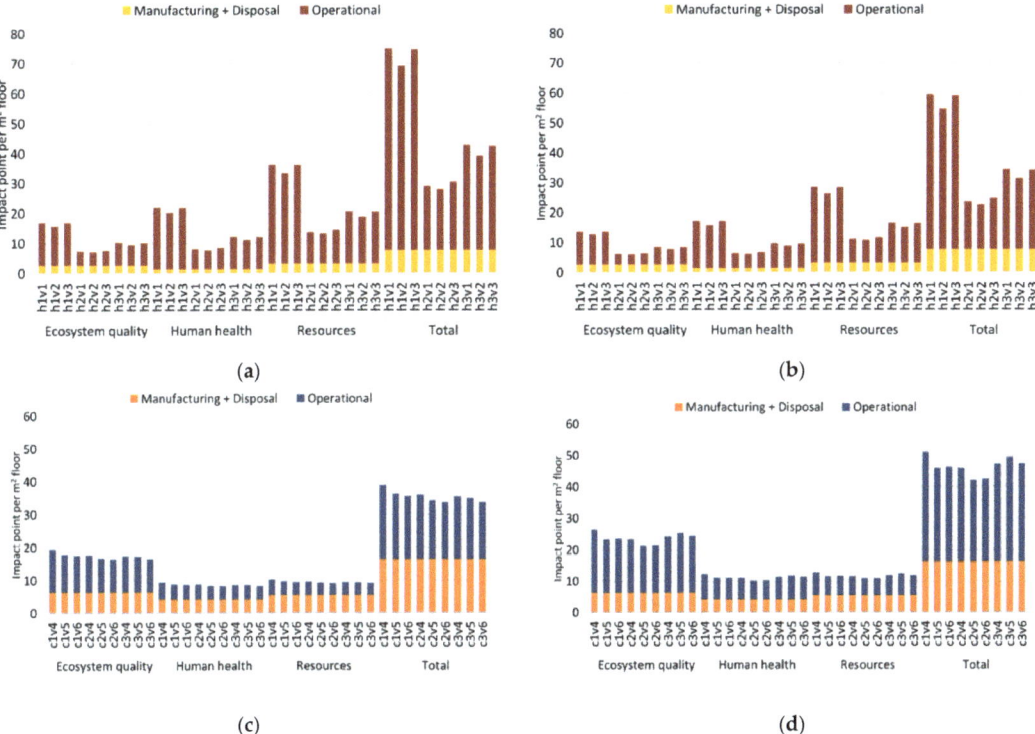

Figure 7. Comparison of the total results per m^2 of floor area (manufacturing and disposal + operational stage) using the indicator ReCiPe calculated: For heating with (**a**) LCA$_b$; and (**b**) LCA$_{cc}$; and for cooling with (**c**) LCA$_b$; and (**d**) LCA$_{cc}$.

Regarding the cooling system, the operational was also the most damaging stage. Only for human health and resources categories, manufacturing appeared to be the most impacting stage. This can be explained by the fact that the impact of the cooling system in these categories was very low if compared with that of the ecosystem quality category.

3.4. Annual Impact of the Two Systems

Figures 8 and 9 show the comparison between the annual results per m^2 of floor area calculated with LCA$_b$ and LCA$_{cc}$ through the indicators GWP and ReCiPe.

In general, LCA$_b$ provided impact values ranging from 1185 to 558 kgCO$_2$-eq/m^2 with GWP 20a (Figure 8a) and from 911 to 419 kgCO$_2$-eq/m^2 with GWP 100a (Figure 8c). The results obtained with LCA$_{cc}$ (Figure 8b,d) were always lower than the previous, in a measure ranging from −3% (with profiles h2v1c2v4 and h3v2c3v5) to −12% (with profiles h1v1c1v4, h1v2c1v5, and h1v3c1v6).

Analysing more in depth the results, for LCA$_b$ the impact due to heating operation was always higher than that caused by cooling. In contrast, the cooling impact appeared to be predominant in three profiles (h2v1c2v4, h2v2c2v5, and h2v3c2v6) with LCA$_{cc}$.

Regarding the annual results obtained with the ReCiPe indicator, it should be noticed that LCA$_b$ provided impacts higher than those of LCA$_{cc}$ only for the first three profiles. Furthermore, LCA$_b$ (Figure 9a) attributed to the heating system the main percentage of the total impact for six out nine profiles (h1v1c1v4, h1v2c1v5, h1v3c1, h3v1c3v4, h3v2c3v5, h3v3c3v6). In the case of LCA$_{cc}$ (Figure 9b), the impact of the heating system is the major contributor of the total impact only for the first three profiles (h1v1c1v4, h1v2c1v5, and h1v3c1v6).

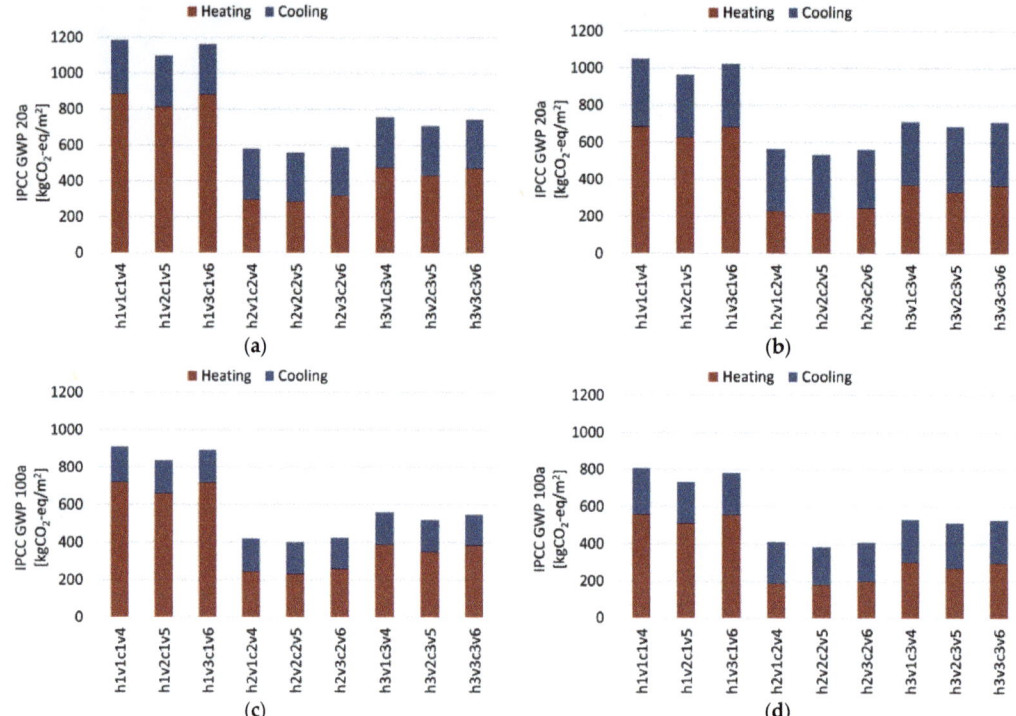

Figure 8. Comparison of the annual results per m^2 of floor area (manufacturing and disposal + operational stage) for heating and cooling calculated: using the indicator IPCC GWP 20a with (**a**) LCA$_b$; and (**b**) LCA$_{cc}$; and using the indicator IPCC GWP 100a with (**c**) LCA$_b$; and with (**d**) LCA$_{cc}$.

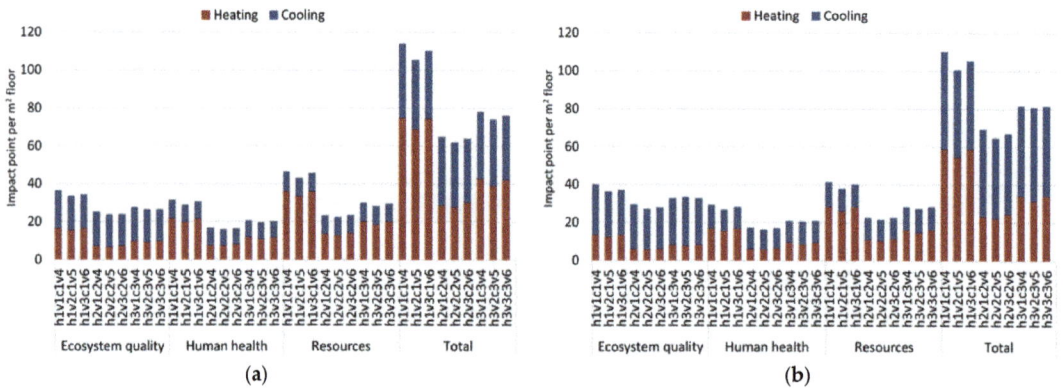

Figure 9. Comparison of the annual results per m^2 of floor area (manufacturing and disposal + operational stage) for heating and cooling using the indicator ReCiPe calculated with: (**a**) LCA$_b$; and (**b**) LCA$_{cc}$.

3.5. Influence of Occupant Behaviour by Varying the Setpoint Temperature

As already mentioned in Section 2.2, energy simulations were also performed by varying the heating and cooling setpoint temperatures of ±2 °C and, hence, considering occupants preferences in thermal comfort. The energy consumption obtained by varying the setpoint temperatures during the three climatic scenarios is shown in Figure 10.

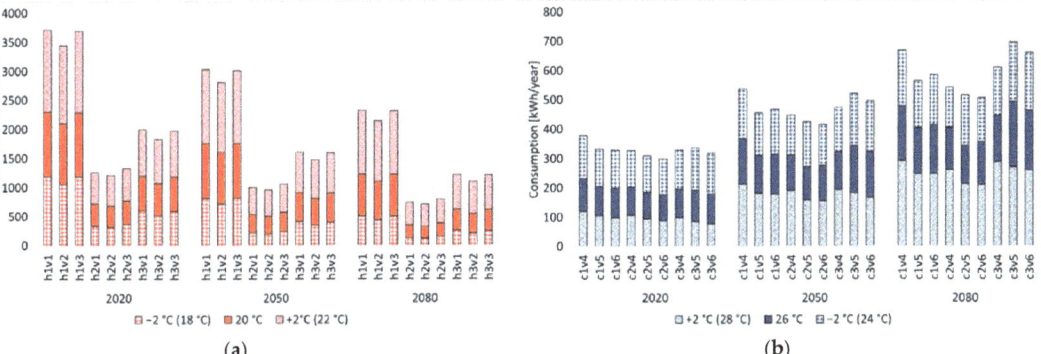

Figure 10. Energy consumption [kWh/year] by varying the setpoint temperatures of ±2 °C for (a) heating; and (b) cooling.

These changes in energy consumption will have a consequence on the environmental impact of the heating and cooling systems. The impact variations, expressed as the percentage difference between the impacts calculated with setpoint temperature at 20 °C (heating) and 26 °C (cooling) and the impacts with a setpoint variation of ±2 °C, are shown in Table 12.

Table 12. Impact variations by varying the setpoint temperatures of ±2 °C.

Season	Profiles	LCA$_b$		LCA$_{cc}$	
		−2 °C	+2 °C	−2 °C	+2 °C
		[Variations (%) of Both GWP and ReCiPe]			
Heating	h1v1	−48	+62	−53	+72
	h1v2	−50	+64	−54	+75
	h1v3	−48	+61	−52	+71
	h2v1	−54	+75	−58	+87
	h2v2	−55	+77	−59	+90
	h2v3	−53	+72	−57	+84
	h3v1	−50	+66	−54	+77
	h3v2	−52	+70	−56	+82
	h3v3	−50	+66	−54	+77
Cooling	c1v4	+62	−49	+47	−42
	c1v5	+62	−48	+47	−42
	c1v6	+66	−50	+49	−43
	c2v4	+62	−47	+44	−40
	c2v5	+68	−49	+57	−42
	c2v6	+67	−51	+51	−44
	c3v4	+68	−50	+47	−40
	c3v5	+75	−56	+51	−48
	c3v6	+77	−57	+53	−48

It can be inferred that LCA$_b$ provided percentage variations of the heating impacts lower than those obtained with LCA$_{cc}$, opposite trends were found for the environmental impact of the cooling system. In general, the reduction of 2 °C of the heating setpoint temperature led to a reduction of the heating impact ranging from −48% to −55% with LCA$_b$ and from −52% to −59% with LCA$_{cc}$. In both cases, the maximum reduction was obtained for the profile h2v2. This profile registered the maximum variation also with the increase of setpoint temperature of +2 °C. In this case, impact variations from +61% to +77% and from +71% to +90% were obtained for LCA_b and LCA_{cc}, respectively.

With cooling, percentage variations of the impact from +62% to +77% and from +44% to +57% were encountered by reducing the setpoint temperature of 2°C with LCA_b and LCA_{cc}. Additionally, an augmentation equal to +2 °C of the cooling setpoint temperature produced an impact drop ranging from −47% to −57% and from −40% to −48%, for LCA_b and LCA_{cc}, respectively.

Unlike heating, the maximum variations were found for different profiles: c3v6 with LCA_b, and for c2v5 (−2 °C) and the profiles c3v5 and c3v6 (+2 °C) with LCA_{cc}.

4. Conclusions

A life cycle assessment was conducted for the heating and cooling systems of a residential building located in southern Italy. The indicators IPCC 2013 Global Warming Potential for short and long-term effect (GWP 20a and GWP 100a) and ReCiPe were adopted to perform the study, for a functional unit of 1 m^2 of usable floor area.

The novelty of the paper consists in the consideration of the effect of two factors usually neglected in such studies. In particular, the analysis was carried out for a lifetime of 90 years by considering the influence of both occupant behaviour and climate change on the operational stage of the systems (LCA_{cc}). The comparison was made with the results of an LCA conducted in a conventional way (LCA_b), namely ignoring the external influences such as climate change. In general, the percentage differences encountered with the GWP during the different stages were also found with the ReCiPe indicator.

Concerning the manufacturing and disposal stage, the results of the LCA_b and LCA_{cc} were the same. For both indicators, the environmental impact of the gas boiler was lower than that of the heat pump. This result is in accordance with that of previous studies (e.g., [17]). In particular, with the ReCiPe indicator, the heat pump appeared to be more degrading for the ecosystem quality, while the gas boiler for the resources category.

Nine usage profiles for the heating season and the same for the cooling were adopted, and significant differences were observed from one profile to another in both LCAs, with similar trends but different magnitude. The impacts caused by the heating operation were highest for LCA_b, while those of cooling resulted greatest for LCA_{cc}. These results can be justified by the fact that the effects of climate change were not considered in LCA_b leading to an overestimation of the heating impacts and an underestimation of those caused by cooling. Regarding heating, the differences were more among the heating profiles (h1, h2, h3) than between one natural ventilation profile to another (v1, v2, v3). Impact values of the order of 800 kgCO$_2$-eq/m^2, 400 kgCO$_2$-eq/m^2, and 200 kgCO$_2$-eq/m^2 were found with LCA_b for h1, h3, and h2, respectively. Maximum and minimum variations were observed between h1v1 and h2v2 (−70%), and between h1v1 and h3v1 (−48%). With LCA_{cc} the maximum and minimum impacts were also found for h1v1 and h2v2, with values 24% and 26% lower than those found with LCA_b, respectively. Unlike the heating system, the environmental impact of the cooling system varied among the three cooling usage profiles and from one ventilation profile to another. Only with LCA_b it was possible to observe a decreasing trend of the impacts moving from c1 to c3 and from v4 to v6. Additionally, while the maximum impact with both LCA_b and LCA_{cc} was encountered for the profile c1v4, the minimum impact was found for the profile c2v6 with LCA_b and the profile c2v5 with LCA_{cc}. In general, the impact caused by the cooling operation with LCA_b was always lower than that calculated with LCA_{cc} in a measure ranging from −33% to −44%.

Regarding the total impact calculated during the entire lifetime, there is a huge difference between the cause of the heating and cooling impacts. In fact, while more than 90% of the heating impact is sourced from the operational stage, that of the cooling is mainly due to the manufacturing and disposal stage for the short-term effect (GWP 20a) and due to operational stage for long-term effect (GWP 100a). With the ReCiPe indicator, the operational stage was predominant for both heating and cooling systems.

The analysis of the annual impact of the system allowed a better understanding of the differences between LCA$_b$ and LCA$_{cc}$ and between the results of GWP and ReCiPe.

According to the GWP indicator, the annual impact provided by LCA$_b$ was always higher than that obtained with LCA$_{cc}$. Impact variations between LCA$_b$ and LCA$_{cc}$ ranging from 3% to 14% were found. Additionally, the environmental impact of the heating was the main contributor of the annual impact for all the nine usage profiles with LCA$_b$, and for six out nine profiles with LCA$_{cc}$. With the ReCiPe indicator, the total impact calculated with LCA$_b$ was higher than LCA$_{cc}$ only with the first three profiles. Additionally, while with LCA$_b$ the impact of the heating was greater than the cooling impact for the first and last three profiles, with LCA$_{cc}$ the heating impact was predominant only for the first three profiles.

This study was the first attempt of including the influence of occupant behaviour and climate change on the assessment of the environmental impact of the heating and cooling systems. Both occupants' behaviour and climate change appeared to highly affect the environmental performance of the systems. With the ongoing concern with climate change, more attention needs to be dedicated to the cooling demand that showed an increasing impact during the LCA analysis. These findings are important and informative for scientists and policymakers for future regulations and design criteria. A limitation of this investigation consists in the fact that one system typology for heating and cooling were considered and the energy simulations were carried out for a building. Moreover, the study was developed on a real case and conducted for its location in terms of climatic conditions. On the other hand, the outcomes of this study can be considered indicative of what could happen in other Mediterranean countries with similar systems. Moreover, the results of this study encourage further studies to analyse more in deep the influence of the occupant behaviour and consider different typologies of heating and cooling systems, the integration of renewable energy sources, and more climatic zones. As a consequence of these results, future studies on the environmental impacts of heating and cooling systems will have to carefully analyse the operational phase, adopting more realistic and contextualised usage profiles and also considering the influence of climate change.

Author Contributions: Conceptualization, G.F., E.B. and L.F.C.; methodology, G.F., E.B. and L.F.C.; software, G.F.; validation, L.F.C.; formal analysis, G.F.; investigation, G.F. and E.B.; resources, L.F.C. and M.D.S.; data curation, L.F.C.; writing—original draft preparation, G.F.; writing—review and editing, E.B., L.F.C., M.D.S. and L.B.; visualization, G.F.; supervision, L.F.C.; project administration, L.F.C. and M.D.S.; funding acquisition, L.F.C. and M.D.S. All authors have read and agreed to the published version of the manuscript.

Funding: This research was funded by the Calabria Region Government with the Gianmarco Fajilla's Ph.D. scholarship (POR Calabria FSE/FESR 2014–2020) grant number H21G18000170006. A part of this publication has received funding from Secretaria Nacional de Ciencia y Tecnologia (SENACYT) under the project code FID18-056. This research was partially funded by the Ministerio de Ciencia, Innovación y Universidades de España (RTI2018-093849-B-C31—MCIU/AEI/FEDER, UE) and by the Ministerio de Ciencia, Innovación y Universidades—Agencia Estatal de Investigación (AEI) (RED2018-102431-T). This work is partially supported by ICREA under the ICREA Academia programme.

Institutional Review Board Statement: Not applicable.

Informed Consent Statement: Informed consent was obtained from all subjects involved in the study.

Data Availability Statement: Data are available upon request to the corresponding author.

Acknowledgments: The authors would like to thank the Catalan Government for the quality accreditation given to their research group (2017 SGR 1537). GREiA is a certified agent TECNIO in the category of technology developers from the Government of Catalonia.

Conflicts of Interest: The authors declare no conflict of interest.

References

1. United Nations. *The Sustainable Development Goals Report 2021*; United Nations Department of Economic and Social Affairs Statistics Division: New York, NY, USA, 2021. [CrossRef]
2. UN General Assembly. *Transforming Our World: The 2030 Agenda for Sustainable Development*; A/RES/70/1; UN General Assembly: New York, NY, USA, 2015.
3. IPCC 2021: Summary for Policymakers. In *Climate Change 2021: The Physical Science Basis. Contribution of Working Group I to the Sixth Assessment Report of the Intergovernmental Panel on Climate Change*; Cambridge University Press: Cambridge, UK, 2021.
4. European Commission. EU Green Deal 2020. Available online: https://ec.europa.eu/info/strategy/priorities-2019-2024/european-green-deal_en (accessed on 29 July 2021).
5. Whitehelm Capital. *Feature Article: The European Heat Sector-Challenges and Opportunities in a Hot Market*; Whitehelm Advisers: London, UK, 2019.
6. Castaño-Rosa, R.; Solís-Guzmán, J.; Marrero, M. Energy poverty goes south? Understanding the costs of energy poverty with the index of vulnerable homes in Spain. *Energy Res. Soc. Sci.* **2020**, *60*, 101325. [CrossRef]
7. Castro, M.d.F.; Colclough, S.; Machado, B.; Andrade, J.; Bragança, L. European legislation and incentives programmes for demand Side management. *Sol. Energy* **2020**, *200*, 114–124. [CrossRef]
8. Araújo, C.; Almeida, M.; Bragança, L.; Barbosa, J.A. Cost-benefit analysis method for building solutions. *Appl. Energy* **2016**, *173*, 124–133. [CrossRef]
9. Rey-Hernández, J.M.; Yousif, C.; Gatt, D.; Velasco-Gómez, E.; San José-Alonso, J.; Rey-Martínez, F.J. Modelling the long-term effect of climate change on a zero energy and carbon dioxide building through energy efficiency and renewables. *Energy Build.* **2018**, *174*, 85–96. [CrossRef]
10. Streimikiene, D.; Balezentis, T.; Alebaite, I. Climate change mitigation in households between market failures and psychological barriers. *Energies* **2020**, *13*, 2797. [CrossRef]
11. Ramesh, T.; Prakash, R.; Shukla, K.K. Life cycle energy analysis of buildings: An overview. *Energy Build.* **2010**, *42*, 1592–1600. [CrossRef]
12. Asdrubali, F.; Baldassarri, C.; Fthenakis, V. Life cycle analysis in the construction sector: Guiding the optimization of conventional Italian buildings. *Energy Build.* **2013**, *64*, 73–89. [CrossRef]
13. Sartori, I.; Hestnes, A.G. Energy use in the life cycle of conventional and low-energy buildings: A review article. *Energy Build.* **2007**, *39*, 249–257. [CrossRef]
14. Cabeza, L.F.; Rincón, L.; Vilariño, V.; Pérez, G.; Castell, A. Life cycle assessment (LCA) and life cycle energy analysis (LCEA) of buildings and the building sector: A review. *Renew. Sustain. Energy Rev.* **2014**, *29*, 394–416. [CrossRef]
15. Shah, V.P.; Debella, D.C.; Ries, R.J. Life cycle assessment of residential heating and cooling systems in four regions in the United States. *Energy Build.* **2008**, *40*, 503–513. [CrossRef]
16. Vignali, G. Environmental assessment of domestic boilers: A comparison of condensing and traditional technology using life cycle assessment methodology. *J. Clean. Prod.* **2017**, *142*, 2493–2508. [CrossRef]
17. Greening, B.; Azapagic, A. Domestic heat pumps: Life cycle environmental impacts and potential implications for the UK. *Energy* **2012**, *39*, 205–217. [CrossRef]
18. Llantoy, N.; Zsembinszki, G.; Palomba, V.; Frazzica, A.; Dallapiccola, M.; Trentin, F.; Cabeza, L.F. Life cycle assessment of an innovative hybrid energy storage system for residential buildings in Continental climates. *Appl. Sci.* **2021**, *11*, 3820. [CrossRef]
19. Zsembinszki, G.; Llantoy, N.; Palomba, V.; Frazzica, A.; Dallapiccola, M.; Trentin, F.; Cabeza, L.F. Life cycle assessment (LCA) of an innovative compact hybrid electrical-thermal storage system for residential buildings in Mediterranean climate. *Sustainability* **2021**, *13*, 5322. [CrossRef]
20. Karkour, S.; Ihara, T.; Kuwayama, T.; Yamaguchi, K.; Itsubo, N. Life cycle assessment of residential air conditioners considering the benefits of their use: A case study in Indonesia. *Energies* **2021**, *14*, 447. [CrossRef]
21. Bazazzadeh, H.; Nadolny, A.; Safaei, S.S.H. Climate change and building energy consumption: A review of the impact of weather parameters influenced by climate change on household heating and cooling demands of buildings. *Eur. J. Sustain. Dev.* **2021**, *10*, 1–12. [CrossRef]
22. Barbosa, R.; Vicente, R.; Santos, R. Climate change and thermal comfort in Southern Europe housing: A case study from Lisbon. *Build. Environ.* **2015**, *92*, 440–451. [CrossRef]
23. Verichev, K.; Zamorano, M.; Carpio, M. Effects of climate change on variations in climatic zones and heating energy consumption of residential buildings in the southern Chile. *Energy Build.* **2020**, *215*, 109874. [CrossRef]
24. Tootkaboni, M.P.; Ballarini, I.; Corrado, V. Analysing the future energy performance of residential buildings in the most populated Italian climatic zone: A study of climate change impacts. *Energy Rep.* **2021**, *7*, 8548–8560. [CrossRef]
25. Ciancio, V.; Falasca, S.; Golasi, I.; de Wilde, P.; Coppi, M.; de Santoli, L.; Salata, F. Resilience of a building to future climate conditions in three European cities. *Energies* **2019**, *12*, 4506. [CrossRef]
26. Berardi, U.; Jafarpur, P. Assessing the impact of climate change on building heating and cooling energy demand in Canada. *Renew. Sustain. Energy Rev.* **2020**, *121*, 109681. [CrossRef]
27. Andric, I.; Al-Ghamdi, S.G. Climate change implications for environmental performance of residential building energy use: The case of Qatar. *Energy Rep.* **2020**, *6*, 587–592. [CrossRef]

28. Yan, D.; Hong, T.; Dong, B.; Mahdavi, A.; D'Oca, S.; Gaetani, I.; Feng, X. IEA EBC Annex 66: Definition and simulation of occupant behavior in buildings. *Energy Build.* **2017**, *156*, 258–270. [CrossRef]
29. O'Brien, W.; Wagner, A.; Schweiker, M.; Mahdavi, A.; Day, J.; Kjærgaard, M.B.; Carlucci, S.; Dong, B.; Tahmasebi, F.; Yan, D.; et al. Introducing IEA EBC Annex 79: Key challenges and opportunities in the field of occupant-centric building design and operation. *Build. Environ.* **2020**, *178*, 106738. [CrossRef]
30. Lucon, O.; Ürge-Vorsatz, D.; Zain Ahmed, A.; Akbari, H.; Bertoldi, P.; Cabeza, L.F.; Eyre, N.; Gadgil, A.; Harvey, L.D.D.; Jiang, Y.; et al. Chapter 9, Buildings. In *Climate Change 2014: Mitigation of Climate Change. Contribution of Working Group III to the Fifth Assessment Report of the Intergovernmental Panel on Climate Change*; Cambridge University Press: Cambridge, UK, 2014; pp. 671–738.
31. Carlucci, S.; De Simone, M.; Firth, S.K.; Kjærgaard, M.B.; Markovic, R.; Rahaman, M.S.; Annaqeeb, M.K.; Biandrate, S.; Das, A.; Dziedzic, J.W.; et al. Modeling occupant behavior in buildings. *Build. Environ.* **2020**, *174*, 106768. [CrossRef]
32. Laskari, M.; de Masi, R.-F.; Karatasou, S.; Santamouris, M.; Assimakopoulos, M. On the impact of user behaviour on heating energy consumption and indoor temperature in residential buildings. *Energy Build.* **2021**, 111657. [CrossRef]
33. Fajilla, G.; De Simone, M.; Cabeza, L.F.; Bragança, L. Assessment of the impact of occupants' behavior and climate change on heating and cooling energy needs of buildings. *Energies* **2020**, *13*, 6468. [CrossRef]
34. Su, S.; Li, X.; Lin, B.; Li, H.; Yuan, J. A comparison of the environmental performance of cooling and heating among different household types in China's hot summer-cold winter zone. *Sustainability* **2019**, *11*, 5724. [CrossRef]
35. Negishi, K.; Tiruta-Barna, L.; Schiopu, N.; Lebert, A.; Chevalier, J. An operational methodology for applying dynamic life cycle assessment to buildings. *Build. Environ.* **2018**, *144*, 611–621. [CrossRef]
36. Su, S.; Li, X.; Zhu, Y.; Lin, B. Dynamic LCA framework for environmental impact assessment of buildings. *Energy Build.* **2017**, *149*, 310–320. [CrossRef]
37. O'Brien, W.; Tahmasebi, F.; Andersen, R.K.; Azar, E.; Barthelmes, V.; Belafi, Z.D.; Berger, C.; Chen, D.; De Simone, M.; D'Oca, S.; et al. An international review of occupant-related aspects of building energy codes and standards. *Build. Environ.* **2020**, *179*, 106906. [CrossRef]
38. Köppen, W.P.; Geiger, R. *Handbuch der Klimatologie*; Gebrüder Borntraeger: Berlin, Germany, 1930. (In German)
39. Italian Standardization Body. *UNI/TS 11300-1 Energy Performance of Buildings, Part 1: Evaluation of Energy Need for Space Heating and Cooling*; Italian Standardization Body: Milan, Italy; Rome, Italy, 2014.
40. Mora, D.; Carpino, C.; De Simone, M. Energy consumption of residential buildings and occupancy profiles. A case study in Mediterranean climatic conditions. *Energy Effic.* **2018**, *11*, 121–145. [CrossRef]
41. ASHRAE ANSI/ASHRAE. *Guideline 14-2002 Measurement of Energy and Demand Savings*; ASHRAE: Peachtree Corners, GA, USA, 2002.
42. DesignBuilder. DesignBuilder Version 6.1.6.008. DesignBuilder Software Ltd. 2020. Available online: http://www.designbuilder.co.uk/ (accessed on 1 February 2020).
43. Meteonorm. *Meteonorm Global Meteorological Database Version 7.1.8.*; METEOTEST: Bern, Switzerland, 2012.
44. Jentsch, M.F.; Bahaj, A.B.S.; James, P.A.B. Climate change future proofing of buildings-Generation and assessment of building simulation weather files. *Energy Build.* **2008**, *40*, 2148–2168. [CrossRef]
45. Ciancio, V.; Salata, F.; Falasca, S.; Curci, G.; Golasi, I.; de Wilde, P. Energy demands of buildings in the framework of climate change: An investigation across Europe. *Sustain. Cities Soc.* **2020**, *60*, 102213. [CrossRef]
46. Met Office. Met office HadCM3: Met Office Climate Prediction Model. Available online: https://www.metoffice.gov.uk/research/approach/modelling-systems/unified-model/climate-models/hadcm3 (accessed on 15 March 2020).
47. Guinée, J.B.; Gorrée, M.; Heijungs, R.; Huppes, G.; Kleijn, R.; de Koning, A.; van Oers, L.; Wegener Sleeswijk, A.; Suh, S.; Udo De Haes, H.a. *Life Cycle Assessment: An Operational Guide to the ISO Standards*; Kluwer Academic Publishers: Dordrecht, Germany, 2001.
48. ISO 14040:2006(en). *Environmental Management-Life Cycle Assessment-Principles and Framework*; ISO: Geneva, Switzerland, 2006.
49. ISO 14044:2006(en). *Environmental Management-Life Cycle Assessment-Requirements and Guidelines*; ISO: Geneva, Switzerland, 2006.
50. Frischknecht, R.; Jungbluth, N.; Althaus, H.; Doka, G.; Dones, R.; Heck, T.; Hellweg, S.; Hischier, R.; Nemecek, T.; Rebitzer, G.; et al. The ecoinvent database: Overview and methodological framework. *Int. J. Life Cycle Assess.* **2005**, *10*, 3–9. [CrossRef]
51. Joint Research Centre European Commission. *International Reference Life Cycle Data System (ILCD) Handbook: Analysing of existing Environmental Impact Assessment methodologies for Use in Life Cycle Assessment*; Publications Office of the European Union: Luxembourg, 2010.
52. Joint Research Centre European Commission. *Recommendations for Life Cycle Impact Assessment in the European Context-Based on Existing Environmental Impact Assessment Models and Factors (International Reference Life Cycle Data System—ILCD Handbook)*; Publications Office of the European Union: Luxembourg, 2011; ISBN 9789279174513.
53. Llantoy, N.; Chàfer, M.; Cabeza, L.F. A comparative life cycle assessment (LCA) of different insulation materials for buildings in the continental Mediterranean climate. *Energy Build.* **2020**, *225*, 110323. [CrossRef]
54. Ben-Alon, L.; Loftness, V.; Harries, K.A.; Cochran Hameen, E. Life cycle assessment (LCA) of natural vs conventional building assemblies. *Renew. Sustain. Energy Rev.* **2021**, *144*, 110951. [CrossRef]

Article

A Case Study on a Stochastic-Based Optimisation Approach towards the Integration of Photovoltaic Panels in Multi-Residential Social Housing

Rui Oliveira [1], Ricardo M.S.F. Almeida [2,3,*], António Figueiredo [1] and Romeu Vicente [1]

[1] Research Center for Risks and Sustainability in Construction—RISCO, Department of Civil Engineering, Campus Universitário de Santiago, University of Aveiro, 3810-193 Aveiro, Portugal; ruioliveira@ua.pt (R.O.); ajfigueiredo@ua.pt (A.F.); romvic@ua.pt (R.V.)
[2] Department of Civil Engineering, Polytechnic Institute of Viseu, Campus Politécnico de Repeses, 3504-510 Viseu, Portugal
[3] CONSTRUCT-LFC, Faculty of Engineering—FEUP, University of Porto, Rua Dr. Roberto Frias s/n, 4200-465 Porto, Portugal
* Correspondence: ralmeida@estv.ipv.pt

Abstract: The socioeconomic reality and the energy retrofit potential of the social housing neighbourhoods in Portugal are stimulating challenges to be addressed by research to pursue suitable energy efficient strategies to be integrated into these buildings. Therefore, this study explored a stochastic-based optimisation approach towards the integration of photovoltaic (PV) panels, considering different scenarios that combine the occupancy rate, the internal gains, the envelope refurbishment and the heating system efficiency. The optimisation approach has as its objective the minimisation of the life cycle cost of the photovoltaic system while using a limited space area on the rooftop for its installation. This study allowed concluding that the use of passive measures such as improving the thermal performance of the building envelope is essential to attain a lower optimal-sizing of a photovoltaic installation. The results reveal a decreasing trend in the PV optimal sizing, attaining a reduction up to 30% of the total number of PV panels installed on the sloped rooftop in several scenarios with 50% of occupancy rate. However, the impact can be greater when passive measures are coupled to more efficient heating systems, with higher COP, which result in a decrease up to 64% of the number of PV panels. Thus, the approach proposed is of paramount importance to aid in the decision-making process of design and sizing of photovoltaic installation, highlighting the practical application potential for social housing and a contribution for mitigation of the energy poverty of low-income families that live in these buildings.

Keywords: social housing; energy demand; energy refurbishment; internal gains; occupation rate; life cycle cost

1. Introduction

The European building stock from the 1970s to the 1990s represent the largest share of the existing buildings and are considered the least energy-efficient, thus presenting enormous potential for intervention. The energy consumption of the built environment over the next 50 years will be mainly dominated by the existing older building stock and their rate of refurbishment and renewal over time. Some projections indicate that without a significant change of practice, the non-retrofitted building stock is estimated by 2050 to represent around 80% of the total energy consumption in the building sector [1]. The massive construction of new buildings and infrastructures has been progressively slowed down, giving priority to the rehabilitation of existing buildings and built heritage. According to this tendency, new challenges and funding programs have emerged across Europe [2–6], focusing on the integration of renewable energy sources into the existing energy systems and new building technologies and more recently the New European Bauhaus Initiative [6].

Building refurbishment has been pointed as a big economic and social challenge that Europe will face over the following decade. In fact, the enormous potential to decrease energy consumption and the advances in renewable energy technologies were two of the main drivers for the implementation of nearly zero energy buildings (nZEB) over the last decade and have led member states to outline new measures and guidelines at the national scale to encourage the refurbishment of existing building stock [7]. The European Commission (EC) published the Directive 2018/844/EU [7] amendment, which includes guidelines and defines strategies that will help accelerate the rate of building renovation towards more energy-efficient systems. This Directive defines that an annual average of 3% renovation is needed to achieve the Union's energy efficiency ambitions cost-effectively, considering that every 1% increase in energy savings reduces gas import by 2.6%. These numbers confirm that the objectives of renovating the existing building are extremely important in the decarbonisation and resource efficiency of cities. Recently, the EC published the Regulation (EU) 2021/1119 [8] of the European Parliament and of the Council of 30 June 2021 establishing the framework for achieving climate neutrality and amending Regulations (EC) n° 401/2009 [9] and (EU) 2018/1999 [10]. In this scope, the EC adopted a comprehensive and interconnected package of proposals for the acceleration of greenhouse gas emission (GGE) reductions in the next decade. Two of those proposals are building-related: increased use of renewable energy and greater energy efficiency [11].

In Portugal, as in other European countries, social housing plays an important role in ensuring adequate living conditions for the population. About 2.5% of the urban population live in social housing complexes, which represent approximately 118,000 dwellings. A large part of these buildings is in need of refurbishment measures [12]. The lack of adequate interior thermal comfort conditions of flats is often reported and the main cause is linked to energy poverty of the families [13]. In a recent publication, Gouveia and Palma [14] gathered information regarding possible approaches to address this issue, namely: development of building management system for social housing [15]; application of energy social tariff for electricity and natural gas [16]; both imply the improvement of measuring and monitoring as well sharing knowledge and best practice on energy poverty mitigation [17].

Some examples of refurbishment interventions in social housing are described in the literature, mostly integrating measures pointing towards a better indoor environmental quality and reduction of the energy consumption, such as increased thermal insulation, new windows and doors with enhanced thermal characteristics, and new energy-efficient HVAC systems [18–22]. The cost-effectiveness of these interventions is important to ensure housing affordability, minimising the risk of tenants being unable to cover energy costs [23]. The lack of financial resources makes the refurbishment interventions in social housing a challenge, either if focused on the improvement of the external envelope thermal properties [24], or in the implementation and operation of more efficient renewable energy-based systems [25,26].

Recent research studies have focused on the relationship between the buildings' occupancy patterns, the energy consumption and indoor thermal comfort, thus concluding that social housing buildings energy consumption is impacted by a significant variation of building occupancy profiles [18,27]. Another key factor with high influence on the buildings' energy consumption and thermal discomfort is the negative impact of the presence of several unoccupied flats in the multi-residential social buildings [28]. Furthermore, the study by Haldi and Robinson [29] point out a non-negligible impact on internal gains as a consequence of the occupant behaviour in the use of lighting and appliances.

Another group of recent research studies have been carried out regarding the incorporation of renewable energy production systems in the context of social housing [30–32], despite all the barriers such as financial, structural, social, and organisational constraints [25]. Lee and Shepley [30] investigated the impact of the installation of photovoltaic panels within the framework of a governmental initiative scheme to support low-income families in Korea. A significant reduction in electricity costs was reported, positively impacting the mitigation of energy poverty. Another study analysed the potential of a photovoltaic installation in a small social housing neighbourhood in Portugal. The results of this latter study pointed to a reduction of energy costs of around 15% and a great potential for energy storage as more than 79% of surplus generated energy [33]. Almeida et al. [34] have been studying several renovation scenarios to be implemented in social housing buildings, based on the effect of embodied energy on cost-effectiveness. The results revealed that a significant reduction of non-renewable primary energy consumption can be achieved using photovoltaic systems, despite the still unattractive installation cost. In fact, several authors [35,36] have reported the impact of the inclusion of PV panels in both new and renovated social housing. Nevertheless, a lack in the literature was identified as no consideration of the variability of the occupation rate was taken into account.

Based on the literature and focusing on the retrofit potential of social housing, a lack of studies regarding the impact of different occupancy rate and consequent uncertainty in quantification of internal gains has been scarce. Combining these issues with the potential for the use of renewable energy sources in social housing context and its practical application potential, constitutes a challenge and is the major aim of this study. Therefore, a novel approach is proposed based on current simulation tools to aid in the decision-making process of design and sizing of photovoltaic installation in the social housing.

The uncertainty in the occupancy of social housing buildings is a well-known issue, which can lead to an oversizing design of the systems with important impact in the operation and maintenance costs, sometimes even jeopardising their use throughout the life cycle. In addition, the proposed methodology can be useful in an early stage of the decision-making process, providing the designer with relevant information. Methodologies that use stochastic design are common in the literature, however, the inclusion of uncertainty in the occupancy-related input parameters are new.

2. Methods: Case Study Definition and Manuscript Organisation

2.1. Building Characterisation

The case study is a multi-residential building located in Aveiro, in the North-Centre region of Portugal. Based on previously conducted studies [28,37], the main characteristics of the building are herein highlighted. The building belongs to a large social housing neighbourhood (see Figure 1) built in the final of 80's, comprising 38 buildings, representing the main public housing complex of the city and more than five thousand habitants. Figure 1 shows the case study building.

Figure 1. General view the social housing neighbourhood with the case study building highlighted in red.

The building is composed by four floors above the ground level. It includes six flat units per floor with an average treated area of 72.4 m² by flat. The total treated floor area of the building is 1737.2 m². Two types of flats configurations exist: T2, composed of two bedrooms, living room, kitchen, and bathroom; and T3, composed of three bedrooms, living room, kitchen, and two bathrooms. The building geometry is depicted in Figure 2.

Figure 2. Geometry of the case study: (**a**) west facade; (**b**) typical floor.

The constructive solutions of the building were designed considering low requirements in respect to energy efficiency, a consequence of the regulations at the time of construction. Two different configurations of the external walls were identified, neither including insulation. The external walls of the ground floor are constituted by double leaf brick masonry while the walls in the elevated floors are concrete prefabricated panels. The roof is pitched and composed of a highly ventilated crawl space above the prefabricated lightweight concrete slab. Windows are single glazed with an aluminium frame with exterior roller-shutters. The ground slab consists of a concrete floor. A detailed description of the building solutions can be found in the previous works [28,37].

Flats are naturally ventilated with the outdoor air admission due to window openings and infiltration through the roller shutter boxes and open balconies. Air extraction grids are installed in the bathrooms. However, the users tend to close the ventilation grids and insufficient ventilation is commonly frequent as reported and discussed by Oliveira et al. [37]. No heating or cooling systems were installed, with rare exceptions in which electric portable devices were seen in some flats and the domestic hot water was supplied by a condensing boiler.

2.2. Climate Regions

To increase the relevance of the findings, the numerical simulations considered two different locations representative of the less and the most severe climates of Portugal. Therefore, besides Aveiro (mild climate conditions), Bragança (severe winter) climate was also simulated [38]. The hourly weather files used in the simulations were extracted from the Portuguese climate database provided by DGEG [39]. The average air temperature, relative humidity, and solar irradiation of the two regions is depicted in Figure 3. Aveiro is characterised by having a heating degree-day (HDD) of 1297 °C while in Bragança is 2015 °C (HDD values are defined with the base temperature of 20 °C). The relative humidity of both regions is balanced in the winter but lower in Bragança during the summer. The solar irradiation values are similar in both regions. Figure 3 shows the main climate details for both regions under the study.

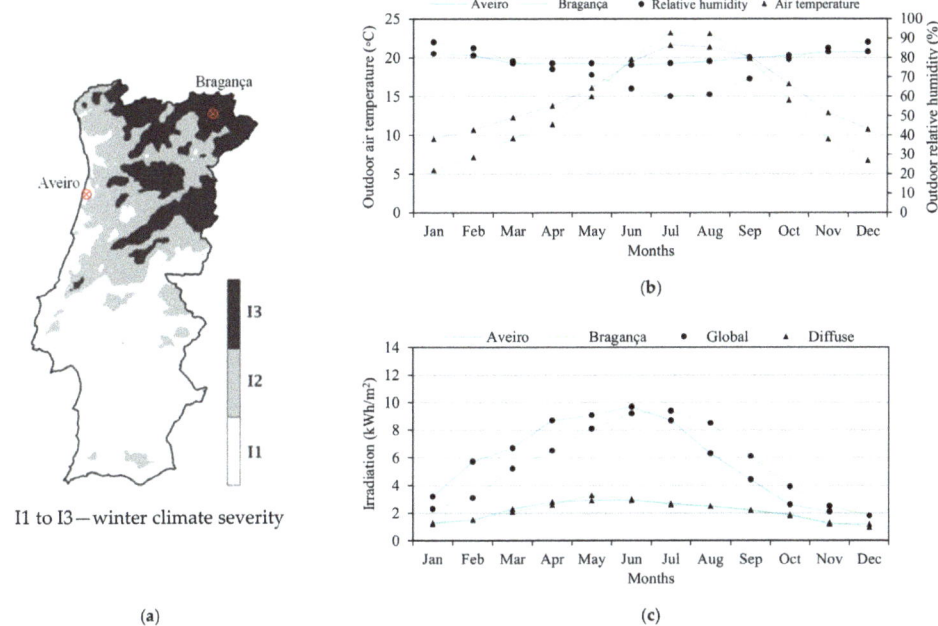

Figure 3. Climate information for Bragança and Aveiro: (**a**) winter map classification; (**b**) outdoor monthly average air temperature and relative humidity [40]; and (**c**) monthly average daily solar horizontal irradiation [40].

2.3. Numerical Design and Sizing Approach

This research work proposes a methodology for the optimisation of the number of photovoltaic panels to be integrated into a multi-residential social housing building taking into account different scenarios. A stochastic approach is implemented to consider the uncertainty of occupancy rate and internal gains in the optimisation procedure and the impact of previous refurbishment actions on the building envelope (opaque and glazed) are also taken into account. The methodology is tested in two different climate regions of Portugal as previously described. The minimisation of the life cycle cost is the main objective. Figure 4 depicts a graphical description of the methodology involving the use of three different software tools (EnergyPlus® (EP) (version 8.9, United States Department of Energy (DOE), Washington, DC, USA), JEPlus® (version 2.0, Energy and Sustainable Development (IESD), United Kingdom, Leicester) and HOMER® (version 3.14.4, National Renewable Energy Laboratory, Boulder, CO, USA)), which can be summarised in the following steps:

- Detailed model definition using the Sketchup software for geometry definition and the Euclides plugin to export for EnergyPlus;
- Definition of the simulation premises to implement in EnergyPlus, such as:
  (i) climatic regions (Aveiro and Bragança);
  (ii) energy refurbishment of the building envelope based on the introduction of thermal insulation and new windows (original envelope and improved envelope);
  (iii) heating system coefficient of performance (COP = 1.0 and COP = 3.4);
  (iv) occupancy rate of the building (100%, 75% and 50% of occupied flats).
- Definition of three levels of internal gains (2.0, 3.0 and 4.0 W/m^2) and random sampling using the Latin Hypercube Sampling (LHS) algorithm of JEPlus. Energy simula-

tion of the different scenarios, using JEPlus as a parametric tool within the EnergyPlus environment. The main output of the simulations to feed into the next step is total energy consumption of the building for the different scenarios.
- Optimisation of the number of PV panels using the HOMER Pro® software by minimising their life cycle cost with a maximum limit 86 photovoltaic panels (rooftop area and shading restrictions).

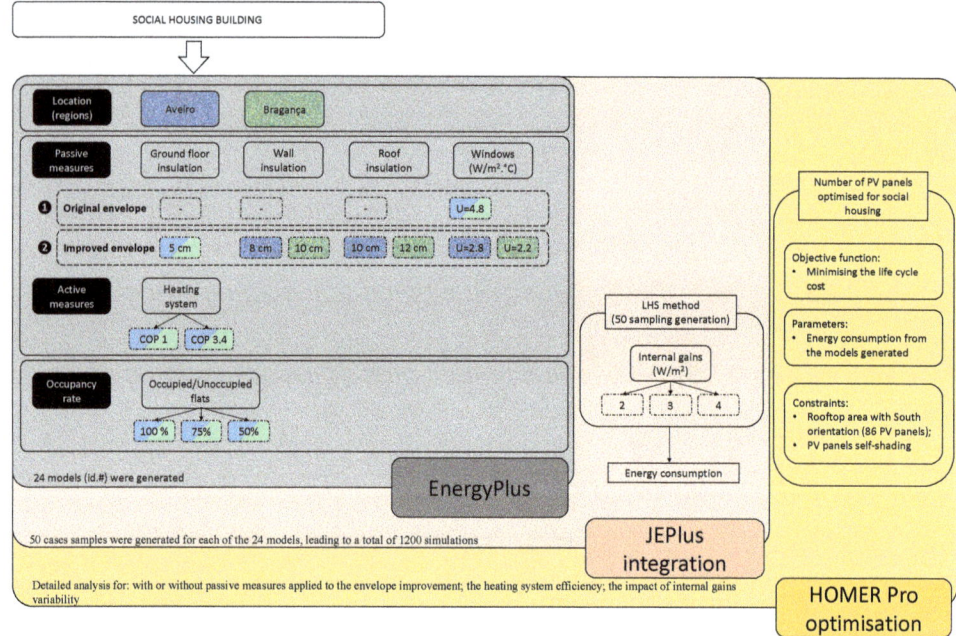

Figure 4. Schematic diagram of the methodology.

2.3.1. Building Energy Model Simulations and Scenarios

The geometry of the building was defined as the 3D view depicted in Figure 5. For modelling purposes, the building is divided into several thermal zones, which correspond to the main compartments of the 24 flats. The building geometry was created in Sketchup® (version 2019, @Last Software and Google Trimble Inc., Boulder, CO, USA) software and automatically converted into EnergyPlus input files for defining further modelling settings. Figure 5 presents the geometry of the building model of the study.

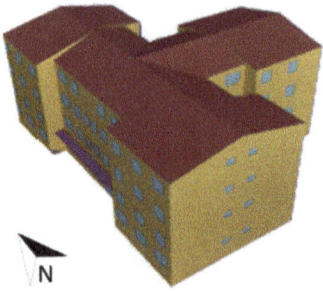

Figure 5. 3D-Geometry of the building model.

The model was previous calibrated by Oliveira et al. [28], resorting to monitoring data collected in-situ, including interior environment data, airtightness evaluation and external envelope characterisation. The model calibration was performed using data from seven flats (14 thermal zones), and the accuracy was evaluated by comparing the goodness of fit (GOF) index (see Figure 6) with the ASHRAE Guideline limits [41].

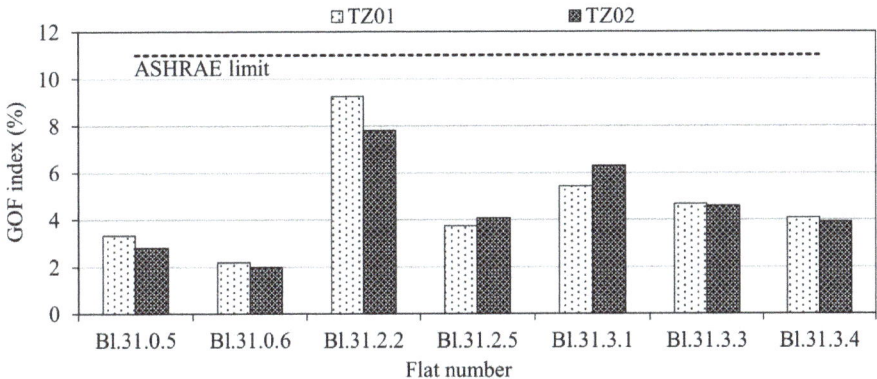

Figure 6. Calibration results of seven flats of the building. Adapted from [28].

To accurately simulate the energy refurbishment of the building envelope, the simulation plan includes two base models, differing in terms of the constructive solutions: (1) original envelope, and (2) improved envelope. In addition, the base models were combined with a heating system varying in terms of COP index and with three occupancy rates, leading to a total of 24 combinations. Figure 7 presents the layout of the building energy models.

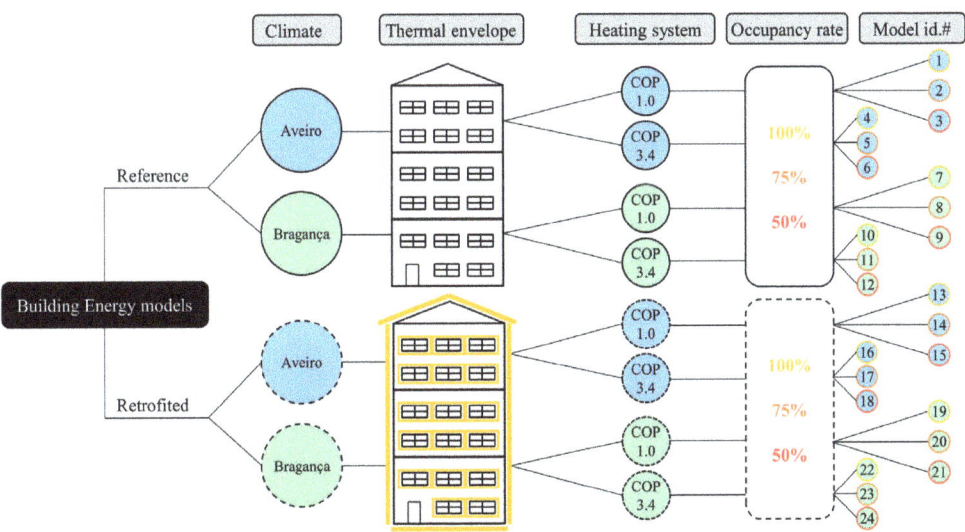

Figure 7. Layout of the building energy models.

The original building characteristics referred to Section 2.1 were considered as the reference scenario (models with original envelope—Id. 1 to 12). The improved envelope (models Id. 13 to 24) includes a set of passive measures, namely: the addition of a thermal insulation layer for vertical and horizontal opaque envelope and the substitution to double-glazing window systems. For the horizontal elements, thermal insulation was applied over the ground floor slab and the roof was thermally isolated on the horizontal slab of the ventilated crawl space. The thermal insulation thickness defined has a thickness range of 5–12 cm, in order to comply with the requirements defined by the Portuguese thermal code for the climate regions [38]. The same criterion was used in the selection of the new double glazing systems. Table 1 lists the main thermal characteristics of the building elements.

Table 1. Thermal characteristics of the improved building envelope.

Thermal Transmittance U (W/m². °C)/Thermal Insulation Thickness (cm)	Aveiro (Zone I1)				Bragança (Zone I3)			
	Ground Floor	External Walls	Roof	Glazing	Ground Floor	External Walls	Roof	Glazing
Original Scenario *	1.24	2.19	2.98	4.80	1.24	2.19	2.98	4.80
Improved scenario	0.49 (5 cm)	0.39 (8 cm)	0.35 (10 cm)	2.80	0.49 (5 cm)	0.33 (10 cm)	0.30 (12 cm)	2.20

* no thermal insulation.

Regarding the active measures, the simulation will consider an improved heating system with a COP of 3.4, which corresponds to an air-conditioning class B (minimum requirement established by the Portuguese regulation [38]). The COP of 1.0 was also considered in the improved envelope scenarios, in order to evaluate the influence of keeping a less efficient heating system. Only the installation of heating systems was considered as typically no cooling systems are used in the Portuguese social housing building stock [37]. Table 2 presents the COP index used in EnergyPlus simulations.

Table 2. COP index of heating systems input.

Heating Systems	COP	Efficiency Classification
Electrical heaters	1.0	G
Air-conditioning	3.4	B

The impact of unoccupied flats can be relevant for the thermal and energy performance of multi-familiar buildings, as shown in a previous study [28]. In the context of social housing, due to their temporary and intermittent occupation rates, this phenomenon can be particularly important. Herein, besides the two fully occupied situations (100% flats occupied), two additional occupancy scenarios of the building were considered (75% flats occupied and 50% flats occupied). No internal gains were considered in the unoccupied flats.

The internal gains are another important source of uncertainty in building simulation. In residential buildings, using a lumped value to simulate all the internal gains is the most common approach. In the framework of this research, three plausible values were defined for the internal gains: 2, 3 and 4 W/m². These values were randomly distributed within the building flats and the LHS algorithm was applied to generate a 50 cases samples for each of 24 models, leading to a total of 1200 simulations in both locations. The sampling algorithm assumes an equal probability for all parameters, i.e., a uniform distribution. Regarding the used software's EnergyPlus and HOMER Pro®, each annual simulation was performed with a computational time associated of approximately 139 s and 70 s respectively, both resourcing to an Intel Core i7 5820 K with eight cores working on a 3.30 GHz processor equipped with 16 GB of RAM.

The model's thermal properties (weather file data, heating system, occupancy rate, etc.) were defined in EnergyPlus and to simplify the automatization of the procedure, the building energy simulation is initiated using the JEPlus add-on for LHS. The samples were thus generated and simulated by JEPlus, using EnergyPlus as the engine. Once determined the energy consumption outputs, these are prepared through a python script to be fed into and processed with HOMER Pro® software (see Section 3.2).

2.3.2. HOMER Pro-Building Energy Production Optimisation

The final goal of the methodology is the optimisation of the number of photovoltaic panels to be integrated into a multi-residential social housing building, using the minimization of the life cycle cost as the ultimate objective. For the optimisation process, the photovoltaic system was configured and modelled in HOMER Pro® software and hourly simulations of its operation were performed to assess the life cycle cost of each solution.

The photovoltaic panels were simulated on the rooftop of the building and the following constraints were defined: not exceeding the sloped rooftop available area (86 panels) with South orientation and avoiding the self-shading of the panels due to their proximity (see Figure 8).

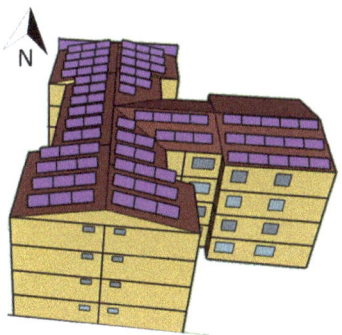

Figure 8. 3-D rooftop view of photovoltaic panels.

The photovoltaic panels were modelled, considering a peak power of 260 Wp, 16.2% efficiency, a tilt angle of 35 degrees and no tracking system. The input values for the inverters were as follows: 1 kW size, 20 years lifetime and 95% inverter efficiency.

Figure 8 shows the rooftop view of full photovoltaic installation and Table 3 the specifications and properties of the PV panels considered in HOMER Pro® software.

Table 3. Technical specifications of photovoltaic panels.

Photovoltaic Properties		Photovoltaic Costs (by Panel)	
Maximum power (Wp)	260	Replacement (€)	200
Open circuit voltage (V)	38.7	Operation & Maintenance (€/year)	1.75
Maximum power point voltage (V)	31	NPV details	
Short circuit current (A)	9.1	Electricity cost (€/kWh)	0.215
Maximum power point current (A)	8.6	Discount rate (%)	3
Module efficiency (%)	16.2	Project lifetime (years)	20
Panel dimensions (m)	1.65 × 0.99		
Cell type	Polycrystalline		

The weather data used in the simulations was collected from the DGEG [39] database, as described in Section 2.2. The input data required by the software are the daily solar irradiation on a horizontal plane, the hourly mean values of outdoor temperature, wind speed, and the hourly energy consumption for the whole year taken from the energy simulations previously carried out in EnergyPlus (see Section 3.1).

In the optimisation procedure, the calculations of the optimal sizing were made using the Net Present Value (NPV) as the base criterion. HOMER Pro® computes the life cycle cost of the photovoltaic system based on the NPV, taking into consideration all costs that occur within the project lifetime, including the effect of a pre-defined discount rate. Therefore, the NPV includes the installation cost of the system and its operating costs and maintenance, which occur during the project lifetime. The project lifetime considered in the analyses of HOMER Pro® was 20 years and it was assumed a unitary cost of 200€ for the photovoltaic panels. A discount rate of 3% was considered and the energy costs were calculated using a unitary electricity price of 0.215 €/kWh.

3. Results and Discussion

3.1. Optimisation Results

The proposed methodology was applied to the entire dataset and thus a large number of results were produced. Therefore, the detailed results of only one model are presented in this section as an example case. Figure 9 shows the results of the 50 energy simulations carried out for the improved envelope scenario, located in Bragança considering a heating system with a COP of 3.4 and with 100% of occupancy rate (Model Id. 22). The effect of the internal gains is evident as the total energy demand varies between 48,000 and 59,000 kWh/year.

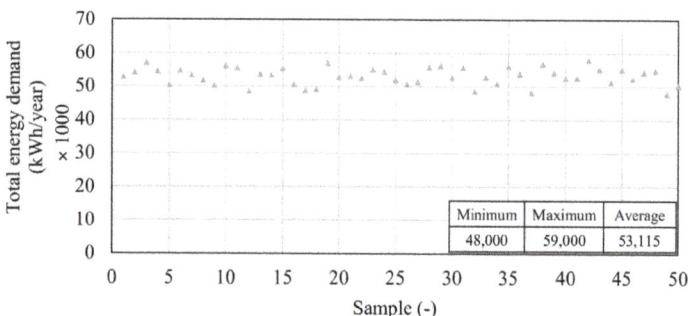

Figure 9. Results of the energy simulation for Model Id. 22.

The total energy demand is then used as input for the calculation of the NPV value associated with the integration of photovoltaic panels. This calculation is carried out in HOMER Pro assuming that the number of photovoltaic panels ranges between 0 and 86. Figure 10 shows the results of this procedure, highlighting the optimum solution of each sample, which corresponds to the minimum NPV value. For this particular set-up (Model Id. 22), the optimum number of photovoltaic panels ranges between 68 and 86.

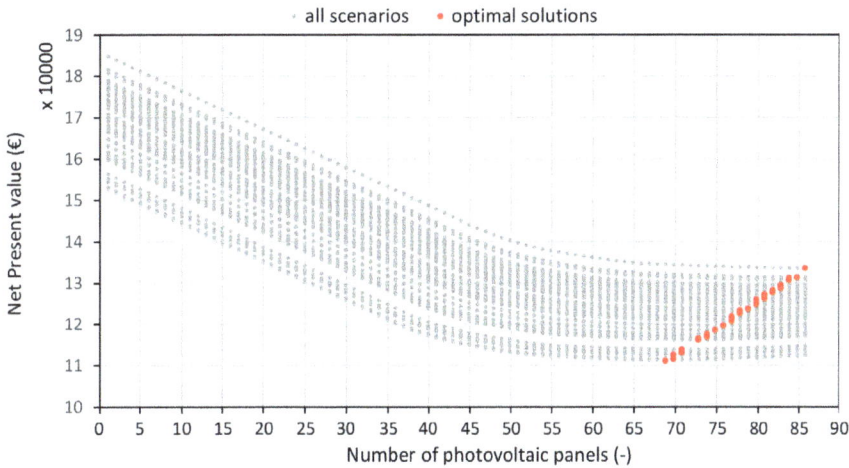

Figure 10. Optimum number of photovoltaic panels for Model Id. 22.

3.2. Impact of the Energy Refurbishment of the Building

To evaluate the impact of the energy refurbishment of the building envelope based on the introduction of thermal insulation and double glazing windows, Figure 11 shows the results of the optimisation procedure for the models with a COP of 1.0, separately for the original envelope and improved envelope scenarios. Moreover, the effect of considering a different number of unoccupied flats is also depicted in the graph. The optimum number of photovoltaic panels are grouped into classes for readability purposes and both figures (Figures 11 and 12) show the share of optimum occurrences (red dots of Figure 10) within each class, considering the entire dataset (24 Models Ids).

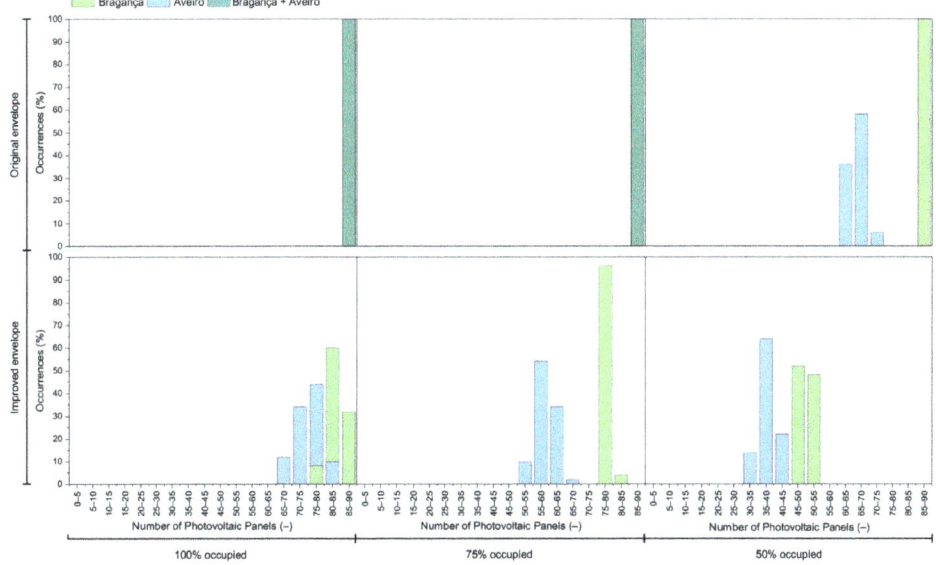

Figure 11. Envelope improvements analysis for heating system with COP 1.

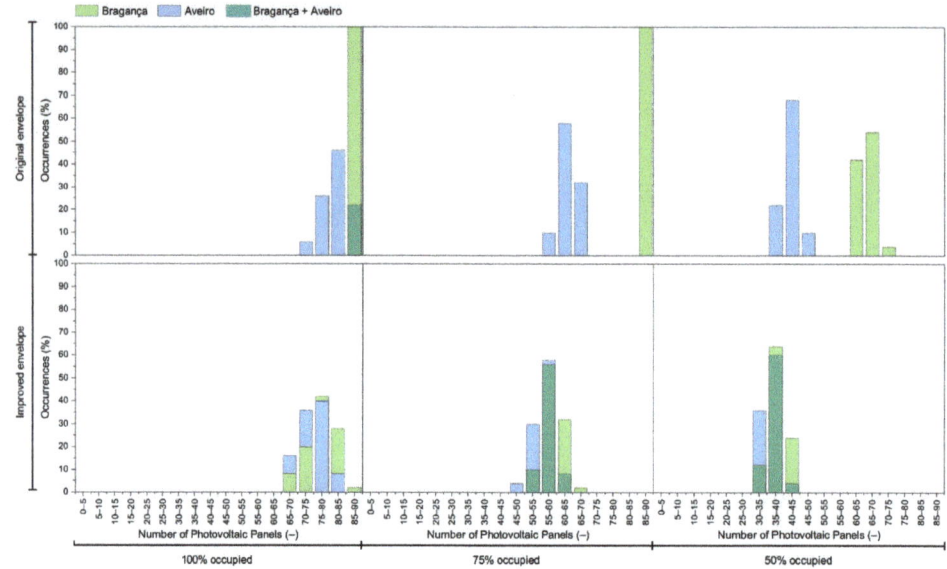

Figure 12. Envelope improvements analysis for heating system with COP of 3.4.

The impact of the energy refurbishment obvious leads to a decreasing trend in the optimum number of photovoltaic panels. In fact, in the scenarios with the original envelope and a COP of 1.0 (see Figure 11), the optimum number of photovoltaic panels results in the maximum (86 panels) in both locations, except for the situation with 50% occupancy rate in Aveiro, where the optimum solution decreases to the range between 60 and 75 panels. On the other hand, in the scenarios with the improved envelope, the optimum number of photovoltaic panels is lower, reaching the minimum range between 30 to 45 panels, once again in the situation with 50% occupancy rate of flats in Aveiro. Regarding the impact of the unoccupied flats, the results show that this situation is even more relevant in the scenarios with the improved envelope. In Aveiro with the improved envelope, the optimum number of photovoltaic panels can vary between 30 and 85, depending on the percent of unoccupied flats and, obviously, on the uncertainty related to the internal gains.

The results for a heating system with a COP of 3.4 are shown in Figure 12. It is noteworthy that the efficiency of the heating system has a high influence on the optimum solutions, generally leading a lower number of photovoltaic panels when compared with the solutions with a COP of 1.0. This situation is more evident in the scenarios with the original envelope.

3.3. Effect of the Uncertainty in the Internal Gains

To highlight the variability associated with the uncertainty in the quantification of the internal gains, Figure 13 shows the box-plot representation of the optimum number of photovoltaic panels, separately for each Model Id. The analysis of the results evidences that the initial premise regarding the available roof area for which the number of PV panels is limited to 86, is predominant in the results. Obviously, in these scenarios, the uncertainty associated with the quantification of internal gains is not felt, since all scenarios lead to an optimal solution of 86 panels. However, when the efficiency of the heating systems is changed, or when the energy performance of the building envelope is improved, or when the two are combined, the relevance of the internal gains is reflected in the results, corresponding to a dispersion of the optimal solutions.

As expected, 100% and 75% occupation rate scenarios are the most influenced by the uncertainty in the internal gains. On the other hand, the impact of the uncertainty is less obvious in the 50% occupation rate scenarios.

In short, the results confirm the importance of internal gains in the optimal design of the number of photovoltaic panels. Considering that in social housing, the variability of internal gains tends to be greater, due to the socio-economic context of the owners/tenants of the buildings, the importance of a probabilistic approach is needed to reduce the performance gap prevision in retrofitting design scenarios. These stochastic approaches can in fact be a powerful tool, supporting the designer in the decision-making process, avoiding oversizing solutions as frequently reported in the literature and that consequently cost and maintenance wise are a long-term issue [42].

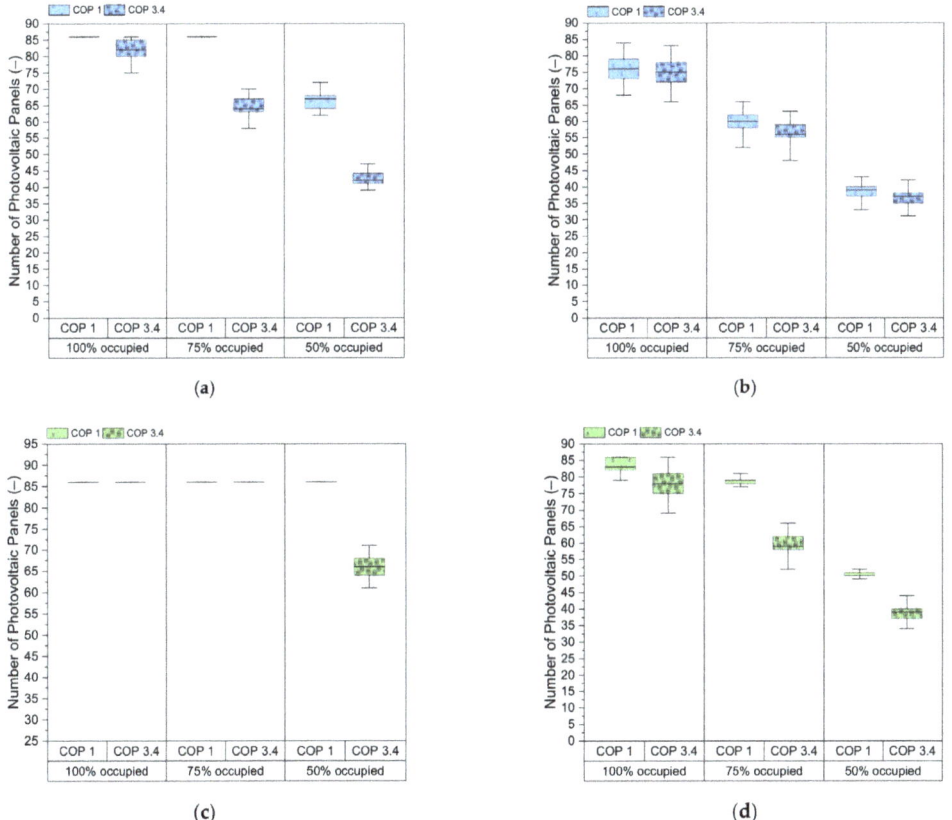

Figure 13. Optimum number of photovoltaic panels: (**a**) Aveiro region with original envelope; (**b**) Aveiro region with improved envelope; (**c**) Bragança region with original envelope; (**d**) Bragança region with improved envelope.

4. Conclusions

This study has highlighted the need to define a robust approach to assess the impact of improvement measures, both active and passive, on the energy efficiency in social housing buildings, specifically tackling the optimisation approach for photovoltaic energy system design, as a contribution for mitigation of the energy poverty of low-income families that live in these neighbourhoods.

Social housing buildings differ from the typical multi-residential building, due to the more pronounced impact of the uncertainty of the occupation and the internal gains that can cause a significant impact on the photovoltaic design for these buildings. Facing the growing trend of photovoltaic systems across the world, this paper develops a novel methodology with a strong practical application potential to minimise the uncertainty in sizing these systems as a paramount issue that has not been thoroughly studied and builds on previous works of the authors [28,37].

The combination of a group of parameters, such as the available rooftop area for PV installation, thermal quality of the external envelope (opaque and glazed), energy efficiency of heating or cooling devices, occupancy rate of the whole buildings, internal gains that implicitly are linked to flat occupation profiles, should be considered in the definition of renewable energy strategies and optimal design of PV installations. A methodology for optimal sizing of photovoltaic energy systems was proposed to support the selection of the optimal area of the number of PV panels, taking into account a number of constraints and variables/parameters based on the minimisation of the life cycle cost. The following main conclusions can be drawn:

- Severe winter regions, as the case of Bragança, can cause space restrictions for an on-site production, since they have major energy consumption for heating and consequently, optimal solutions can led to a maximum area (86 panels) allowed in the building;
- Renovation and improvement measures over the building external envelope reduce the total energy consumption, herein this study referred as improved envelope, leading to lower optimal-sizing PV solutions. In some cases a reduction up to 30% of the total number of photovoltaic panels allowed on the rooftop is attained. However, the impact can be greater when coupled to more efficient heating systems, with higher COP;
- The occupancy rate has a significant impact on the energy consumption for heating, having a more significant consequence in the scenarios with the improved envelope leading a lower PV design (number of panels). As an example, a reduction up to 64% of the number of PV panels was attained in the scenario with 50% occupancy rate;
- To the previous findings the variability associated with the uncertainty in the quantification of the internal gain is more notable in the cases of higher occupation rates (75 and 100%) revealing the importance of the definition of internal gains in the optimal design of the number of photovoltaic panels.
- Renewable energy production is growing continuously, and consequently supported by the investments and funding sources to intensify the implementation of renewable energy for 2030 horizon and also by the decreasing costs of renewable energy technology, namely in the case of photovoltaics and their components.
- At this stage of its development, the proposed methodology includes some limitations, namely: the inclusion of uncertainty in the definition of the economic scenarios, the definition of the range of variation for internal gains, and the compatibility with HVAC systems, including the possibility of cooling in summer, responding to future scenarios arising from climate change.
- In future developments, the optimisation of PV should be complemented with battery storage systems scenarios, in order to optimise the surplus energy production and off-peak demand, as well as the potential increasing energy costs.

Author Contributions: Conceptualization, R.M.S.F.A.; Data curation, R.O.; Formal analysis, R.O.; Funding acquisition, R.V.; Investigation, R.O. and A.F.; Methodology, R.M.S.F.A. and R.V.; Resources, R.V.; Supervision, R.M.S.F.A., A.F. and R.V.; Validation, A.F.; Writing—original draft, R.O. and R.M.S.F.A.; Writing—review & editing, R.M.S.F.A, R.V. and A.F. All authors have read and agreed to the published version of the manuscript.

Funding: This work was supported by the Foundation for Science and Technology (FCT)—Research Centre for Risks and Sustainability in Construction (RISCO), University of Aveiro, Portugal [FCT/UIDB/ECI/04450/2020].

Institutional Review Board Statement: Not applicable.

Informed Consent Statement: Not applicable.

Data Availability Statement: Data sharing not applicable.

Conflicts of Interest: The authors declare no conflict of interest.

References

1. Pfeiffer, A.; Koschenz, M.; Wokaun, A. Energy and building technology for the 2000 W society—Potential of residential buildings in Switzerland. *Energy Build.* **2005**, *37*, 1158–1174. [CrossRef]
2. Euroconstruct. *European Construction: Market Trends Until 2019*; Euroconstruct: Amsterdam, The Netherlands, 2017.
3. Hermelink, A.; Schimschar, S.; Boermans, T.; Pagliano, L.; Zangheri, P.; Armani, R.; Voss, K.; Musall, E. *Towards Nearly Zero-Energy Buildings: Definition of Common Principles under the EPBD*; Europäische Kommission: Köln, Germany, 2013.
4. European Comission. *A Policy Framework for Climate and Energy in the Period from 2020 to 2030*; European Comission: Brussels, Belgium, 2014.
5. European Union. *A Renovation Wave for Europe—Greening Our Buildings, Creating Jobs, Improving Lives*; European Union: Brussels, Belgium, 2020; Volume 53, pp. 1689–1699.
6. European Union. *New European Bauhaus*; European Union: Brussels, Belgium, 2020; p. 2.
7. Directive 2018/844/EU of the European Parliament and of the Council of 30 May 2018. *Off. J. Eur. Union* **2018**, *19*, 75–91.
8. European Union. Regulation 2021/1119. *Off. J. Eur. Union* **2021**, *9*, 12–25.
9. European Commission. Regulation 401/2009. *Off. J. Eur. Union* **2009**, *21*, 13–22.
10. European Union. Regulation 2018/1999. *Off. J. Eur. Union* **2018**, *328*, 1–77.
11. European Commission. European Green Deal: Commission proposes transformation of EU economy and society to meet climate ambitions. European Comission: Brussels, Belgium, 2021.
12. IHRU. Levantamento Nacional das Necessidades de Realojamento Habitacional. Instituto da Habitação e Reabilitação Urbana: Lisboa, Portugal, 2018. (In Portuguese)
13. Gouveia, J.P.; Seixas, J.; Long, G. Mining households' energy data to disclose fuel poverty: Lessons for Southern Europe. *J. Clean. Prod.* **2018**, *178*, 534–550. [CrossRef]
14. Gouveia, J.P.; Palma, P. Perspective on Energy Poverty in the Public Debate and in Research in Portugal. *EP-pedia ENGAGER COST Action.* **2021**, 1–8.
15. European Union. Interreg Sudoe Sudoe Energy Push. *Sudoe Effic. Energy Public Soc.* **2021**.
16. Silva, P.P.; Martins, R.; Antunes, M.; Furtonato, A. Estudo sobre a aplicação da tarifa social de energia em portugal. *Energy Obs.* **2019**. (In Portuguese)
17. European Commission. Member State Reports on Energy Poverty 2019. *Energy Poverty Obs.* **2020**, 1–118.
18. Escandón, R.; Suárez, R.; Sendra, J.J. Field assessment of thermal comfort conditions and energy performance of social housing: The case of hot summers in the Mediterranean climate. *Energy Policy* **2019**, *128*, 377–392. [CrossRef]
19. Ramos, N.M.M.; Almeida, R.M.S.F.; Simões, M.L.; Delgado, J.M.P.Q.; Pereira, P.F.; Curado, A.; Soares, S.; Fraga, S. Indoor hygrothermal conditions and quality of life in social housing: A comparison between two neighbourhoods. *Sustain. Cities Soc.* **2018**, *38*, 80–90. [CrossRef]
20. Curado, A.; de Freitas, V.P. Influence of thermal insulation of facades on the performance of retrofitted social housing buildings in Southern European countries. *Sustain. Cities Soc.* **2019**, *48*, 101534. [CrossRef]
21. Agliardi, E.; Cattani, E.; Ferrante, A. Deep energy renovation strategies: A real option approach for add-ons in a social housing case study. *Energy Build.* **2018**, *161*, 1–9. [CrossRef]
22. Rosso, F.; Peduzzi, A.; Diana, L.; Cascone, S.; Cecere, C. A Sustainable Approach towards the Retrofit of the Public Housing Building Stock: Energy-Architectural Experimental and Numerical Analysis. *Sustainability* **2021**, *13*, 2881. [CrossRef]
23. Sojkova, K.; Volf, M.; Lupisek, A.; Bolliger, R.; Vachal, T. Selection of Favourable Concept of Energy Retrofitting Solution for Social Housing in the Czech Republic Based on Economic Parameters, Greenhouse Gases, and Primary Energy Consumption. *Sustainability* **2019**, *11*, 6482. [CrossRef]
24. D'Oca, S.; Ferrante, A.; Ferrer, C.; Pernetti, R.; Gralka, A.; Sebastian, R.; op 't Veld, P. Technical, Financial, and Social Barriers and Challenges in Deep Building Renovation: Integration of Lessons Learned from the H2020 Cluster Projects. *Buildings* **2018**, *8*, 174. [CrossRef]
25. McCabe, A.; Pojani, D.; van Groenou, A.B. The application of renewable energy to social housing: A systematic review. *Energy Policy* **2018**, *114*, 549–557. [CrossRef]
26. IRENA. Future of solar photovoltaic: Deployment, investment, technology, grid integration and socio-economic aspects. In *International Renewable Energy Agency*; International Renewable Energy Agency: Abu Dhabi, United Arab Emirates, 2019; pp. 1–88.
27. Guerra-Santin, O.; Boess, S.; Konstantinou, T.; Romero Herrera, N.; Klein, T.; Silvester, S. Designing for residents: Building monitoring and co-creation in social housing renovation in the Netherlands. *Energy Res. Soc. Sci.* **2017**, *32*, 164–179. [CrossRef]
28. Oliveira, R.; Figueiredo, A.; Vicente, R.; Almeida, R.M.S.F. Impact of unoccupied flats on the thermal discomfort and energy demand: Case of a multi-residential building. *Energy Build.* **2020**, *209*, 109704. [CrossRef]

29. Haldi, F.; Robinson, D. The impact of occupants' behaviour on building energy demand. *J. Build. Perform. Simul.* **2011**, *4*, 323–338. [CrossRef]
30. Lee, J.; Shepley, M.M. Benefits of solar photovoltaic systems for low-income families in social housing of Korea: Renewable energy applications as solutions to energy poverty. *J. Build. Eng.* **2020**, *28*, 101016. [CrossRef]
31. Guerra-Santin, O.; Bosch, H.; Budde, P.; Konstantinou, T.; Boess, S.; Klein, T.; Silvester, S. Considering user profiles and occupants' behaviour on a zero energy renovation strategy for multi-family housing in the Netherlands. *Energy Effic.* **2018**, *11*, 1847–1870. [CrossRef]
32. Pinto, J.T.M.; Amaral, K.J.; Janissek, P.R. Deployment of photovoltaics in Brazil: Scenarios, perspectives and policies for low-income housing. *Sol. Energy* **2016**, *133*, 73–84. [CrossRef]
33. Seabra, B.; Pereira, P.; Corvacho, H.; Pires, C.; Ramos, N. Low Energy Renovation of Social Housing: Recommendations on Monitoring and Renewable Energies Use. *Sustainability* **2021**, *13*, 2718. [CrossRef]
34. Almeida, M.; Barbosa, R.; Malheiro, R. Effect of Embodied Energy on Cost-Effectiveness of a Prefabricated Modular Solution on Renovation Scenarios in Social Housing in Porto, Portugal. *Sustainability* **2020**, *12*, 1631. [CrossRef]
35. Carpino, C.; Bruno, R.; Arcuri, N. Social housing refurbishment for the improvement of city sustainability: Identification of targeted interventions based on a disaggregated cost-optimal approach. *Sustain. Cities Soc.* **2020**, *60*, 102223. [CrossRef]
36. Domingos, R.M.A.; Pereira, F.O.R. Comparative cost-benefit analysis of the energy efficiency measures and photovoltaic generation in houses of social interest in Brazil. *Energy Build.* **2021**, *243*, 111013. [CrossRef]
37. Oliveira, R.; Vicente, R.; Almeida, R.M.S.F.; Figueiredo, A. The Importance of In Situ Characterisation for the Mitigation of Poor Indoor Environmental Conditions in Social Housing. *Sustainability* **2021**, *13*, 9836. [CrossRef]
38. REH. *Regulamento de Desempenho Energético de Edifícios de Habitação (Portuguese regulation on the energy performance of residential buildings)*; Decree-Law No. 118/2013. Diário da República—1.ª Série—N° 159; Ministério da Economia e do Emprego: Lisboa, Portugal, 2013; 159, pp. 4988–5005.
39. DGEG. Directorate-General for Energy and Geology. Available online: http://www.dgeg.gov.pt/ (accessed on 26 April 2021).
40. DGEG. Portuguese Climate Data for Building Simulation. *DEIR Stud. Port. Energy Syst. 006. Dir. Energy Geol. Div. Res. Renewables.* **2021**, 29.
41. ASHRAE Guideline. ASHRAE Guideline 14-2002 Measurement of Energy and Demand Savings. *Am. Soc. Heat. Refrig. Air-Cond. Eng.* **2002**, *8400*, 170.
42. Decreto-Lei no 162/2019. *Diário da República, N° 206, 1ª Série* **2019**, 45–62. (In Portuguese)

Article

Effect of HVAC's Management on Indoor Thermo-Hygrometric Comfort and Energy Balance: In Situ Assessments on a Real nZEB

Rosa Francesca De Masi [1,*], Antonio Gigante [1], Valentino Festa [1], Silvia Ruggiero [1] and Giuseppe Peter Vanoli [2]

[1] DING—Department of Engineering, University of Sannio, 82100 Benevento, Italy; gigante@unisannio.it (A.G.); valefesta@unisannio.it (V.F.); sruggiero@unisannio.it (S.R.)
[2] Department of Medicine and Health Sciences—Vincenzo Tiberio, University of Molise, 86100 Campobasso, Italy; giuseppe.vanoli@unimol.it
* Correspondence: rfdemasi@unisannio.it; Tel.: +39-0824-305577

Abstract: This paper proposes the analysis of real monitored data for evaluating the relationship between occupants' comfort conditions and the energy balance inside an existing, nearly zero-energy building under different operational strategies for the heating, ventilation, and air-conditioning system. During the wintertime, the adaptive comfort approach is applied for choosing the temperature setpoint when an air-to-air heat pump provides both heating and ventilation. The results indicate that in very insulated buildings with high solar gains, the setpoint should be decided taking into consideration both the solar radiation and the outdoor temperature. Indeed, when the room has large glazed surfaces, the solar radiation can also guarantee acceptable indoor conditions when a low setpoint (e.g., 18.7 °C) is considered. The electricity consumption can be reduced from 17% to 43% compared to a conventional setpoint (e.g., 20 °C). For the summertime, the analysis suggests the adoption of a dynamic approach that should be based on the outdoor conditions and differentiated according to room characteristics. Considering the indoor comfort and the maximization of renewable integration, the direct expansion system has better performance than the heat pump; this last system should be integrated with a pre-handling unit to be energy convenient.

Keywords: nZEB; HVAC management; monitoring campaign; indoor comfort; load matching; computational fluid dynamics

1. Introduction

A key element in order to achieve energy-efficient buildings with high integration of renewable sources is the adopted strategy of nearly zero-energy buildings (nZEBs). According to the European Directive 2010/31 [1], starting from 31 December 2020, all new buildings must be nZEB, and the goal of a zero energy balance is also promoted [2] for the refurbishment of existing stock. The design of high-performance buildings requires us to fully understand the connection between indoor comfort and energy use [3], because the low energy demand should not compromise the occupants' wellness. At the same time, the analysis of operational behavior is important for understanding the connection between indoor comfort and energy use, as well as between on-site generation and the building's self-consumption. Therefore, as was underlined by Butera [4], the comfort conditions and the real-time energy balance are two key research fields.

With regard to the first aspect, some numerical analyses indicated that energy use could be reduced by choosing appropriate setpoints [5,6]; conversely, the adoption of only one thermostat may be deficient in certain occupancy cases [7,8]. Controller tuning by setpoint regulation often provides relatively low- or no-extra-cost solutions for existing buildings [9]. With the aim of reducing energy demand and mitigating energy poverty, adaptive thermal comfort should be considered rather than fixed or tightly specified parameters [10].

Most scheduling methods lack a systematic approach to ensuring consumption reduction and occupant comfort [11]. As shown by Guillén et al. [12], the heating energy saving that can be achieved by moving the set temperature from 20 °C to 19 °C is between 30% and 46% for a nZEB and between 13% and 23% for a traditional dwelling. Reda et al. [13] have found that the adoption of information and communication technologies for setting heating setback and indoor setpoint temperatures and for controlling mechanical ventilation can ensure consistent reductions of energy demand.

Thermal comfort is always affected by the energy efficiency of buildings and vice versa [14], but it is also indisputable that the adoption of extended comfort ranges can cause significant energy saving [15,16]. For an office building, Sánchez-García et al. [17] have found that daily setpoint temperatures based on the adaptive thermal comfort approach can cause a 74.6% reduction in energy demand and a 59.7% drop in energy consumption. For different climatic conditions, Ming et al. [18] have shown how the use of a dynamic comfort range can reduce the cooling demand by around 34% compared with a static temperature setpoint. Similarly, Mui et al. [19] have found that demand-controlled ventilation and an adaptive comfort temperature setpoint have energy-saving potential between 21.4% and 24.3% in a subtropical climate zone. With surveys in mechanically ventilated offices, Roussac et al. [20] have found a reduction of daily heating, ventilation, and air-conditioning energy use of about 6% when a static control strategy is applied; on the other hand, a dynamic control approach causes an energy saving of around 6.3%. Data analytics applied to an office building have found that an alternative HVAC (Heating, Ventilation, and Air-Conditioning system) schedule that can reduce energy consumption by, on average, 2.7% [21]. A new controller based on occupancy prediction makes energy savings up to 75% possible [22]. Traylor et al. [23], by means of EnergyPlus simulations, have found that savings up to 5–15% could be achieved by modulating indoor temperatures in cooling applications. The results of Kim et al. [24] have demonstrated that with a controller based on the temperature and humidity, the use of the HVAC system can be reduced due to the human capability to thermally adapt after reaching the comfort range of temperature. Jazizadehet et al. [25] have shown the potentials of integration of diffuser-level adaptive actuation with improvements in thermal comfort satisfaction and energy savings of around 25%. Recently, the role of the key enabling technologies has been experimentally verified by Lourenço et al. [26]. Moreover, Kalaimani et al. [27] have shown how energy use could be reduced during periods of low occupancy by choosing appropriate thermal setpoints.

Briefly, the interest in adaptive models has led to the definition of different theories for the management of the heating and cooling system based on the relation of the operative indoor temperature [28] with the mean outdoor temperature, also considering an extended monitoring of buildings with high performance [29]. However, these studies are mainly focused on tertiary buildings during the cooling period. Although wider acceptable temperature ranges have been included in international standards, adoption of the adaptive principles into design practice is still limited, mainly with reference to the winter period [30] and for residential high-efficiency buildings. Some in situ analyses [31] have shown that the performance of natural night ventilation depends highly on the external weather conditions, and especially on the outdoor temperature. However, some monitored data indicate that active and passive ventilation and shading systems, if manually controlled, cannot guarantee a high-quality indoor environment [32].

Findings from O'Donovan et al. [33] show that passive control strategies maintained comfortable internal conditions between 57% and 95% of the occupied hours, without the need for mechanical cooling. However, variable setpoints could favor a match between the load and productivity curves [34] with minimal loss in comfort [35].

Perhaps the gaps in the current literature suggest that one of the most important omissions in the current literature is the limited availability of real applications that testify the implications and improvements of an adaptive and dynamic control of the heating, cooling, and ventilation system in a nearly zero-energy building, as well as the evaluation of occupant's perceptions. Indeed, with reference to this aspect, there are no studies based on in situ monitoring campaigns that evaluate the influence of a comfort control strategy on the building energy balance and the effect on local renewable sources adoption, or in other words, on the matching between energy demand and renewable production. More generally, there is a lack of data regarding the operational behavior and the energy performance of nZEBs in a real context. The available scientific literature is mainly focused on design strategies and not on the post-occupancy performance evaluation of this type of building. Instead, the occupants' behavior is widely recognized as a major factor in the disparity between predicted and measured building performance [36]. Therefore, hourly and sub-hourly analysis can give more accurate information compared with measured peak values [37]. The comparative analysis of O'Donovan et al. [38] has highlighted the need to consider diverse occupancy schedules and opening control strategies for evaluating the performance of nearly zero-energy buildings. In another study [39], the same authors have found that grey box modelling with an automatic model-calibration technique can reduce the human labor input for simulating the internal air temperature of a naturally ventilated nZEB by approximately 90%.

Considering that the integration of renewable energy sources is the key factor for achieving zero-energy performance, the second important issue is the evaluation of the impact of the uncertainty of renewable sources also in consideration of occupants' behavior. Indeed, the large-scale diffusion of nZEBs can affect the stability of the existing power grid with consequences on operational costs and environmental impacts. Real-time energy monitoring devices are significant tools for improving the load-matching between production and consumption [40]. Tumminia et al. [41], through some indicators, have underlined how the designing of a photovoltaic system able to cover the yearly estimated energy use has negative implications on the power grid. Aelenei et al. [42] have investigated the potential of increasing the load-matching by means of battery energy systems for improving the grid interaction. Demand-side management could bring various benefits such as: reduction in electricity cost, reduction in peak demand, and improvement in load factor [43]. Monitoring data from a real case study investigated by Stasi et al. [44] indicated that the energy performance gap concerning energy production on a yearly basis is equal to 9.1%, while on a monthly basis, the performance gap ranges from 3.5% to 27.1%. Demand-side management is one possible approach for reducing the electricity cost and peak demand, and for improving the load factor [43,45]. Moreover, because of extensive variations in occupancy patterns and their use of electrical equipment, accurate occupants' behavior detection is valuable for reducing a building's energy demand and carbon emissions [46].

This paper analyzes the interaction between these two aspects, and it proposes a critical post-occupancy analysis of a building designed to be nearly zero-energy, considering both the aspects connected to indoor thermal-hygrometric comfort and to the hourly energy balance between consumption and renewable production. With this aim, the results of a monitoring campaign and some numerical analysis are discussed, considering different HVAC settings. This kind of discussion is not available for other existing studies because research and national or international legislation are usually focused on long-term energy balances (based on annual or monthly time step), without considering that the energy exchange at a short time scale is often more critical. This case study is a single-story dwelling built in Benevento (South Italy, Mediterranean climate), named BNZEB (Benevento Nearly Zero-Energy Building).

2. Case Study and Method

2.1. Case Study Building: BNZEB

Figure 1A shows the BNZEB, outcome of an Italian project named "SMARTCASE" developed under the umbrella of the European Regional Development Fund. The net conditioned area is 70 m^2, the window/wall ratio is 22.5%, and the surface-to-volume ratio is equal to 1.03 m^{-1}.

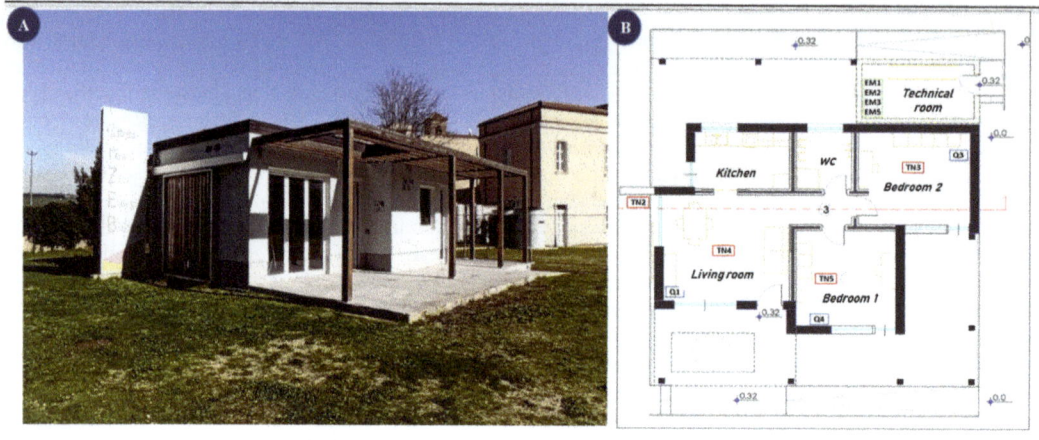

Code	Variable	Model	Accuracy
EM	Energy meter	Schneider electric A9MEM3250	0.5%
Q	CO_2 concentration	Siemens QPA2002	≤ ±(50 ppm +2%)
TN	Air temperature / relative humidity	Testo 177-H1	±0.5°C / ± 2%
AN	Digital Anemometer	AS65	±(2%+ 1dgt)

Figure 1. BNZEB: (**A**) external view, (**B**) layout, and (**C**) accuracy of main sensors.

The BNZEB is, at the same time, a research laboratory, suitable for testing and measuring energy, and indoor and outdoor conditions, but also a dwelling in which a comfortable life is allowed. In addition, as it has a small size, the BNZEB can be considered representative of a new design method aimed at reducing, as much as possible, the adoption of active systems and covering all of its energy needs entirely with available on-site renewable sources. The size does not influence the type of proposed analysis, which is contextualized to the type of occupation and external conditions.

The detailed design phase [47] and in situ characterization [48] have already been described by the authors, but some details are summarized here to improve the readability. Herein, it was also demonstrated that the building is classified as a nearly zero-energy one according to Italian standards; thus, it has the best energy class, and more than 50% of the energy consumed in the building is covered by on-site renewable production.

The BNZEB has been built in a west-facing area of the University of Sannio, in Benevento (lat. 41°7′55″, long. 14°46′40″, elevation 135 m), which is inside the Italian climatic zone "C", characterized by 1315 heating degree-days (baseline 20 °C). Benevento has a typical Mediterranean climate, characterized by warm to hot, dry summers and mild to cool, wet winters.

Cross-laminated wood makes up the structural frame, and it is coupled with two layers of fiber-wood insulation, with an overall thickness of 33 cm for the external walls, and about 50 cm for the roof. The measured thermal transmittance is equal to 0.19 W/m^{2}K and 0.22 W/m^{2}K, respectively, for the wall and the roof. Double-glazing systems with low-emissivity treatment and polyvinyl chloride frame are installed. The measured thermal transmittance for the window glass is 1.5 W/m^{2}K.

The layout is almost a square (Figure 1B), with the main entrance in the living room (22 m^2), a kitchen with openings to the north-west, and west exposures and two bedrooms. The living room, oriented south-west, has two large windows, one of which (5.8 m^2), on the south exposure, is permanently shaded by the wooden porch. The second window (5.3 m^2), on the west exposure, has an external shading system (Figure 1A) made of vertical wooden slats that can be automatically moved by means of a temporal program or by means of the monitored incident radiation. Bedroom 1 has two windows with an internal white curtain. One of these is a smart window (1.7 m^2), but during the monitoring, it was clear all of the time; the other one has a net surface area of 2.1 m^2. Bedroom 2 (Figure 1B) has two windows (2.4 m^2 and 2.1 m^2) shaded by a white curtain and by the porch; this room borders the technical room on the north side.

For the heating and cooling needs, there are two systems. The first one is an aerothermal heat pump with a nominal heating power of 3.18 kW and cooling power of 2.14 kW. It provides hot water, heating, cooling, dehumidification, and mechanical ventilation, with an active thermo-dynamic heat recovery. The second one is a direct expansion heating/cooling system with two internal units. Moreover, to improve the overall performance of the HVAC system, a pre-treatment section for the outdoor ventilation air has been added. In detail, a heat exchanger may pre-cool the ventilation air during the summer period and pre-heat it during the winter before it reaches the aerothermal heat pump. The intermediate water-to-water heat exchanger is linked alternately to a horizontal ground-to-water heat exchanger or a solar collector. The horizontal geothermal probes (earth-to-water heat exchanger) are positioned at a depth of 2 m, with a total linear length of 100 m. The solar thermal collector (2.16 m^2) is also used for the domestic hot water stored in a tank (196 L). Finally, a photovoltaic system is installed on the roof. This is composed of 16 monocrystalline silicon panels, each with an area about 1.63 m^2 and a peak power of 330 Wp. The photovoltaic (PV) modules are oriented to the south (i.e., azimuth angle of 0°) with a tilt angle of 5°. Moreover, there is a lithium battery of 6.5 kWh for electricity storage.

2.2. Methodology

This study is organized in two main sections aimed at evaluating the indoor comfort condition with different approaches. The wintertime investigation is based only on the results of a monitoring campaign performed from November 2019 to January 2020, and it is aimed at evaluating the effect of different setpoint temperatures when the heat pump is operating in heating and ventilation mode with the intermediate heat exchanger turned off. On the other hand, during the summertime, different possible configurations for the HVAC systems are monitored during July in order to calibrate a numerical model of the building; this energy model is used for establishing the most adequate configuration to meet the comfort conditions. In both seasons, the building energy balance will be discussed considering the monitored values.

All information about the energy consumption, the generation from renewable sources, and the indoor microclimate conditions, like air temperature and relative humidity, have been collected with a time step of 15 min. The sensors available in the building have been adopted (see Figure 1B), and their accuracy is reported in Figure 1C.

To carry out the measurement campaign, air temperature and relative humidity sensors were installed in each room at 1.1 m (the head-height for a man sitting for typical studying work) near the table of the living room and kitchen and near the desks in the bedrooms. This does not always mean they are in the center of the room, but rather in a representative position with respect to the position mainly occupied by the students. At the center of the room, a globe-thermometer (emission equal to 0.95 and diameter of 15 cm) is positioned, and also an air temperature and relative humidity sensor (1.1 m), as well as the anemometer. The measurement of globe temperature with knowledge of the air temperature allows us to obtain the mean radiant temperature starting from a non-linear equation.

The monitoring campaign is performed in real operational conditions, when the building is occupied during the weekdays by two students from 9:00 to 18:00. For a complete analysis, the carbon dioxide (CO_2) concentration has been continuously monitored by the sensors located in each room. However, it is not discussed, because it was verified that for all configurations and setups the CO_2 concentration is 500 ppm above the outdoor concentration, as required by considering the category II of comfort in residential buildings according to the EN 15251 standard [49].

2.2.1. Wintertime Investigation

During the wintertime, the monitoring campaign is based on the application of the adaptive theory for establishing the setpoint value for the air temperature starting from the monitored value of the outdoor temperature. The base rule for the adaptive comfort is a linear equation, $T_c = a \cdot T_{ext} + b$, where T_c is the expected indoor comfort temperature, T_{ext} is the outdoor reference temperature, a indicates the degree of adaption, and b is the y-intercept [50]. The a and b values are different for each adaptive thermal comfort model, and these are usually found by means of experimental data. Starting from the study of Dear and Brager [51], the ANSI/ASHRAE 55:2004 has included the adaptive model in the reference comfort standard; the standard was updated in 2020. In the same year, ISSO 74 was adopted in Netherlands [52], and revised in 2014. Later, in 2007, the new European standard EN 15251 was introduced, and updated by UNI EN 16798-1 [53]. Most recently, it was introduced in China [54]. The following table summarizes the main equations proposed in the mentioned standard. In the table, T_{ext} is usually calculated starting from the daily mean external air temperature (t_{ed}) for a time d of a series of equal intervals (day). In its equation, α is a constant ranging between 0 and 1, and in some cases, it is recommend to use 0.8. For the standard ANSI/ASHRAEE and in the case of UNI EN 16798-1, only the optimal expected indoor comfort temperature is reported; instead, the first standard provides two ranges for T_c, namely 80% and 90% acceptability, and in the second case, an upper and lower limit of category are provided. For the other two models, equations of the normal level of expectation are considered (this category refers to the design of new buildings or to substantial renovation).

According to the reference standard, the proposed equations can be applied only to occupant-controlled naturally conditioned spaces. Only in the case of beta spaces (ISSO 74) can the model be used for centrally controlled environments in the summer period.

Thus, the application of the adaptive theory for the wintertime is worthy of investigation. Moreover, it is also proposed for the selected case study because the building is characterized by a high level of insulation, and it is interesting to evaluate if the thermal-hygrometric conditions are comfortable when the setpoint temperature is different from the threshold value (20 °C) indicated by the Italian standards [55]. The proposed approach is based on a linear equation, as seen in the models summarized in Table 1. In particular, the setpoint temperature (T_{set}) has been fixed according to an empirical relation reported in Equation (1). It is calculated taking into account the reference outdoor temperature (T_{ext}):

$$T_{set} = 20 \pm a * T_{ext} \quad (1)$$

Herein, a is the adaptation coefficient that ranges, arbitrarily, between 0 and 0.15, and T_{ext} is the average value of the external temperature recorded during the seven days preceding the one to be heated. The equation indicates that for finding the setpoint, a factor that is a function of external conditions can be added or subtracted to the reference value of 20 °C. This approach is different from the models proposed in Table 1, where starting from the reference external temperature, a minimum comfort temperature is added or sometimes subtracted.

Table 1. Equations of adaptive models from literature.

Standard	Equations
ANSI/ASHRAE 55: 2010 [51]	$T_c = 0.31 \cdot T_{ext} + 17.8$ $T_{ext} = (1-\alpha) \cdot [t_{ed-1} + \alpha \cdot t_{ed-2} + \alpha^2 \cdot t_{ed-3} + \alpha^3 \cdot t_{ed-4}]$
UNI EN 16798-1:2019 [53]	$T_c = 0.33 \cdot T_{ext} + 18.8$ $T_{ext} = (1-\alpha) \cdot [t_{ed-1} + \alpha \cdot t_{ed-2} + \alpha^2 \cdot t_{ed-3} + \alpha^3 \cdot t_{ed-4} + ..]$
ISSO 74:2014 [52]	**ALFA spaces** Upper limit: $\begin{cases} T_c = 0.33 \cdot T_{ext} + 18.8 + 2 & 10\,°C \leq T_{ext} \leq 25\,°C \\ T_c = 24 & -5\,°C \leq T_{ext} < 10\,°C \end{cases}$ Lower limit: $\begin{cases} T_c = 0.2 \cdot T_{ext} + 18 & 10\,°C \leq T_{ext} \leq 25\,°C \\ T_c = 20 & -5\,°C \leq T_{ext} < 10\,°C \end{cases}$ $T_{ext} = 0.2 \cdot [t_{ed-1} + 0.8 \cdot t_{ed-2} + 0.8^2 \cdot t_{ed-3} + 0.8^3 \cdot t_{ed-4} + ..]$ **BETA spaces** Upper limit: $\begin{cases} T_c = 24 & -5\,°C \leq T_{ext} < 10\,°C \\ T_c = 0.33 \cdot T_{ext} + 18.8 + 2 & 10\,°C \leq T_{ext} \leq 16\,°C \\ T_c = 26 & 16\,°C < T_{ext} \leq 25\,°C \end{cases}$ Lower limit: $\begin{cases} T_c = 0.2 \cdot T_{ext} + 18 & 10\,°C \leq T_{ext} \leq 25\,°C \\ T_c = 20 & -5\,°C \leq T_{ext} < 10\,°C \end{cases}$ $T_{ext} = 0.2 \cdot [t_{ed-1} + 0.8 \cdot t_{ed-2} + 0.8^2 \cdot t_{ed-3} + 0.8^3 \cdot t_{ed-4} + ..]$
GB/T 50785:2012 [54]	Cold climates: $\begin{cases} T_c = 0.73 \cdot T_{ext} + 15.28 & 18\,°C \leq T_{ext} \leq 28\,°C \\ T_c = 0.91 \cdot T_{ext} - 0.48 & 16\,°C \leq T_{ext} \leq 28\,°C \end{cases}$ Hot and mild climates: $\begin{cases} T_c = 0.73 \cdot T_{ext} + 12.72 & 18\,°C \leq T_{ext} \leq 30\,°C \\ T_c = 0.91 \cdot T_{ext} - 3.69 & 16\,°C \leq T_{ext} \leq 28\,°C \end{cases}$ $T_{ext} = (1-\alpha) \cdot [t_{ed-1} + \alpha \cdot t_{ed-2} + \alpha^2 \cdot t_{ed-3} + \alpha^3 \cdot t_{ed-4} + ..]$

However, it should be noted that this relation has not been found in the literature or in the international standards, and it is not validated. The proposed analysis is the first attempt to verify the effectiveness of changing the setpoint temperature in a building designed to be very insulated and with a high contribution of solar gains during the winter. After this step, a deeper work will be done with the aim of validating the equation, while also considering different types of occupation.

Based on the monitored data, the most used thermal comfort indices have been calculated with the aim of comparing a quantitative variable with the occupants' subjective judgment. In particular, the predicted mean vote (PMV) and the predicted percentage of dissatisfied (PPD) are evaluated according to the definition proposed in [56] in two representative rooms at different times: 10:00, 13:00, and 17:00. The analysis is reported for the living room, representative of high solar gains during the winter, and Bedroom 1.

Moreover, an hourly energy balance is performed, and some assessment indices are discussed for evaluating the incidence of the renewable electricity source. The first index (RenEl) is the ratio between the amount of electricity from a renewable source used to satisfy the request and the total daily consumption of the building. The second one (PV$_{in}$) is the ratio between the generation from the PV system and the daily consumption. The last one is the Load Match Index, defined in Equation (2). It quantifies the ability to temporally match a building's load and the in situ energy generation, and thus it indicates the ability to work beneficially with respect to the needs of the grid infrastructure.

$$F_{loadmatch} = \frac{1}{N} * \sum_p \min\left[1, \frac{g(t)}{l(t)}\right] \qquad (2)$$

In more detail, $g(t)$ is the on-site electricity generation, $l(t)$ is the load, and t is the time interval (e.g., hour, day, or month). Meanwhile, p is the evaluation period, and N is the number of data samples (i.e., if p is equal to 1 year, this value is 12 for a monthly time interval and 8760 for an hourly time interval, respectively).

2.2.2. Summertime Investigation

The summertime investigation starts from a previous work of the authors, where it was demonstrated that the coupling of a ground-to-water heat exchanger with the two available cooling systems had great potential to achieve the nZEB target, with reference to the energy requirement for space cooling [57]. Instead, in this paper, the effect of the HVAC configurations, reported in Table 2, on indoor comfort conditions is evaluated by means of numerical simulations with EnergyPlus [58] though the DesignBuilder v.6.0 interface [59].

Table 2. Information about tested HVAC configurations.

Conf.	Monitoring Period	People Presence and Thermal Load	Ventilation System	Pre-Cooling Activation	Cooling System	Indoor Setpoint
C1	8–11 July	Yes	Packaged heat pump	Off	Packaged heat pump	25 °C
	12 July	No				
C2	15–18 July	Yes	Packaged heat pump	On	Packaged heat pump	25 °C
C3	22–25 July	Yes	Packaged heat pump	On	Direct Expansion heat pump	25 °C
C4	27–28 July	No	Packaged heat pump	Off	Direct Expansion heat pump	25 °C
	29–30 July	Yes				

The numerical model of the BNZEB has been already created and calibrated in another work [48], but for the purpose of the present study, it has been upgraded, also using a new version of the software interface. The numerical simulation needs a weather file as input. In this case study, it was defined with outdoor data monitored on the roof of the building during the same days during which the energy balance is evaluated and the indoor conditions are monitored. This procedure allows evaluation of the calibration indices for reliable internal and external forcing. In more detail, the meteorological datasets contain climatic data with hourly frequency based on measurements of the external air temperature and relative humidity, the global horizontal radiation, the wind speed, and direction, and so on.

The calibration of the numerical model has been performed according to the methodology proposed by M&G Guidelines [60]. The statistical indices are the Mean Bias Error (MBE), Equation (3), and the Coefficient of Variation of Root Mean Square Error (CvRMSE), Equations (4)–(6).

$$MBE = \frac{\Sigma_p (S-M)_{daily}}{\Sigma_p M_{daily}} \quad (3)$$

$$RMSE = \sqrt{\Sigma \frac{(S-M)^2_{daily}}{N_{interval}}} \quad (4)$$

$$A_{hourly} = \frac{\Sigma_p M_{interval}}{N_{interval}} \quad (5)$$

$$CvRMSE\ (\%) = \frac{RMSE_{period}}{A_{period}} * 100 \quad (6)$$

In the equations: S is the simulated data, M is the measured data, and N is the number of sampling. Obviously, the lower the value of these indicators, the better the quality and the predictive capacity of the numerical model.

The calibrated model is used for a CFD (Computational Fluid Dynamics) analysis aimed at verifying the thermal comfort indicators at some critical points. The CFD technique is used for the analysis of the heat transfer by following these two steps:

- discretization of the governing differential equation using numerical methods;
- solving of the discretized version of equation with high-performance computers.

Unlike an energy simulation that gives results in which the investigated variables assume different values over time (i.e., hourly results), before running a CFD simulation, a precise time instant must be fixed. In this study, CFD simulations are carried out in the days reported in Table 1, and the results will be shown for three different times, that is, 10:00, 14:00, and 18:00.

CFD analysis is a space-variant analysis. The calculation method requires that the geometric space across which the calculations are conducted is firstly divided into a number of non-overlapping adjoining cells, which are collectively known as the finite volume grid (or mesh). The grid used by DesignBuilder CFD is a non-uniform rectilinear Cartesian grid, and the adopted grid has been generated with 0.15 m grid spacing.

The equations are solved considering the 2-k-ε turbulence model that involves replacing the instantaneous velocity in the Navier–Stokes and energy equations with a mean and fluctuating component. The calculation process involves replacing the defining set of partial differential equations with a set of finite difference equations. The upwind scheme has been used. It allows the convective term to be calculated assuming that the value of the dependent variable at a cell interface is equal to the value at the cell on the upwind side of the interface.

In more detail, 2D distributions of indoor air temperature and 3D distributions of PPD and PMV inside the building will be shown.

In this season, the energy balance between the generation from the PV system and the overall energy consumption allows us to understand the potential daily energy saving and the best management strategy for the trade-off between the energy requirements and production.

3. Results

3.1. Wintertime Assessment Results

In this section, firstly, the characterization of thermo-hygrometric conditions inside the building during the wintertime is discussed; subsequently, the energy consumption and the energy balance are shown. During this period, the air-to-air heat pump provides temperature control and mechanical ventilation with an internal filter that purifies the air before supplying it into the building. That system is also provided with active thermodynamic heat recovery, and this means that part of the exhaust air interacts with the outdoor heat exchanger (evaporator in winter), so that the thermal level of the ambient air is more suitable to discharge or supply energy. Another part of return air is recirculated, mixed with primary (outdoor) air, filtered again, and handled in the packaged heat pump before the new supply into the rooms.

3.1.1. Indoor Comfort Analysis by Means of Measured Variables

Figure 2 shows the monitored values of relative humidity inside and outside the building for the considered periods with variable setpoint temperatures (indicated in the orange boxes), taking into consideration the living room and Bedroom 1 for the first week. These trends allow us to conclude that the setpoint temperature does not influence the value of relative humidity, which varies between 30% and 60% in both rooms; it seems mainly influenced by the outdoor value and its interaction with the recirculated flow. In more detail, considering 12 January and 8 December, when the setpoint is 20 °C and the outdoor relative humidity varies respectively from 48% to 87% and 62% to 96%, in the living room, the recorded values vary between 35% and 42% in January (with lower outdoor values) and between 46% and 56% in December.

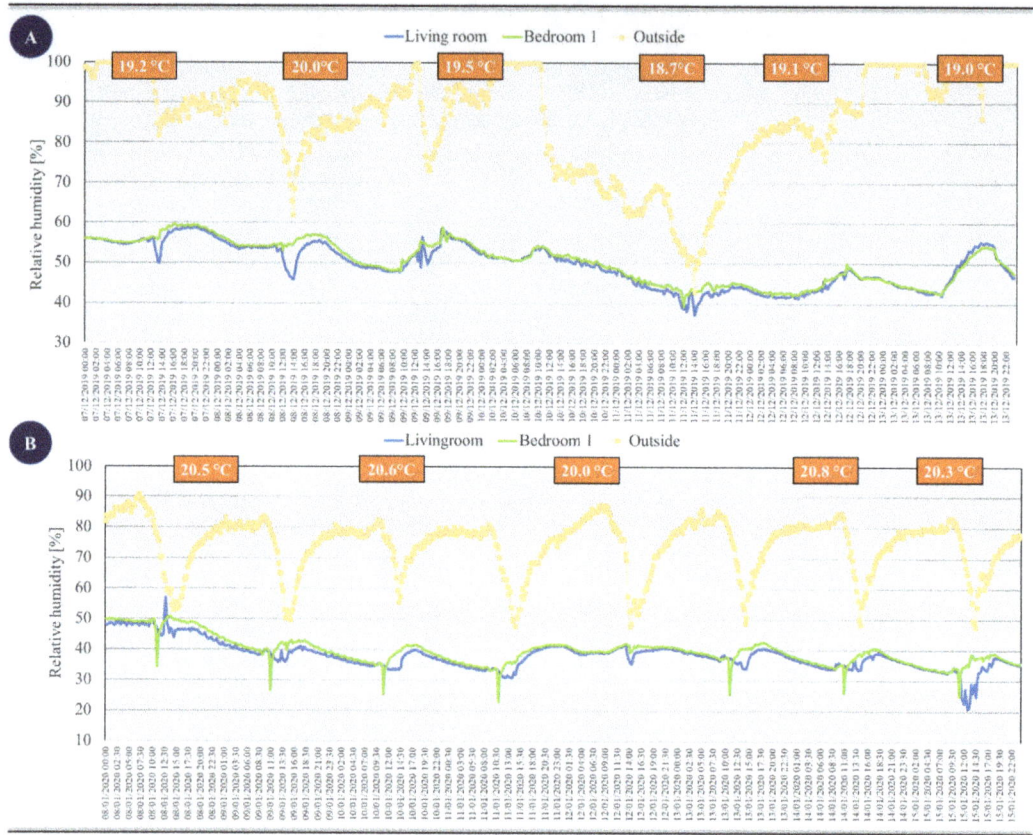

Figure 2. Relative humidity trend: (**A**) setpoint lower that 20 °C and (**B**) upper 20 °C.

The influence of solar gains on the humidity value is also notable. Indeed, considering the same days between 12:00 and 15:00, in the living room there is a sensible increment of the indoor temperature and a decrement of the relative humidity. In this time interval, the maximum solar gains are achieved, and during 12 January, the temperature passes from 18.3 °C to 21.4 °C, and the indoor humidity from 41% to 36%. Similarly, during 8 December, the temperature goes from 19.8 °C to 21.4 °C, and the indoor humidity from 54% to 50%.

Moreover, when the setpoint is increased (14 January, 20.8 °C) and the outdoor relative humidity is comparable to the 12 January (48–85%), the indoor value varies from 33% to 39%, as in the case of a setpoint of 20 °C. The same observation can be made for 12 December, with a lower setpoint.

Figure 3A reports the monitored air temperature during 7–13 December. In this period, the external temperature varies in the range between 4 °C and 14 °C, and in all days, the temperature monitored in the bedroom is more uniform. It is some tenths higher than the living room value, and thus closer to the setpoint. It is due to the higher percentage of glazed surfaces in the living room. Even by using a setpoint lower than 20 °C, the temperature inside the building is within 17.5 °C and 19.5 °C. Only in some conditions, when the solar radiation gives an important contribution, does the air temperature rise up to the setpoint value. For instance, during 11 December, the setpoint is 18.7 °C, but inside the living room, during the afternoon, the sensor recorded 21.6 °C. However, the occupant perception is not positive for this management strategy, because they have described the indoor condition as "slightly cold" for all days.

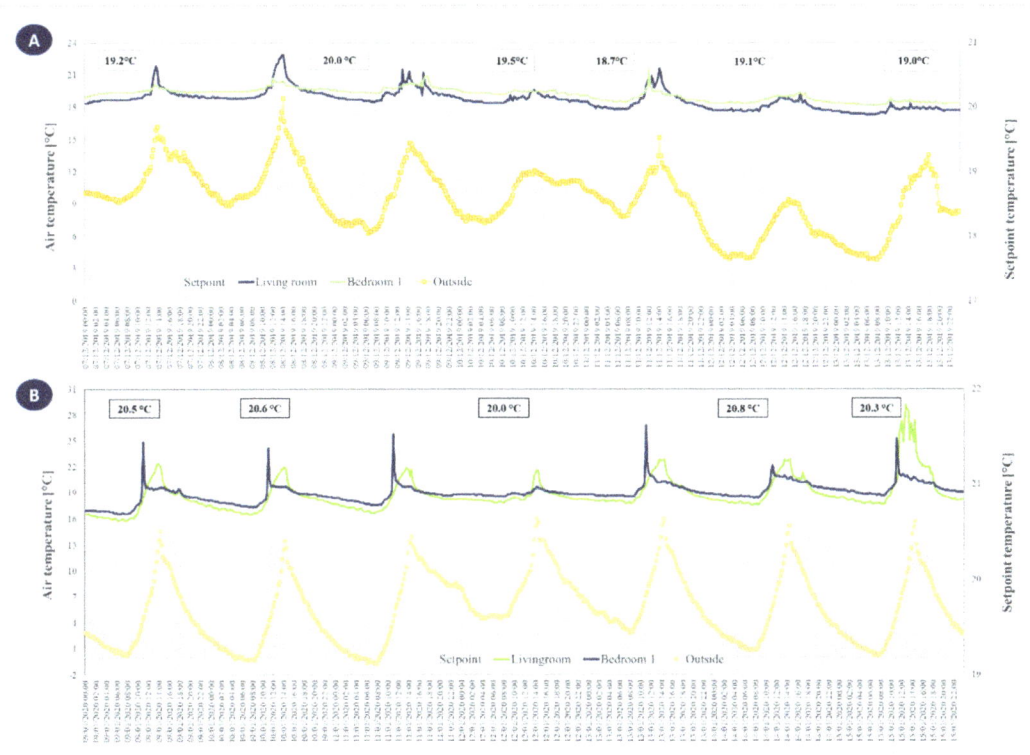

Figure 3. Air temperature trend: (**A**) setpoint lower that 20 °C and (**B**) upper 20 °C.

In the period 9–15 January, the setpoint was increased; meanwhile, the outdoor temperature varied between 0 °C and 14 °C. For all days, the value of the air temperature is usually lower than the setpoint, except during the afternoon, when in the living room the temperature rises up 20 °C, with a maximum value of 28 °C during 15 January that is characterized by a solar radiation (monitored on the roof) of around 400 W/m².

It is interesting that when the outside temperature decreases during the night, the temperature inside the building drops about two degrees, even if the setpoint is settled at 20 °C. For these days, the occupants have affirmed they are in comfort conditions. This is a notable conclusion that underlines how the occupants are influenced by the knowledge of the test to which they are subjected, and how the comfort has a significant psychological component. Indeed, it can be objectively observed that the indoor trends of the two periods are comparable.

This conclusion is also supported by comparison of the temperature trend in three days with extreme setpoints: 9 December, 11 December, and 14 January. Indeed, with 20 °C and mild external conditions (09/01, outdoor temperature from 6.3 °C to 14.7 °C) in the living room, the temperature ranges between 18.5 and 21.5 °C, with the maximum value at 14:00, when there is the maximum outdoor temperature and the solar radiation on south wall; in Bedroom 1, this range is 19.3–20.9 °C (maximum at 11:45). For comparable outdoor conditions (11/01, outdoor temperature from 5.6 °C to 15.2 °C) and the lowest setpoint, the temperature varies between 17.6 °C and 21.6 °C, with maximum value at 14:00 in the living room, and between 18.4 °C and 21.6 °C (maximum at 12:00) in Bedroom 1. Finally, with the higher setpoint and extreme variable conditions (14/01, outdoor temperature from 0.9 °C to 15.2 °C), in the living room the recorded temperature varies between 17.6 °C and 22.9 °C (maximum at 13:30), and in Bedroom 1 between 18.4 °C and 22.2 °C (maximum at 11:15).

Two main conclusions can be found. Firstly, the design of glazed elements assures comparable peak values of indoor temperature during the early afternoon when the setpoint temperature is different, and thus it seems to be the main force of the energy balance. Secondly, for the case study, the solar contribution mainly regards the window on the south-west exposure of the living room (in the early afternoon); meanwhile, the designed porch reduces the effect of solar gains on the south window during the late morning. Instead, Bedroom 1 receives the benefits of solar radiation during the late morning.

Table 3 shows the value of the comfort indices calculated with the monitored values during the days with occupants. These have been calculated according to the model proposed in [56] with the implementation of the equations for PMV and PPD in a mathematical sheet. Herein, the measured parameters are used as input data as explained in the previous section. The metabolic rate of individuals in this study was assumed to be uniform at 1.1 met (corresponding to work office); winter clothing resistance of 1 clo was also assumed.

Table 3. PMV and PPD calculated indices.

		PMV (%)	PPD (%)	PMV (%)	PPD (%)
		Living Room		Bedroom	
9 December 2019	10:00	−0.5	10.2	−0.3	6.9
	13:00	−0.3	6.9	−0.3	6.9
	17:00	−0.2	5.8	−0.1	5.2
10 December 2020	10:00	−0.6	12.5	−0.4	8.3
	13:00	−0.5	10.2	−0.4	8.3
	17:00	−0.6	12.5	−0.4	8.3
11 December 2019	10:00	−0.7	15.3	−0.5	10.2
	13:00	−0.4	8.3	−0.5	10.2
	17:00	−0.5	12.5	−0.5	10.2
12 December 2019	10:00	−0.8	18.5	−0.6	12.5
	13:00	−0.7	15.3	−0.5	10.2
	17:00	−0.6	12.5	−0.5	10.2
13 December 2019	10:00	−0.8	18.5	−0.6	12.5
	13:00	−0.8	18.5	−0.6	12.5
	17:00	−0.8	18.5	−0.6	12.5
9 January 2020	10:00	−1.3	40.3	−1.3	40.3
	13:00	−0.6	12.5	−0.9	22.1
	17:00	−0.9	22.1	−0.8	18.5
10 January 2020	10:00	−0.9	22.1	−0.8	18.5
	13:00	−0.2	5.8	−0.5	10.2
	17:00	−0.6	12.5	−0.5	10.2
13 January 2020	10:00	−0.9	22.1	−0.7	15.3
	13:00	−0.2	5.8	−0.5	10.2
	17:00	−0.7	15.3	−0.6	12.5
14 January 2020	10:00	−0.5	10.2	−0.4	8.3
	13:00	0.0	5.0	−0.3	6.9
	17:00	−0.4	8.3	−0.4	8.3
15 January 2020	10:00	−0.6	12.5	−0.4	8.3
	13:00	0.0	5.0	−0.2	5.8
	17:00	−0.3	6.9	−0.3	6.9

Considering the obtained results, for the bedroom, the PMV values are not really different in the considered weeks, with a presumable opinion of "slightly cold", except during 9 January, which is really cold according to the calculated indices. Considering the living room, during 9 December, 11 December, and 14 January, the analytical calculation of the PMV and PPD indicates comparable conditions, with a setpoint of 20 °C or higher (14 January). The main difference is at 13:00, but this is attributable to the direct solar radiation and not to the heating system; indeed, at this time, the HVAC is turned off, because in both days the setpoint has been exceeded. In Bedroom 1, the indices for 9 December are even better than 14 January. Instead, when the lower setpoint is set, the conditions are more critical in both rooms, but these are quite acceptable according to [51,52,57], since the PMV is 0.5. The values in the living room, once again, underline the important role of solar gains. Indeed, in the morning the heating system is not able to meet comfort conditions, but at 13:00, the comfort indices indicate pleasant conditions, comparable to the other days. When the sun has set, the living room is colder, and the PPD is higher than the threshold value.

When the setpoint is 18.7 °C, at 10:00, the living room seems to be more comfortable according to PMV, and this happens also at 17:00; on the other hand, during the early afternoon at 13:00, the higher setpoint causes only 5.8% of people to be dissatisfied and results in a lower value of 8.3%. However, in both cases the indoor conditions could be considered acceptable. Regarding the external conditions, the solar radiation monitored both on vertical and horizontal surfaces was higher during 13 January. This condition may have contributed to different occupants' perception.

More generally, the proposed analysis suggests that the occupants, in their environmental assessment, are affected both by the external weather conditions and by the knowledge of the test being carried out. Indeed, the second week was characterized by an average value of solar radiation higher than the considered week in December, and the sky was usually clear. Probably, this condition led the occupants to consider the rooms more comfortable, due to the solar gains, and also because the indoor temperatures were comparable. Moreover, it can be remarked that when the room is characterized by large glazed surfaces, the contribution of solar radiation also allows one to balance the indoor conditions when a low setpoint is considered. For the case study, the temperature inside the room can exceed the setpoint by even more than 3 °C. The proposed data suggest that for reducing the heating demand in highly insulated buildings, the setpoint can be decided starting from the external conditions, but it could be better to take into account the solar radiation and not only the air temperature for the formulation of an analytical relation for the setting.

3.1.2. Building Energy Balance from Monitored Data

The impact of the selected setpoint on the energy balance can be found in Figure 4; it reports the monitored daily energy consumption for both selected weeks. The overall energy consumption and the separated contributions of the main loads are plotted. The slight difference is attributable to the energy consumption for the monitoring system (sensors and computer) with which the BNZEB is equipped.

The results show that the HVAC system covers around 63–72% and 72–85% of the total building energy consumption, respectively, for the first and second week. The artificial lighting accounts for around 2% of the overall energy consumption. The analysis is not easy, and it needs to consider the global external conditions. For 11 December and 13 January, with comparable outdoor temperature and relative humidity (e.g., average daily temperature respectively of 7.2 °C and 6.9 °C), the adoption of a lower setpoint assures a reduction of heating consumption of 43%, while also maintaining comfortable conditions according to the previous analysis. On the other hand, when the comparison is done with 9 December, characterized by milder external conditions, the reduction is 17%. In this case, the heating demand is influenced by lower heat losses and high solar gains during the day, with a setpoint of 20 °C, and thus the benefit of the adoption of 18.7 °C is lower. Considering 8 January, with an average daily temperature of 6.7 °C and minimum and maximum peak values of 3 °C and 14 °C, the heating consumption increases by 65%, passing from 20 °C to 20.3 °C, and it is more than three times higher than the case with 18.7 °C.

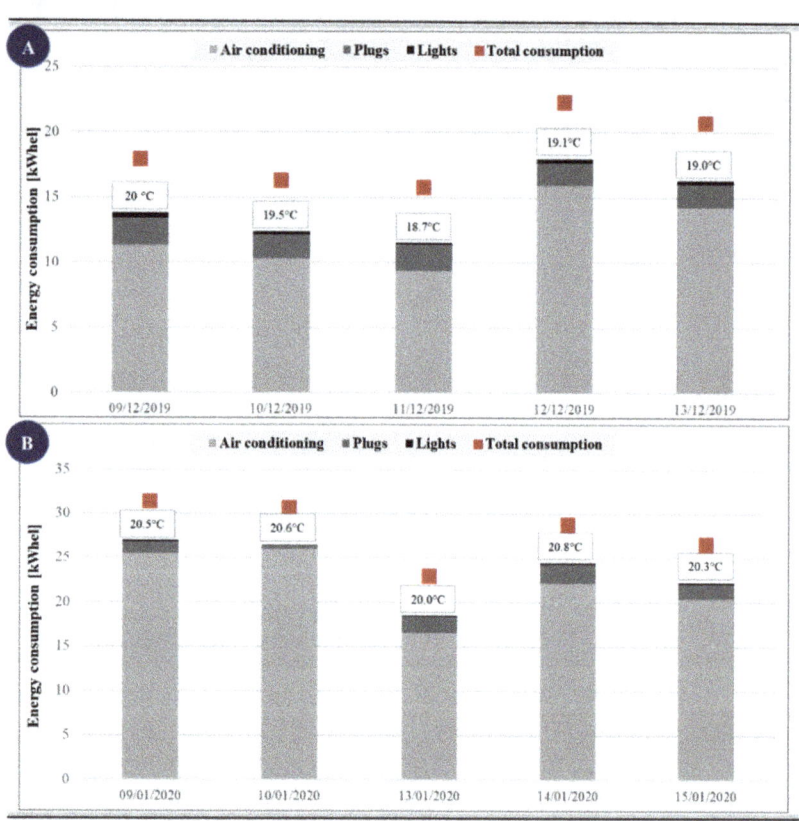

Figure 4. Energy consumption trend: (A) setpoint lower that 20 °C and (B) upper 20 °C.

The variation of the energy consumption cannot be charged only to the adopted setpoint; the evaluation of external conditions that can positively (solar gains) or negatively (heat losses due to colder conditions) influence the heating request is also important. For instance, considering 13 December, with 19.0 °C as setpoint and an average daily temperature of 7.3 °C, if the comparison is done with 9 December (mild climate, average daily temperature of 9.7 °C), it is found that the heating request increases by 26%; if it is compared with 13 January instead, the heating request is lower by around 14%.

These data, based not on simulations but on monitoring of real conditions, confirm that the management strategy of the heating system has a great influence on the building's energy balance. However, the adopted strategy can be decided taking into consideration, dynamically, the external conditions for reaching the energy-saving and the thermal comfort objectives.

Figure 5 shows the daily energy balance considering the electric energy consumption and the generation from the PV system. Table 4 reports the introduced hourly indices.

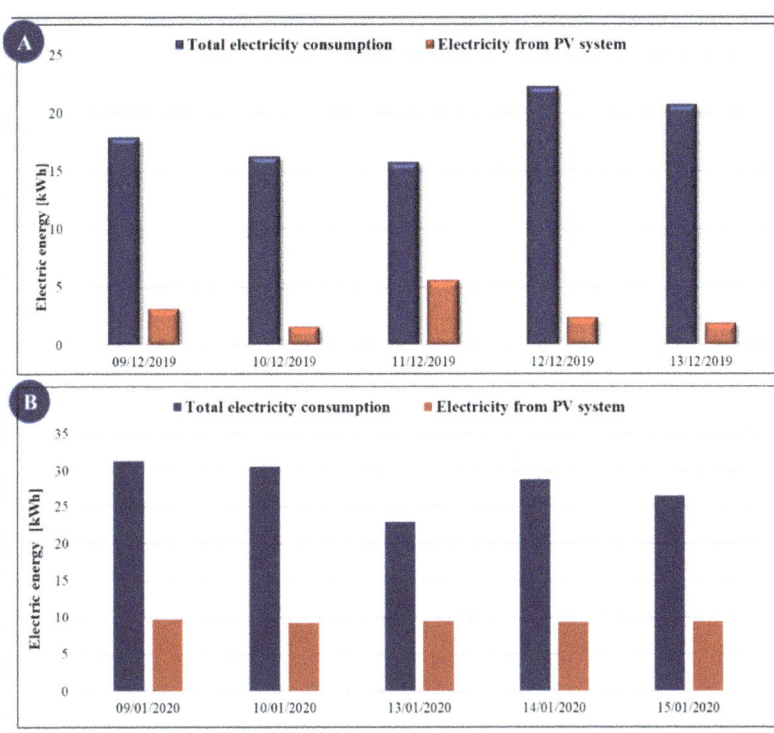

Figure 5. Daily energy balance between energy consumption and PV generation: (**A**) setpoint lower that 20 °C and (**B**) upper 20 °C.

Table 4. Hourly indices regarding energy balance.

	RenEl	PV$_{in}$	F$_{loadmatch}$
9 December 2019	17.8%	17.8%	17.4%
10 December 2019	10.1%	10.7%	11.21%
11 December 2019	35.8%	35.8%	34.8%
12 December 2019	12.0%	11.0%	10.9%
13 December 2019	9.28%	9.28%	8.73%
9 January 2020	30.9%	30.9%	35.7%
10 January 2020	30.1%	30.1%	36.2%
13 January 2020	41.2%	41.2%	52.0%
14 January 2020	32.3%	32.3%	39.1%
15 January 2020	35.4%	35.4%	48.4%

During the winter period, the renewable electricity generation cannot satisfy the energy need of the building, and the selected setpoint does not significantly change the balance. Indeed, the RenEl indicates that the renewable energy used for the electricity request cannot reach 50% during the analyzed days. More generally, the comparison of RenEl and PV$_{in}$ indicates that thanks to the energy storage, the used renewable energy coincides with the production in almost all days. This testifies that the BNZEB has been designed to maximize self-consumption, but the production is usually lower than the expected, and for this reason the building requires for more time using electricity from the energy grid. The Load Match Index results are within 11% and 56%, and thus the percentage of electrical demand covered by on-site generation at the hourly level is very low.

There is not a unique interpretation for the incidence of the adopted setpoint. Indeed, the effect on the renewable integration requires us to take into consideration the global external conditions. Globally, the first considered week was cloudier that the second one, and this justifies the lower value of the indices if the adopted setpoints were lower. In more detail, for 11 December and 13 January, also with comparable conditions, it is not possible to compare the results. Indeed, during both days, all produced electricity has been used by the building (RenEl is equal to PV_{in}), but thanks to higher solar radiation on the roof, the production during 13 January is almost double compared to 11 December; thus, the results seem better than the case with the lower setpoint. However, 11 December, with the lowest setpoint, is characterized by a higher percentage of integration during the first selected week, and this is due to both the lower consumption and the higher production of all other days.

In the second considered week, the producibility of the PV system is comparable in all days; thus, some comparisons can be made. Looking at 13 January, with the lower setpoint, it is characterized by the maximum integration, since the renewable production (9450 Wh) can satisfy 41.2% of the consumption, and the match between the request and the production is higher than in the other cases (52%). With a comparable production (9415 Wh) and a setpoint of 20.3 °C, the integration becomes 35.4%.

The proposed data indicate that it is difficult to establish a relation between the utilization of renewable in situ production and the adopted setpoint, because the production is related to the solar radiation captured by the panels, which is a variable that does not influence the needs of the building. However, when the renewable production is comparable, a decimal variation of the adopted setpoint greatly influences the integration rate.

Based on the obtained results, it can be concluded that the external temperature monitored in the days before the setting could be used for deciding the setpoint in a high-performance building. However, the solar radiation in the most glazed room should be used for selecting the management strategy. The analytical evaluation of comfort parameters demonstrates that there is not a notable variation in the people dissatisfied, and the integration of renewable adoption can be increased if the external conditions are not favorable.

Other tests are needed for evaluating the incidence of external conditions on the occupants' perception.

3.2. Summertime Assessment Results

As explained in the methodological section, four different configurations of HVAC system were tested and compared by means of simulations as well as monitoring results.

3.2.1. Simulation Analysis of Indoor Comfort Conditions

The simulation model has been verified for the variables that are used during the CFD investigations. The output of simulations performed in all different configurations were compared in terms of indoor air temperature and relative humidity with the measured data. Table 5 shows the evaluated indices for the air temperature and the relative humidity in Bedroom 1 (Figure 1B). The results are always within the admitted tolerances; in particular, the M&G Guidelines [60] require for the MBE a value $\leq 10\%$ and for the CvRMSE $\leq 30\%$ when the validation is performed with hourly data. All indices are below the threshold value provided by [60]; thus, the model can be considered well calibrated and well representative of the real conditions inside the building for all configurations.

Table 5. HVAC system configurations and corresponding calibration indices for one building room.

	HVAC System		MBE (%)	CvRMSE (%)	MBE (%)	CvRMSE (%)	Evaluation Period
	Cooling System	Pre-Cooling Activation	Air Temperature		Relative Humidity		
C1	Heat pump	Off	−1.55	3.11	5.94	11.56	8–12 July
C2	Heat pump	On	1.86	−1.91	−7.79	10.74	15–18 July
C3	DX system	On	−3.66	−1.87	−6.38	8.95	22–25 July
C4	DX system	Off	3.04	3.08	0.23	5.91	27–30 July

The calibrated numerical model of the building has been used with the purpose of investigating the air temperature and the distribution of global comfort indices by means of the CFD tool of DesignBuilder. In more detail, before running the simulation, all boundary conditions have been fixed by using the measured values for surface and window temperatures, HVAC supplies, extracts, and so on. In the various rooms, the cooled air is introduced by means of vents, which are reproduced by boundary-defining conditions such as the minimum velocity of the air jet, the supply temperature, the supply angle, and the flow rate; then, the air is extracted by extraction vents. Moreover, CFD calculations have been performed by selecting the values for metabolic rate and clothing thermal resistance of 1 met (light-intensity activity) and 0.5 clo for the clothing insulation (a typical summer value), respectively.

Table 6 shows the selected days and the indication of the HVAC configuration. During these days, the setpoint temperature inside the building is 25 °C, and the simulations are carried out at three different times, that is, 10:00, 14:00, and 18:00. Table 5 shows the values of external temperature and the average indoor temperature for each day.

Table 6. Investigated hours: outdoor temperature and average indoor temperature.

HVAC Configuration and Investigated Day	Investigated Hours	Outdoor Temperature (°C)	Average Indoor Temperature (°C)
C1—12 July	10:00	27.9	24.5
	14:00	31.9	24.9
	18:00	24.2	24.1
C2—17 July	10:00	26.6	25.1
	14:00	38.7	24.6
	18:00	33.6	24.3
C3—24 July	10:00	31.3	24.3
	14:00	37.9	24.8
	18:00	39.5	24.5
C4—30 July	10:00	27.6	24.2
	14:00	33.6	24.4
	18:00	30.7	24.1

Considering 12 July, at 10:00, Figure 6A shows the 2D distribution of the indoor air temperature; the 3D distributions of PPD and PMV inside the building are reported in Figure 6B,C, respectively. The distribution of indoor air temperature varies around the real average indoor value indicated in Table 6.

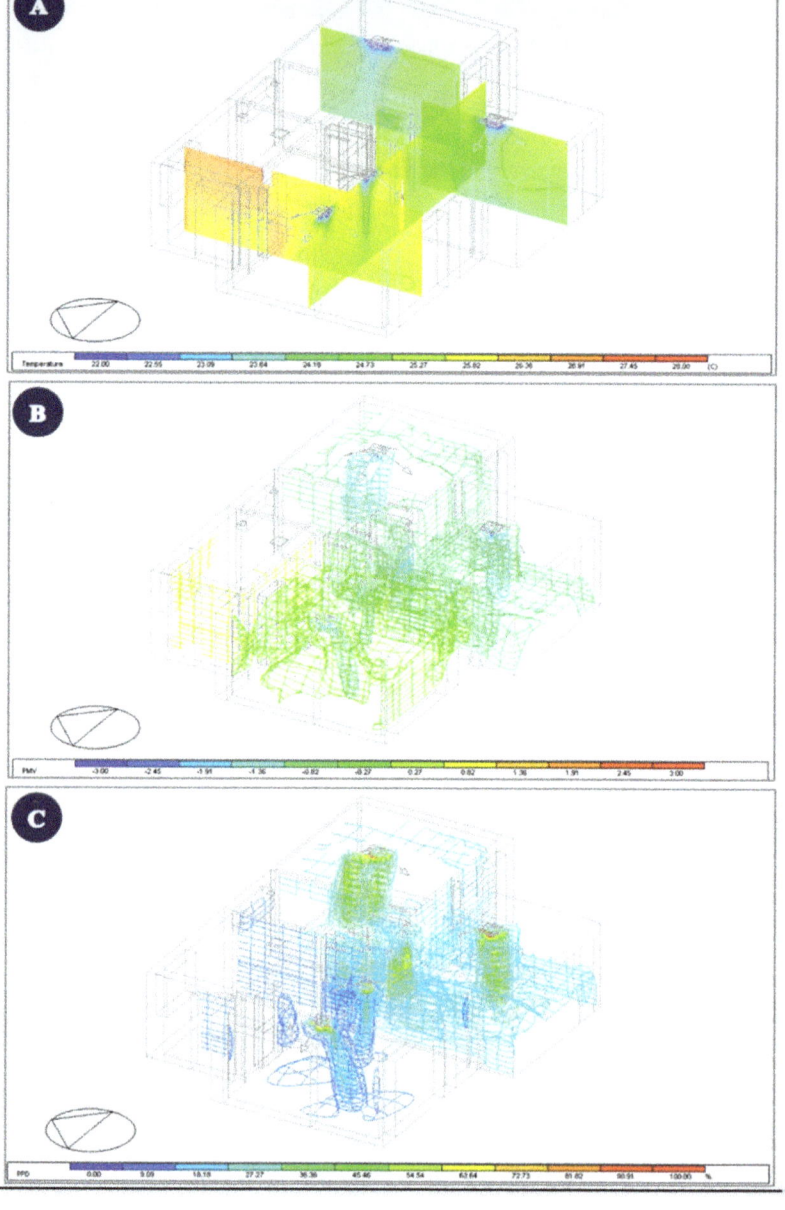

Figure 6. CFD results on 12 July at 10:00: (**A**) air temperature 2D distribution; (**B**) PMV and (**C**) PPD 3D distribution.

In particular, the bedrooms are characterized by rapid temperature equalization, mainly in Bedroom 2, which is disadvantaged in terms of solar gains due to the prevalent exposure. As indicated in Figure 6B, the temperature distribution in Bedroom 1 determines a "slightly cold" sensation, according to PMV distribution, except at the vent where the air volume is associated with a cold flow, which corresponds (Figure 6C) to a PPD higher than the threshold value of 10% [59]. The punctual value of these indices in the room center at a height of 1.1 m (man sitting at a table) is reported in Table 7 for the coldest (Bedroom 2) and the hottest (kitchen) rooms and for the living room. The PPD is 20%, and this indicates that without the contribution of inner and solar gains, the temperature of the supply air is low for Bedroom 2. This could be maintained in comfort conditions with a higher setpoint and thus lower cooling consumption.

Table 7. PMV and PPD values calculated on 12 July.

12 July	Bedroom 2			Kitchen			Living Room		
	10:00	14:00	18:00	10:00	14:00	18:00	10:00	14:00	18:00
Operative temperature (°C)	24	23.7	23.6	26.4	26.2	25.5	25.2	26.3	25.3
PMV (%)	−0.83	−0.94	−0.98	0.05	−0.03	−0.28	−0.39	0.01	−0.36
PPD (%)	20	24	25	5	5	7	8	5	8

In the living room, the cold flow is neutralized by the solar gains and the volume assumes a higher temperature mainly near the windows, despite the presence of two vents. Figure 6B also indicates that the PMV is near a judgment of "slightly warm" in correspondence of the vents; indeed, as reported in Table 7, the PMV is 8%. The selected setpoint and the diffusion strategy are adequate for balancing the solar gains in the morning.

The kitchen is characterized by a higher level of temperature due to the absence of cooling vents, which also causes a stratification of the air due to a lack of significant convective flows and air speed. In this room, according to PMV and PPD (Table 6), the conditions are acceptable and more adequate than the other rooms.

Figure 7 shows the 2D distribution of air temperature inside the BNZEB, resulting from two CFD simulations concerning 12 July at 14:00 and 18:00, when the outside mean temperature was 31.9 °C and 24.2 °C, respectively.

At both times, the bedrooms are characterized by lower and more uniform values of temperature, and the path of the cold air introduced by the HVAC to the floor is also evident, mainly in Bedroom 2. The solar radiation, due to the two south-facing windows in Bedroom 1 and in the living room, enters inside the room, and it causes an increment of the air temperature. In the living room, the temperature is even higher than in Bedroom 1, despite the presence of two vents that provide cooled air. The kitchen is the hottest room of the dwelling, due to the absence of cooling vents, which also causes air stratification. However, according to values reported in Table 6, the indoor sensation complies with the comfort standard, except in the morning.

The values of PPD and PMV (Table 7) confirm that Bedroom 1 should be managed with a different setpoint, because the percentage of dissatisfied persons increases until it reaches 25% in the evening, and thus the room could also assure comfortable conditions without the active system due to effect of thermal inertia and lower inner solar gains during the night. On the other hand, the supply conditions (flow rates and temperature) are optimal for the living room, where the PPD is also always lower than 10% during the early afternoon when the solar radiation is incoming. The shading system is able to intercept the radiation; thus, the design of BNZEB seems to be adequate for reducing the overheating problem.

Figure 8 reports the 2D distributions of air temperature during 17 July. These values are the result of CFD simulations when the HVAC configuration was C2, and thus the heat pump worked to provide cooled air, and the ventilation air was pre-cooled by crossing the water-to-air heat exchanger.

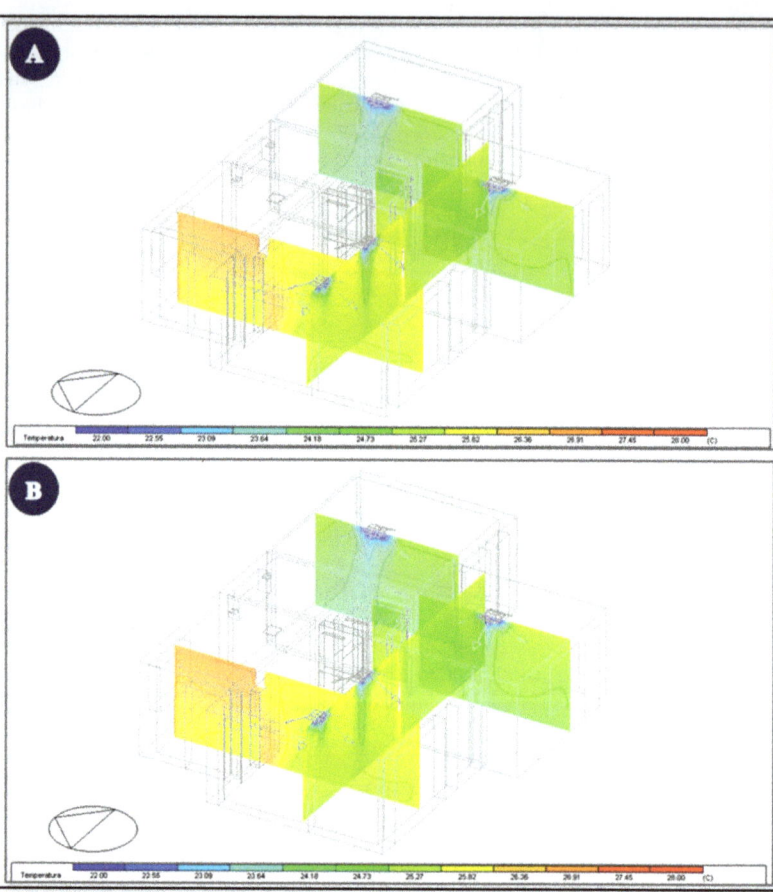

Figure 7. 2D distribution of air temperature on 12 July: (**A**) 14:00 and (**B**) 18:00.

The temperature on 17 July was warmer than 12 July, with a maximum temperature of 38.7 °C compared to 32 °C, but with comparable global solar radiation of 670 W/m² and 617 W/m², respectively. These conditions have led to comparable temperature distributions for both HVAC configurations. However, in this case, it can be noted that the bedrooms have more comparable conditions, and assuming as reference the PMV and PPD calculated for Bedroom 2 (see Table 8), a cold feeling could be experienced in the morning, with perfect conditions in the other investigated hours. In comparison with configuration C1, the living room is characterized by higher values of temperature; the temperature difference between the center surface and the window is around 6 °C. Thus, if the PPD is also lower than 10% in the afternoon and evening, considering the center of the room, some local discomfort phenomena could occur near the glazed surfaces. The kitchen, during the three considered hours, is characterized by a higher level of temperature compared to 12 July, due to both the absence of cooling ceiling diffusers and to the adjacency to the living room. Even in these cases, there is a stratification of the air in the kitchen. However, the comfort indices (Table 8) indicate that the room has acceptable conditions in the afternoon and evening.

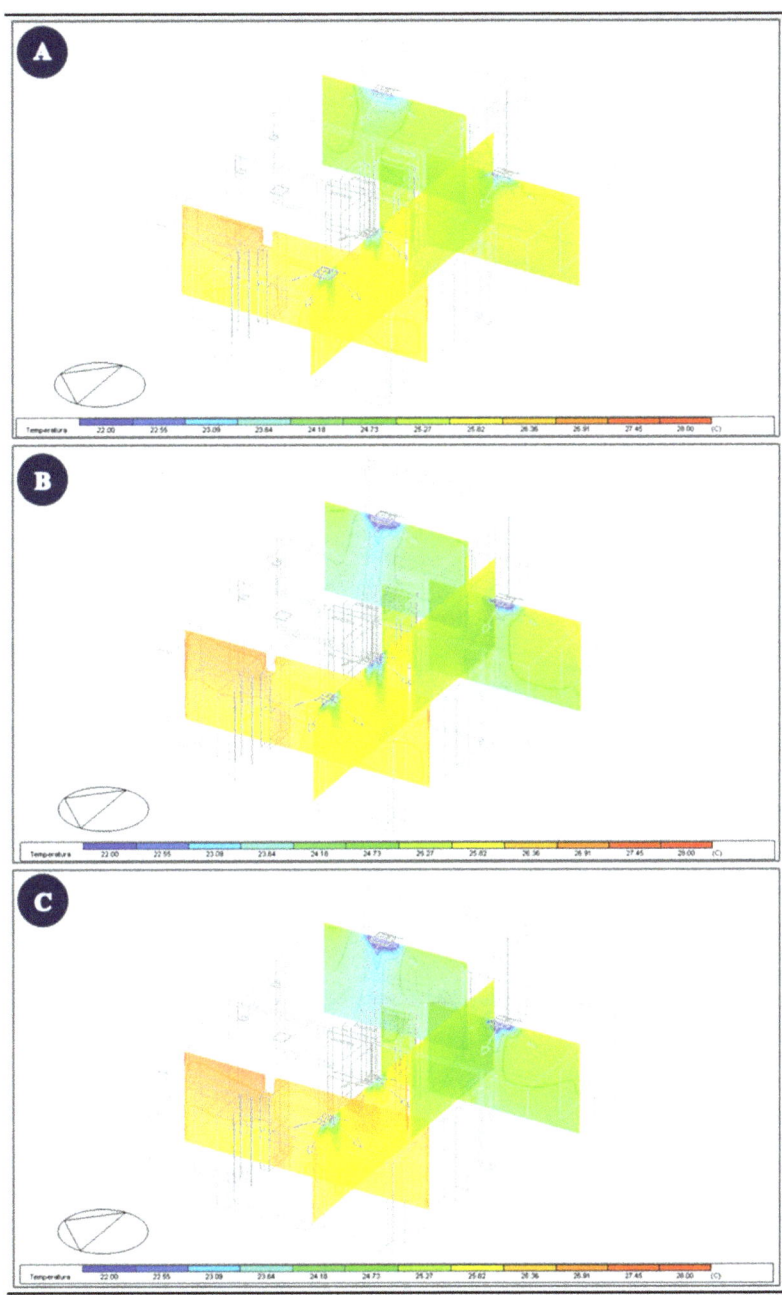

Figure 8. 2D distribution of air temperature on 17 July at (**A**) 10:00, (**B**) 14:00, and (**C**) 18:00.

Table 8. PMV and PPD values calculated on 17 July.

17 July	Bedroom 2			Kitchen			Living Room		
	10:00	14:00	18:00	10:00	14:00	18:00	10:00	14:00	18:00
Operative temperature (°C)	24.3	26.4	26.0	24.0	26.7	26.3	24.1	26.8	26.4
PMV (%)	−0.72	0.05	−0.1	−0.83	0.16	0.01	−0.80	0.19	0.05
PPD (%)	16	5	7	20	6	5	18	6	5

Briefly, the temperature profiles and the calculated indices are comparable to the previous case, and thus it can be concluded that considering thermal comfort, the pre-handling of the ventilation air does not make significant differences. What happens in terms of energy consumption has to be evaluated.

Moreover, it can be underlined that the selected setpoint as well as the temperature and flow rate of HVAC are suitable for the management of indoor conditions when the outdoor conditions are very warm. Starting from the temperature distribution until 10:00, for reducing the consumption without compromising the indoor comfort, it is probably possible to operate with a higher setpoint or supply conditions. The findings about the kitchen are also interesting; indeed, thanks to the designed architectural distribution, the selected glazed surfaces and the building's thermal mass assure comfortable conditions without direct cooling.

In Figure 9, the 2D distributions of air temperature inside the BNZEB are shown during 24 July for the selected hours. In these cases, the HVAC configuration was C3, and thus the heat pump worked only to provide ventilation, while the backup DX multi-split system worked to supply cooled air, and the outdoor air was pre-cooled by crossing the water-to-air heat exchanger.

At 14:00, the living room is the warmer room, but globally, the temperature distribution indicates lower values than 17 July, and the PPD is 6%. In this case, the supplied cold air is able to neutralize the effect of solar gains, and the more uniform distribution suggests that local discomfort phenomena should not occur. Bedroom 2 is the coldest room, and considering that the PPD is 25% both in the morning and in the evening, it is clear that less cold supply air could be used, or a lower flow rate.

According to Figure 9, Bedroom 1 is less affected by solar gains compared to the living room, and the temperature distributions are comparable in the three hours as reported in Table 9.

Table 9. PMV and PPD values calculated on 24 July.

24 July	Bedroom 2			Kitchen			Kitchen		
	10:00	14:00	18:00	10:00	14:00	18:00	10:00	14:00	18:00
Operative temperature [°C]	23.6	25.2	23.6	24.1	25.4	24.8	24.3	25.7	25.0
PMV [%]	−0.98	−0.39	−0.98	−0.80	−0.32	−0.54	−0.72	−0.21	−0.47
PPD [%]	25	8	25	18	7	11	16	6	10

Finally, in Figure 10 the 2D distributions of air temperature inside the BNZEB are shown for 30 July when the heat pump worked only to provide ventilation, while the backup DX multi-split system worked to supply cooled air. The outdoor conditions are similar to 12 July, with temperature variation between 20.0 °C and 33.6 °C, and maximum solar radiation of 622 W/m^2. In addition, during this day, it is clear that the average indoor temperature is lower compared to those obtained with C1 and C2 (12 and 17 July).

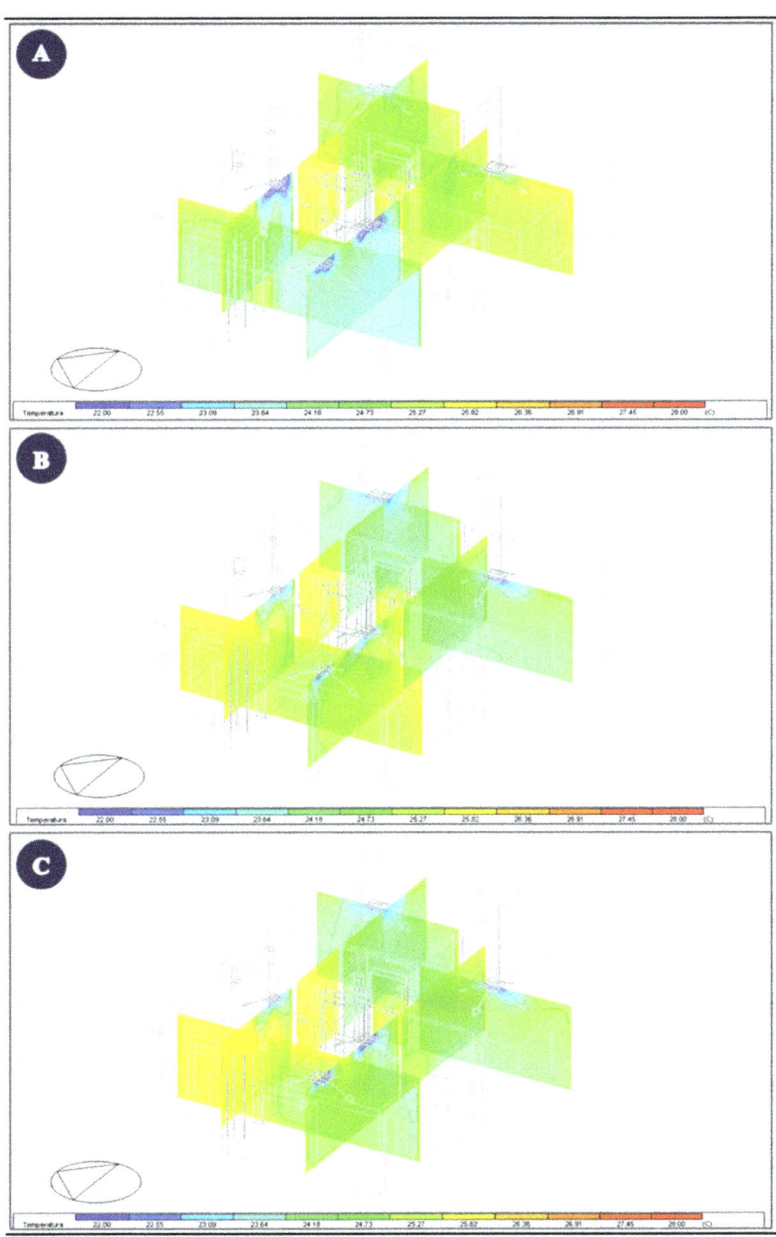

Figure 9. 2D distribution of air temperature on 24 July at (**A**) 10:00, (**B**) 14:00, and (**C**) 18:00.

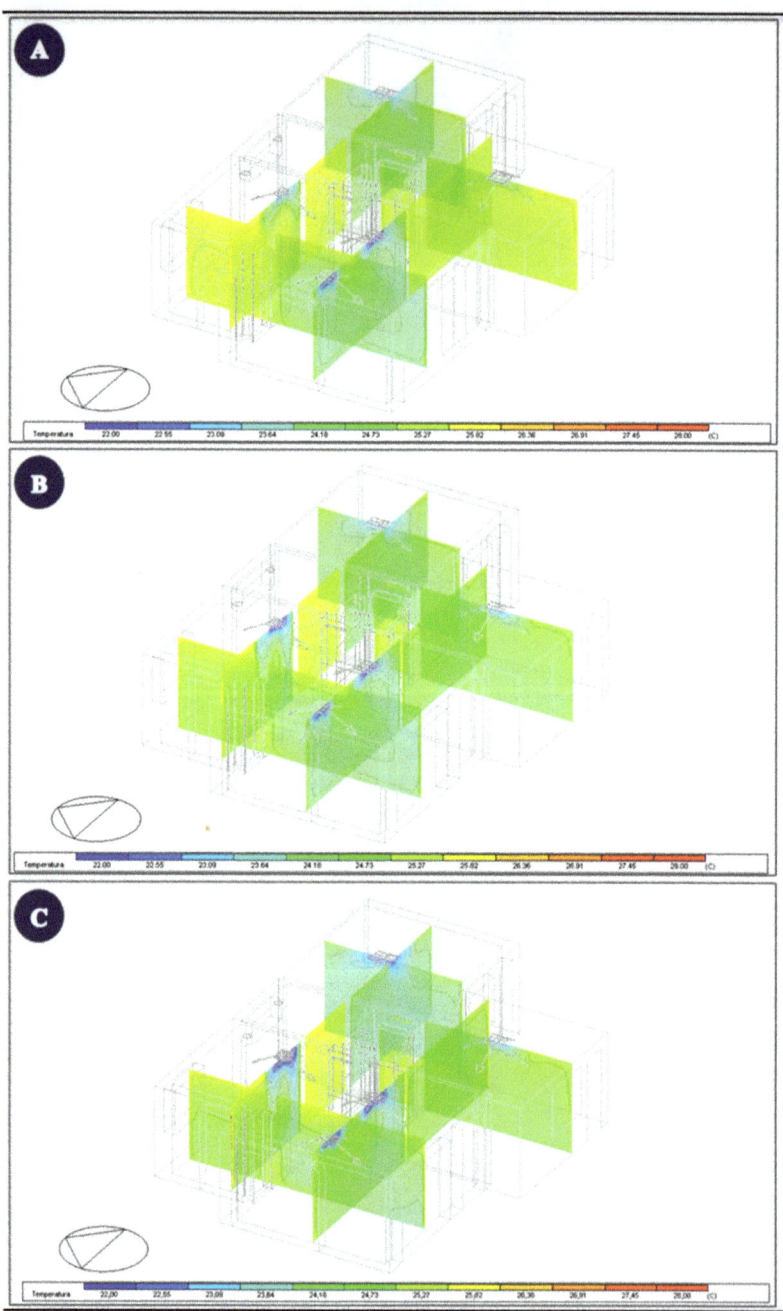

Figure 10. 2D distribution of air temperature on 30 July at: (**A**) 10:00, (**B**) 14:00, and (**C**) 18:00.

In more detail, Bedroom 2 is always the coldest room, due to its exposure and supplied air. In the living room, the management of HVAC system assures uniform and slightly cold conditions, which is also due to the induced cooling effect by the vent of the kitchen. The hottest room in these cases is also the kitchen, where due to the cooling vent, there is no stratification of the air; however, at the most critical time (14:00), the PPD is 8%. Really, the indoor conditions are quite uniform in the whole building, but at the most favorable times, the perceived sensation would be slightly cool, and this means that the cooling system provided too much cooled air. Detailed results are reported in Table 10.

Table 10. PMV and PPD values calculated on 30 July.

30 July	Bedroom 2			Kitchen			Living Room		
	10:00	14:00	18:00	10:00	14:00	18:00	10:00	14:00	18:00
Operative temperature (°C)	24	25.4	24	24.2	25.2	24.2	24	25.3	24
PMV (%)	−0.83	−0.32	−0.83	−0.76	−0.39	−0.76	−0.83	−0.36	−0.83
PPD (%)	20	7	20	17	8	17	20	8	20

In general, the monitoring results coupled with the CFD analysis have indicated that in the summertime as well, the management of the HVAC system could be decided with a dynamic approach based on the outdoor conditions. Indeed, for buildings such as the BNZEB with good passive control of indoor conditions, the supply variables and the setpoint could be changed during the day following the variation of outdoor temperature and solar radiation. Moreover, the analysis suggests that in residential buildings as well, the indoor comfort could benefit from adoption of differentiated microclimate control in each room. When the building envelope is designed with bioclimatic criteria, each room is characterized by different conditions during the day, and thus, higher energy savings could be obtained with an adaptive regulation system.

The CFD analysis performed according to the four different configurations allowed us to verify that the pre-handling system has no effect on the indoor comfort. Indeed, it only guarantees a lower temperature of the air supplied to the heat pump; therefore, a saving of the energy demand and costs are expected, while the supply temperature, operated by the A/C vents, does not change. Moreover, it can be noticed that during the summertime, the pre-handling of the ventilation air, based on air pre-treatment, does not make substantial differences in terms of indoor comfort, or in terms of energy consumption.

3.2.2. Building Energy Balance from Monitored Data

Figure 11 shows the daily energy balance, and thus the electricity consumption, the generation from PV system, and the electricity available in the battery. Table 11 reports the calculation of the introduced hourly indices for evaluating the incidence of production from the photovoltaic system at a smaller time scale.

Globally, it can be found that the most convenient configuration is C4, because in each day the energy consumptions are lower than the other configurations. A further optimization is also possible according to the CFD analysis, because the DX system could be managed with a higher setpoint or higher supply temperature with a reduction of the energy consumption. Instead, the comparison between C1 and C2 highlights that the adoption of the heat pump requires the integration of a pre-handling unit to be convenient in terms of energy request.

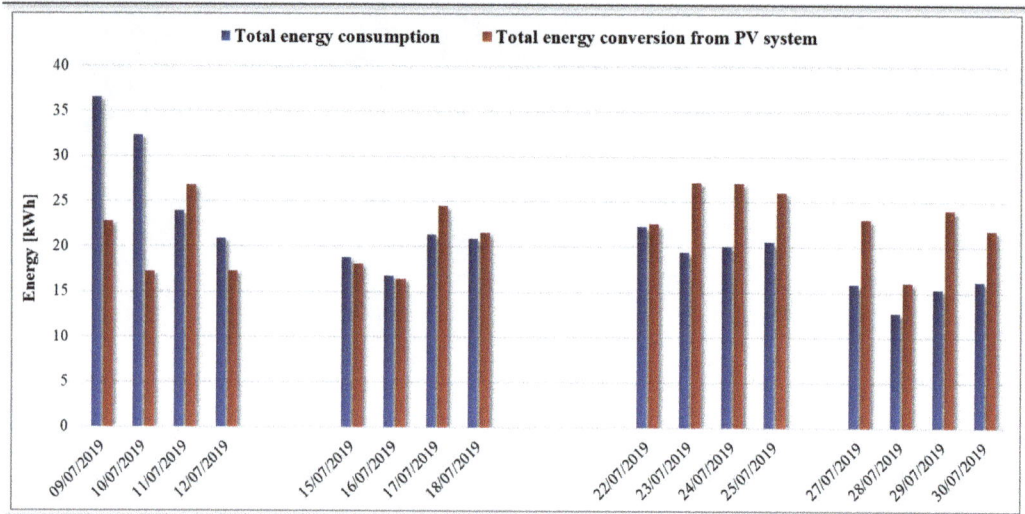

Figure 11. Daily energy balance between energy consumption and PV generation.

Table 11. Hourly indices concerning energy balance in the building.

	RenEl (%)	PV (%)	$F_{load\ match}$ (%)
9 July	63.0	62.5	60.9
10 July	57.0	53.3	50.4
11 July	85.7	111.8	61.8
12 July	80.4	82.60	54.2
15 July	81.9	96.3	60.5
16 July	86.4	97.8	73.6
17 July	95.5	114.8	90.8
18 July	80.4	103.0	56.8
22 July	80.9	101.3	72.2
23 July	81.6	139.8	72.4
24 July	89.2	135.2	81.1
25 July	85.3	126.3	75.0
27 July	87.2	145.0	79.5
28 July	85.8	126.3	84.2
29 July	95.4	156.6	91.8
30 July	95.7	134.8	92.1

With regard to the renewable production, the PV_{in} index is always high and, starting from 17 July, it is always more than 100%, meaning that the PV production, from 16 to 27 kWh/day, exceeds the daily electricity demand of the building. This value does not indicate that the energy request is completely covered by the generation from the PV system as represented through the RenEl index, which is between 56% to 97%. This finding is important. Indeed, with a monthly or daily approach, it could be said that the renewable production is able to satisfy the entire energy demand of the building within the considered period. Instead, with a short timestep, the unbalance between the demand and the production and the importance of working on a more adequate management strategy for maximizing the matching on an hourly basis is clear. The indication of the load match for the case study by means of the $F_{load\ match}$ is quite satisfactory, being between 50% and 92%.

For a better understanding of these considerations, the days analyzed in terms of thermal comfort can be compared. The measures indicate that for 12 July, the energy consumption for air conditioning is 15.7 kWh, while, on 17 July, this is equal to 12.9 kWh, and thus about 18% lower. This result confirms that with the same indoor conditions and comparable thermal comfort indices, the pre-handling unit allows one to reduce the energy consumption. This finding is even more interesting taking into consideration that 17 July is characterized by warmer conditions. The configuration C2 also increases the matching between the renewable production and the energy demand, with the $F_{load\ match}$ that passes from 54% to 91%.

The comparison between 24 and 17 July, with comparable outdoor conditions, indicates that the C3 configuration is characterized by lower consumption (around 6%). The energy saving could be higher taking into consideration the findings of the CFD analysis. Indeed, with the DX system, the supply conditions or the setpoint can be set at a higher level without compromising the comfort, and this management action could assure lower requests. However, it is also important to understand that the configuration C2 guarantees higher integration of the renewable source. Indeed, despite the higher production of 24 July (PV equal to 135%), the $F_{load\ match}$ is 81%.

Moreover, the daily energy consumption of the configuration C3 is higher by around 46% compared to C4, considering 24 and 30 July. Although this result seems predictive of the inconvenience of the pre-handling system, it must be considered that 24 July has been characterized by more heavy external weather conditions than 30 July. This last day assures a higher load match (92%), and almost the entire energy demand (95.7%) is covered by in situ production.

The comparison between C1 and C4 suggests, again, that with comparable external conditions, the energy request of the DX system is lower (23%). Finally, for this case study, the configuration C4 is the best one from the points of view of both energy consumption and the load-matching problem. Different management of the DX system could increase this convenience.

4. Conclusions

Starting from monitored data in an nZEB in a Mediterranean climate during the wintertime, how the variation of the setpoint temperature affects the comfort conditions and the energy consumption was evaluated. The results are encouraging for further analysis. Indeed, it was found that energy savings (from 17% to 43%) are achievable by lowering the setpoint with respect to the Italian legislation threshold and through the utilization of on-site renewable electricity, which can be maximized during the day with the lowest productivity. The comfort indices suggest that the indoor conditions are also acceptable with lower setpoints, mainly when the room is characterized by a large glazed surface. However, the perception of people does not always agree with analytical evaluations; the judgment seems closely related to the external conditions, mainly in term of sky clearness.

Moreover, for the summertime, the CFD analysis indicates that the pre-handling system does not affect indoor comfort. It only guarantees a lower temperature of the air supplied to the heat pump, and therefore a saving of the energy demand and costs, but the supply temperature does not change. Globally, considering all available data, the most convenient configuration is the operation of the DX system with a reduction of energy consumption near 20%. In this case, management with a higher setpoint or higher supply temperature is also possible; this strategy can determine a further reduction of the energy consumption and the increment of load matching. Moreover, it was shown that although currently, the energy balances on nZEBs are usually based on annual or monthly data, what happens at lower scales of time (daily or hourly) should be considered, in order to have profitable designs oriented towards maximizing the self-consumption of energy production. This would assure economic profitability and a better interaction with the local grid infrastructure.

Obviously, all presented data are strictly related to the building application and to the climate zone under investigation. However, the adopted solutions and the learned results could be useful guidelines for building and system design in areas with similar boundary conditions, and thus climates, living styles, and construction types.

Author Contributions: Conceptualization, methodology, R.F.D.M. and G.P.V.; software, validation, data curation, experimental analysis, V.F., A.G. and S.R.; writing—original draft preparation, R.F.D.M. and G.P.V.; supervision, R.F.D.M. and G.P.V. All authors have read and agreed to the published version of the manuscript.

Funding: This research received no external funding.

Institutional Review Board Statement: Not applicable.

Informed Consent Statement: Not applicable.

Data Availability Statement: The datasets generated during and analyzed during the current study are available from the corresponding author on reasonable request.

Conflicts of Interest: The authors declare no conflict of interest.

References

1. European Parliament. Directive 2010/31/EU of the European Parliament and of Council of 19 May 2010 on the energy performance of buildings (recast). *Off. J. Eur. Union* **2010**, *L153*, 13–35.
2. European Parliament. Directive 2018/844 of the European Parliament and of Council of 30 May 2018 on amending Directive 2010/31/EU on the energy performance of buildings and Directive 2012/27/EU on energy efficiency. *Off. J. Eur. Union* **2018**, *L156*, 75–91.
3. Chen, Y.; Lin, G.; Crowe, E.; Granderson, L. Development of a Unified Taxonomy for HVAC System Faults. *Energies* **2021**, *14*, 5581. [CrossRef]
4. Butera, F.M. Zero-energy buildings: The challenges. *Adv. Build. Energy Res.* **2013**, *7*, 51–65. [CrossRef]
5. Mor, G.; Cipriano, J.; Gabaldon, E.; Grillone, B.; Tur, M.; Chemisana, D. Data-Driven Virtual Replication of Thermostatically Controlled Domestic Heating Systems. *Energies* **2021**, *14*, 5430. [CrossRef]
6. Moon, J.W.; Han, S.H. Thermostat strategies impact on energy consumption in residential buildings. *Energy Build.* **2011**, *43*, 338–346. [CrossRef]
7. Ben-Nakhi, A.E.; Mahmoud, M.A. Application of building-dynamics-based control strategies to improve air-conditioning performance in educational buildings. *Adv. Build. Energy Res.* **2017**, *11*, 153–179. [CrossRef]
8. Kazanci, O.B.; Olesen, B.W. Beyond nearly-zero energy buildings: Experimental investigation of the thermal indoor environment and energy performance of a single-family house designed for plus-energy targets. *Sci. Tech. Built Environ.* **2016**, *22*, 1024–1038. [CrossRef]
9. Zalewska, M.E. The Impact of Incentives on Employees to Change Thermostat Settings—A Field Study. *Energies* **2021**, *14*, 5315. [CrossRef]
10. De Feijter, F.J.; van Vliet, B.J. Housing retrofit as an intervention in thermal comfort practices: Chinese and Dutch householder perspectives. *Energy Effic.* **2021**, *14*, 1–18. [CrossRef]
11. Haniff, M.F.; Selamat, H.; Khamis, N.; Almin, A.J. Optimized scheduling for an air-conditioning system based on indoor thermal comfort using the multi-objective improved global particle swarm optimization. *Energy Effic.* **2019**, *12*, 1183–1201. [CrossRef]
12. Guillén-Lambea, S.; Rodríguez-Soria, B.; Marín, J.M. Comfort settings and energy demand for residential nZEB in warm climates. *Appl. Energy* **2017**, *202*, 471–486. [CrossRef]
13. Reda, F.; Pasini, D.; Laitinen, A.; Teemu, V. ICT intelligent support solutions toward the reduction of heating demand in cold and mild European climate conditions. *Energy Effic.* **2019**, *12*, 1443–1471. [CrossRef]
14. Katafygiotou, M.C.; Serghides, D.K. Indoor comfort and energy performance of buildings in relation to occupants' satisfaction: Investigation in secondary schools of Cyprus. *Adv. Build. Energy Res.* **2014**, *8*, 216–240. [CrossRef]
15. Ghahramani, A.; Zhang, K.; Dutta, L.; Yang, Z.; Becerik-Gerber, B. Energy savings from temperature setpoints and deadband: Quantifying the influence of building and system properties on savings. *Appl. Energy* **2016**, *165*, 930–942. [CrossRef]
16. Tzivanidis, C.; Antonopoulus, K.A.; Gioti, F. Numerical simulation of cooling energy consumption in connection with thermostat operation mode and comfort requirements for Athens buildings. *Appl. Energy* **2011**, *88*, 2871–2884. [CrossRef]
17. Sánchez-García, D.; Rubio-Bellido, C.; Martín del Río, J.J.; Pérez-Fargallo, A. Towards the quantification of energy demand and consumption through the adaptive comfort approach in mixed mode office buildings considering climate change. *Energy Build.* **2019**, *187*, 173–185. [CrossRef]
18. Ming, R.; Yu, W.; Zhao, X.; Liu, Y.; Li, B.; Essah, E.; Yao, R. Assessing energy saving potentials of office buildings based on adaptive thermal comfort using a tracking-based method. *Energy Build.* **2020**, *208*, 109611. [CrossRef]

19. Mui, K.W.; Wong, L.T.; Cheung, C.T.; Yu, H.C. Improving cooling energy efficiency in Hong Kong offices using demand-controlled ventilation (DCV) and adaptive comfort temperature (ACT) systems to provide indoor environmental quality (IEQ) acceptance. *HKIE Trans.* **2017**, *24*, 78–87. [CrossRef]
20. Roussac, A.C.; Steinfeld, J.; de Dear, R. A preliminary evaluation of two strategies for raising indoor air temperature setpoints in office buildings. *Archit. Sci. Rev.* **2011**, *54*, 148–156. [CrossRef]
21. Akram Hossain, M.; Khalilnejad, A.; Haddadian, R.; Pickering, E.M.; French, R.H.; Abramson, A.R. Data analytics applied to the electricity consumption of office buildings to reveal building operational characteristics. *Adv. Build. Energy Res.* **2020**. [CrossRef]
22. Franco, A.; Miserocchi, L.; Testi, D. HVAC Energy Saving Strategies for Public Buildings Based on Heat Pumps and Demand Controlled Ventilation. *Energies* **2021**, *14*, 5541. [CrossRef]
23. Traylor, C.; Zhao, W.; Tao, Y.X. Utilizing modulating-temperature setpoints to save energy and maintain alliesthesia-based comfort. *Build. Res. Inf.* **2019**, *47*, 190–201. [CrossRef]
24. Kim, S.K.; Hong, W.H.; Hwang, J.H.; Jung, M.S.; Park, Y.S. Optimal Control Method for HVAC Systems in Offices with a Control Algorithm Based on Thermal Environment. *Buildings* **2020**, *10*, 95. [CrossRef]
25. Jazizadeh, F.; Joshi, V.; Battaglia, F. Adaptive and distributed operation of HVAC systems: Energy and comfort implications of active diffusers as new adaptation capacities. *Build. Environ.* **2020**, *186*, 107089. [CrossRef]
26. Lourenço, J.M.; Aelenei, L.; Facão, J.; Gonçalves, H.; Aelenei, D.; Pina, J.M. The Use of Key Enabling Technologies in the Nearly Zero Energy Buildings Monitoring, Control and Intelligent Management. *Energies* **2021**, *14*, 5524. [CrossRef]
27. Kalaimani, R.; Jain, M.; Keshav, S.; Rosenberg, C. On the interaction between personal comfort systems and centralized HVAC systems in office buildings. *Adv. Build. Energy Res.* **2020**, *14*, 129–157. [CrossRef]
28. Van der Linden, A.A.; Boerstra, A.C.; Raue, A.K.; Kurvers, A.R.; de Dear, R.J. Adaptive temperature limits: A new guidelines in The Netherlands A new approach for the assessment of building performance with respect to thermal indoor climate. *Energy Build.* **2006**, *38*, 8–17. [CrossRef]
29. Rijal, H.B.; Yoshida, K.; Humphreys, M.A.F.; Nicol, J. Development of an adaptive thermal comfort model for energy-saving building design in Japan. *Archit. Sci. Rev.* **2021**, *64*, 109–122. [CrossRef]
30. Hellwin, R.T.; Teli, D.; Schweiker, M.; Choi, J.H.; Lee, M.C.J.; Mora, R.; Rawal, R.; Wang, Z.; Al-Atrash, F. A framework for adopting adaptive thermal comfort principles in design and operation of buildings. *Energy Build.* **2019**, *205*, 109476. [CrossRef]
31. Breesch, H.; Janssens, A. Performance evaluation of passive cooling in office buildings based on uncertainty and sensitivity analysis. *Sol. Energy* **2010**, *84*, 1453–1467. [CrossRef]
32. Psomas, T.; Heiselberg, P.; Lyme, T.; Duer, K. Automated roof window control system to address overheating on renovated houses: Summertime assessment and intercomparison. *Energy Build.* **2017**, *138*, 35–46. [CrossRef]
33. O'Donovan, A.; Murphy, M.D.; O'Sullivan, P.D. Passive control strategies for cooling a non-residential nearly zero energy office: Simulated comfort resilience now and in the future. *Energy Build.* **2021**, *231*, 110607. [CrossRef]
34. Fratean, A.; Dobra, P. Control strategies for decreasing energy costs and increasing self-consumption in nearly zero-energy buildings. *Sustain. Cities Soc.* **2018**, *39*, 459–475. [CrossRef]
35. Gonzalez, R.M.; Torbaghan, S.S.; Gibescu, M.; Cobben, S. Harnessing the flexibility of thermostatic loads in microgrids with solar power generation. *Energies* **2016**, *9*, 547. [CrossRef]
36. Darakdjian, Q.; Billé, S.; Inard, C. Data mining of building performance simulations comprising occupant behaviour modelling. *Adv. Build. Energy Res.* **2019**, *13*, 57–173. [CrossRef]
37. Salom, J.; Widén, J.; Candanedo, J.; Byskov, K. Lindberg, Analysis of grid interaction indicators in net zero-energy buildings with sub-hourly collected data. *Adv. Build. Energy Res.* **2015**, *9*, 89–106. [CrossRef]
38. O'Donovan, A.; O'Sullivan, P.D.; Murphy, M.D. Predicting air temperatures in a naturally ventilated nearly zero energy building: Calibration, validation, analysis and approaches. *Appl. Energy* **2019**, *250*, 991–1010. [CrossRef]
39. Murphy, M.D.; O'Sullivan, P.D.; da Graça, G.C.; O'Donovan, A. Development, Calibration and Validation of an Internal Air Temperature Model for a Naturally Ventilated Nearly Zero Energy Building: Comparison of Model Types and Calibration Methods. *Energies* **2021**, *14*, 871. [CrossRef]
40. Minarčík, P.; Procházka, H.; Gulan, M. Advanced Supervision of Smart Buildings Using a Novel Open-Source Control Platform. *Sensors* **2021**, *21*, 160. [CrossRef]
41. Tumminia, G.; Guarino, F.; Longo, S.; Aloisio, D.; Cellura, S.; Sergi, F.; Brunaccini, G.; Antonucci, V.; Ferraro, M. Grid interaction and environmental impact of a net zero energy building. *Energy Convers. Manag.* **2020**, *203*, 112228. [CrossRef]
42. Aelenei, D.; Lopes, R.A.; Aelenei, L.; Gonçalvesc, H. Investigating the potential for energy flexibility in an office building with a vertical BIPV and PV roof system. *Renew. Energy* **2019**, *137*, 189–197. [CrossRef]
43. Hemanth, G.R.; Raja, S.C.; Nesamalar, J.J.D.; Kumar, J.S. Cost effective energy consumption in a residential building by implementing demand side management in the presence of different classes of power loads. *Adv. Build. Energy Res.* **2020**. [CrossRef]
44. Stasi, R.; Liuzzi, S.; Paterno, S.; Ruggiero, F.; Stefanizzi, P.; Stragapede, A. Combining bioclimatic strategies with efficient HVAC plants to reach nearly-zero energy building goals in Mediterranean climate. *Sustain. Cities Soc.* **2020**, *63*, 102479. [CrossRef]
45. Franco, A.; Fantozzi, F. Experimental analysis of a self-consumption strategy for residential building: The integration of PV system and geothermal heat pump. *Renew. Energy* **2016**, *86*, 1075–1085. [CrossRef]

46. Tien, P.W.; Wei, S.; Calautit, J. A Computer Vision-Based Occupancy and Equipment Usage Detection Approach for Reducing Building Energy Demand. *Energies* **2021**, *14*, 156. [CrossRef]
47. Ascione, F.; Bianco, N.; De Masi, R.F.; de Rossi, F.; Vanoli, G.P. Concept, Design and Energy Performance of a Net Zero-Energy Building in Mediterranean Climate. *Proc. Eng.* **2016**, *169*, 26–37. [CrossRef]
48. Ascione, F.; Borrelli, M.; De Masi, R.F.; de Rossi, F.; Vanoli, G.P. A framework for NZEB design in Mediterranean climate: Design, building and set-up of a lab-small villa. *Sol. Energy* **2019**, *184*, 11–29. [CrossRef]
49. UNI EN 15251: 2008 Indoor Environmental Input Parameters for Design and Assessment of Energy Performance of Buildings Addressing Indoor Air Quality, Thermal Environment, Lighting and Acoustics; UNI: Milan, Italy, 2008.
50. Carlucci, S.; Bai, L.; de Dear, R.; Yang, L. Review of adaptive thermal comfort models in built environmental regulatory documents. *Build. Environ.* **2018**, *137*, 73–89. [CrossRef]
51. De Dear, R.J.; Brager, G.S. Developing an adaptive model of thermal comfort and preference. *ASHRAE Trans.* **1998**, *104*, 145–167.
52. ISSO 74:2014, Termische behaaglijkheid. *Eisen en Achtergronden Betreffende het Thermisch Binnenklimaat in Kantoren, Scholen en Vergelijkbare Utiliteitsbouw*; ISSO: Rotterdam, The Netherlands, 2014. (In Dutch)
53. UNI EN 16798-1:2019, Energy Performance of Buildings—Ventilation for Buildings—Part 1: INDOOR Environmental Input Parameters for Design and Assessment of Energy Performance of Buildings Addressing Indoor Air Quality, Thermal Environment, Lighting and Acoustics—Module M1-6; UNI: Milan, Italy, 2019.
54. Li, B.; Yao, R.; Wang, Q.; Pan, Y. An introduction to the Chinese Evaluation Standard for the indoor thermal environment. *Energy Build.* **2014**, *82*, 27–36. [CrossRef]
55. Decree of President of Italian Republic, D.P.R. 412/1993—Requisiti e Dimensionamento Degli Impianti Termici. Available online: https://www.apeapz.it/images/PDF/CaldaiaSicura/Leggi/DPR_412_93.pdf (accessed on 15 June 2021). (In Italian).
56. UNI EN ISO 7730:2006, Ergonomics of the Thermal Environment—Analytical Determination and Interpretation of Thermal Comfort Using Calculation of the PMV and PPD Indices and Local Thermal Comfort Criteria; UNI: Milan, Italy, 2006.
57. Ascione, F.; Borrelli, M.; De Masi, R.F.; de Rossi, F.; Vanoli, G.P. Hourly operational assessment of HVAC systems in Mediterranean Nearly Zero-Energy Buildings: Experimental evaluation of the potential of ground cooling of ventilation air. *Renew. Energy* **2020**, *155*, 950–968. [CrossRef]
58. U.S. Department of Energy. *Energy Plus Simulation Software, Version 8.1.0*; U.S. Department of Energy: Washington, DC, USA, 2020.
59. Design Builder. v. 6.0. 2018. Available online: https://doi.org/10.1016/j.renene.2020.03.180 (accessed on 25 June 2020).
60. U.S. Department of Energy. Federal Energy Management Program. M&G Guidelines: Measurement and Verification of Performance-Based Contacts Version 4.0. 2015. Available online: https://www.academia.edu/32378615/M_and_V_Guidelines_Measurement_and_Verification_for_Performance_Based_Contracts_Version_4_0 (accessed on 15 June 2021).

Article

Energy Performance and Benchmarking for University Classrooms in Hot and Humid Climates

Jaqueline Litardo [1,*], Ruben Hidalgo-Leon [2] and Guillermo Soriano [2]

1. Department of Architecture, Built, Environment and Construction Engineering (DABC), Politecnico di Milano, Via Ponzio 31, 20133 Milan, Italy
2. Centro de Energías Renovables y Alternativas CERA, Escuela Superior Politécnica del Litoral ESPOL, Km. 30.5 Vía Perimetral, Guayaquil 090902, Ecuador; rhidalgo@espol.edu.ec (R.H-L.); gsorian@espol.edu.ec (G.S.)
* Correspondence: jaqueline.litardo@polimi.it

Abstract: In this paper, the energy performance of a university campus in a tropical climate is assessed, and four mixed classroom buildings are compared using benchmarking methods based on simple normalization: the classic Energy Use Intensity (EUI), end-used based EUI, and people-based EUI. To estimate the energy consumption of the case studies, building energy simulations were carried out in EnergyPlus using custom inputs. The analysis found that buildings with more classroom spaces presented higher energy consumption for cooling and lighting than others. In comparison, buildings with a greater percentage of laboratories and offices exhibited higher energy consumption for plug loads. Nevertheless, differences were identified when using the people-based EUI since buildings with larger floor areas showed the highest values, highlighting the impact of occupant behavior on energy consumption. Given the fact that little is known about a benchmark range for university campuses and academic buildings in hot and humid climates, this paper also provides a comparison against the EUIs reported in the literature for both cases. In this sense, the identified range for campuses was 49–367 kWh/m^2/year, while for academic buildings, the range was 47–628 kWh/m^2/year. Overall, the findings of this study could contribute to identifying better-targeted energy efficiency strategies for the studied buildings in the future by assessing their performance under different indicators and drawing a benchmark to compare similar buildings in hot and humid climates.

Keywords: Energy Use Intensity; higher education buildings; energy consumption; benchmarking; hot and humid climates; EnergyPlus

Citation: Litardo, J.; Hidalgo-Leon, R.; Soriano, G. Energy Performance and Benchmarking for University Classrooms in Hot and Humid Climates. *Energies* **2021**, *14*, 7013. https://doi.org/10.3390/en14217013

Academic Editor: Paulo Santos

Received: 28 September 2021
Accepted: 19 October 2021
Published: 26 October 2021

Publisher's Note: MDPI stays neutral with regard to jurisdictional claims in published maps and institutional affiliations.

Copyright: © 2021 by the authors. Licensee MDPI, Basel, Switzerland. This article is an open access article distributed under the terms and conditions of the Creative Commons Attribution (CC BY) license (https://creativecommons.org/licenses/by/4.0/).

1. Introduction

The assessment of the energy performance of existing and new buildings is of paramount importance for minimizing the energy consumption of this sector. This is due to the fact that buildings and their related sectors consume about 35% of the global energy and are responsible for about 38% of global greenhouse gas (GHG) emissions [1]. The latter makes this sector the largest source of carbon dioxide emissions [2]. The energy use in a building is directly influenced by its physical characteristics such as geometry, envelope, and systems [3]. Several studies have shown that about 70% of energy consumption in buildings comes from HVAC systems (around 50%) and artificial lighting (around 20%) [4–8]. In hot and humid cities, the use of air conditioners considerably increases the energy consumption of buildings, and this can also be exacerbated by the urban heat island effect [3,9].

Within the building sector, educational buildings worldwide have evidenced high energy consumption. For instance, university buildings in the USA account for about 13% of the total building energy consumption, with teaching buildings being key drivers of this due to their schedules and occupancy densities [10]. Similarly, in China, Liu and Ren reported that colleges and universities use 8% of the total energy consumed by Chinese society [11]. They also mentioned that university students consume four times more energy than the average Chinese citizen. To overcome this issue, higher education institutes are investing in improving the energy efficiency (EE) of their campuses through implementing sustainability programs, pursuing a low carbon economy [12], and enhancing their prestige in the national and international context [13].

Among the strategies that higher education institutes are implementing to improve the sustainability of their campuses are those related to EE and those for energy conservation [14]. Nevertheless, measures should not only involve technical improvements such as those mentioned above but should also focus on scheduling and occupancy, which vary from campus to campus. In this sense, changing the academic calendar from semester to trimester resulted in a reduction in annual energy consumption of about 5%, as observed at Griffith University [15]. However, regardless of the selected strategies to improve energy use in buildings, indicators are required to measure building performance against a reference.

Energy Use Intensity (EUI) is one of the most used indicators to evaluate the energy performance of buildings [16,17]. It results from the ratio between the annual energy consumption of the building and its total floor area [18]. Using the EUI, it is possible to perform benchmarking analysis, which refers to comparing buildings from the same uses and located in similar climate zones [19]. In this sense, building energy benchmarking is a reference point for how efficient the buildings are, enabling the possibility of proposing energy efficiency strategies. In the context of educational buildings, there are several references worldwide. In the USA, the mean EUI for educational buildings from climates 1A (very hot and humid) and 2A (hot and humid) is about 420 kWh/m^2/year [20]. If only electricity use is considered, they account for a mean index of about 130 kWh/m^2/year [20]. In Europe, the EUI of these buildings ranges between 150 and 250 kWh/m^2/year [21]. In the Ecuadorian Coast, a study carried out in 123 primary schools determined a median EUI of about 14 kWh/m^2/year [19].

Several methodologies on building energy benchmarking have been proposed in the pursuit of finding better alternatives for comparison among buildings [17,22]. For instance, Li and Chen investigated the correlation between the EUI of 24 higher education buildings and the percentages of the areas destined for different uses [23]. Through a regression model, the authors found that laboratory spaces were major contributors to energy consumption compared to public and school spaces. A similar approach was performed by Khoshbakht et al. [13], where the authors compared 80 higher education buildings using an EUI based on their different academic activities. Their findings indicated that research buildings were more energy-consuming than others, presenting a maximum EUI of more than 200 kWh/m^2/year. Furthermore, other benchmarking methods have focused on comparing buildings by the disaggregation of their EUIs [2] or normalizing the annual energy consumption by people instead of floor area [24,25].

In this study, the energy performance of the ESPOL campus located in the tropical climate of Guayaquil, Ecuador was evaluated, and the results from the energy modeling of four existing classroom buildings were introduced. The research aims to compare these buildings using different Energy Use Intensity (EUI)-based indicators. The conclusions of this paper are relevant to establishing a benchmark for university buildings since, in general, little is known about this topic in hot and humid climates, particularly for the case of the Ecuadorian Coast. Hence, better-targeted energy efficiency measures could be proposed for these buildings in the future, considering the results that emerge from the evaluation with the different indicators.

This paper is structured as follows. In Section 2, the case study and the analyzed buildings are described, the procedure to perform the building energy models in EnergyPlus is explained, and the benchmarking methods to compare the buildings under study are introduced. In Section 3, the obtained results are presented, including the assessment of the energy performance of the case study, the estimation of the annual energy consumption, and the benchmarking analysis. In Section 4, the results are briefly discussed. Finally, in Section 5, some conclusions about this work are drawn.

2. Methodology

In this section, the case study is described, and the inputs for assessing the energy performance of the selected buildings are reported in more detail. Furthermore, the three methods used for the building energy benchmarking analysis are presented.

2.1. Case Study

ESPOL is an Ecuadorian Public University located on the south coast of Ecuador [26], in the city of Guayaquil (2°8'51.08" S, 79°57'52.21" W). Guayaquil has a very hot and humid climate, presenting more than 5000 cooling degree days (CDD 10 °C) [27]. This climate is also included within the group Aw of the Köppen–Geiger classification [28]. The latter corresponds to a tropical savannah climate characterized by having two seasons: wet (January–April) and dry (May–December). The typical meteorological conditions in the surroundings of the case study are depicted in Figure 1. The monthly minimum temperatures are greater than 18 °C, the monthly average oscillates between 23.4 °C and 26.5 °C, and the monthly maximum is above 30 °C. Monthly average humidity values range from 63% to 76%. The monthly average solar radiation values are over 3.4 kWh/m² and below 5.4 kWh/m².

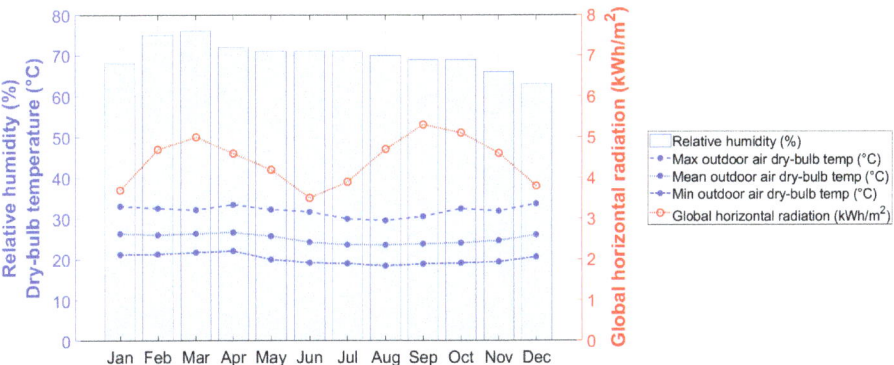

Figure 1. Climate conditions in Guayaquil, Ecuador for the typical meteorological year (TMY, source: Meteonorm 7.3.4).

The ESPOL campus spreads over an area of about 703 ha (see Figure 2). It consists of 106 buildings of different types, including classrooms, laboratories, restaurants, and other spaces, as listed in Table 1, accounting for a total building area of about 154,564 m². The primary energy source used on campus is electricity. On this basis, the monthly electricity consumption of the last four years is summarized in Table 2 and includes the energy for street lighting (around 2% of the total) and buildings. In 2020, the total campus electricity consumption decreased by about 40% due to the COVID-19 pandemic; therefore, this year was not considered for the analysis.

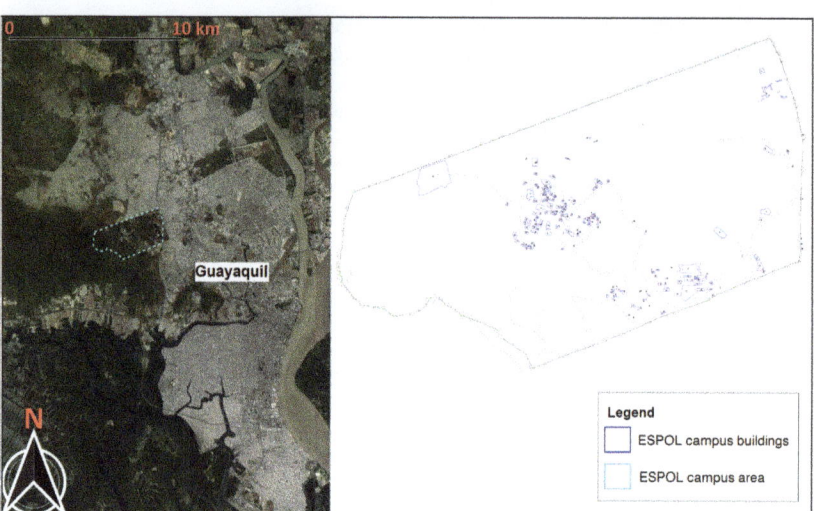

Figure 2. ESPOL campus area in Guayaquil and its buildings.

Table 1. Number of buildings in ESPOL campus according to their type and total built-up area.

Building Type	Number of Buildings	Building Area (m^2)
Classrooms and teaching areas	36	60,956
Laboratories	28	26,749
Sport facilities	5	27,698
Administrative offices	18	27,304
Residences	3	1123
Restaurants	8	4901
Others (storage and small spaces)	8	5832

Table 2. Monthly electricity consumption of ESPOL campus from 2017 to 2020.

	Energy Consumption (kWh)			
	2017	2018	2019	2020
January	1,058,148	1,143,828	1,050,000	1,385,857
February	1,199,520	968,184	1,163,400	976,681
March	1,053,864	1,079,400	982,800	819,613
April	921,060	932,400	1,062,600	563,711
May	1,216,656	1,184,400	1,226,400	584,262
June	1,190,952	1,012,200	1,226,400	552,691
July	1,130,976	1,071,000	1,134,000	559,621
August	1,109,556	1,058,400	1,071,000	588,479
September	1,019,592	861,000	970,200	288,305
October	1,053,864	1,045,800	907,200	626,674
November	1,152,396	1,104,600	1,134,000	577,420
December	1,216,656	1,157,000	1,155,000	357,298
TOTAL	13,323,240	12,618,212	13,083,000	7,880,611

To assess the energy performance of the ESPOL campus, available data from 2017 to 2020 were collected. In this sense, Table 3 lists the total number of occupants, the electricity consumed by buildings, and the CDD per year. The annual CDD was calculated using Equation (1). The plus symbol (+) indicates that only positive values are considered for the calculations of the CDD. In this case, T_b was defined as 18.3 °C, which agrees with the base temperature recommended for different hot climates [29] and by IEA [30].

$$\text{CDD} = \sum_{i=1}^{N}(T_d - T_b)^+ \qquad (1)$$

where T_d is the daily mean temperature; T_b is the reference or base temperature, which indicates that cooling is required when outdoor temperatures are above it; and N is the number of days in a year.

Table 3. Annual cooling needs, building electricity consumption, and number of occupants of ESPOL campus from 2017 to 2020.

Year	CDD for T_b = 18.3 °C	Number of Students	Number of Professors and Administrative Staff	Number of Occupants	Building Electricity Consumption (kWh)	% Savings
2017	2904	12,323	1470	13,793	13,143,660	Base year
2018	2815	11,949	1462	13,411	12,438,632	5.4
2019	2935	11,599	1453	13,052	12,903,420	1.8
2020	3264	11,595	1387	12,982	7,701,031	41.4

ESPOL started its sustainability program in 2018, intending to contribute to the UN's Sustainable Development Goals (SDGs) [31]. The action plan is focused on reducing the carbon footprint of the campus through the implementation of diverse initiatives such as the construction of a bike lane, the replacement of fluorescent lighting with LED systems, and the use of renewable energies [32]. According to Criollo et al. [33], 66% of the GHG emissions source in the ESPOL campus comes from electricity, which is mostly used to supply air conditioners in buildings. Currently, the ESPOL administration has started the installation of energy meters in every building since before there was only a general meter for the whole campus; therefore, the consumption of each building is unknown yet. In this regard, the development of energy models of the most energy-intensive buildings on campus is key to determining rapid actions to improve their energy efficiency.

2.2. Energy Models: Classroom Buildings

In order to evaluate the performance of the campus buildings in more detail, four mixed classroom buildings were selected on the basis that these are among the most energy-consuming buildings that include teaching areas. Figure 3 shows the four classrooms that were modeled and compared in this study. Building 1 (Figure 3a) has an area of about 983 m², with 70% being classrooms and the remaining area corresponding to public spaces such as corridors and bathrooms. Building 2 (Figure 3b) has an area about 1912 m², of which 40% is classrooms, 15% is laboratories, 8% is offices, 31% is public spaces, and 6% is others (storage rooms). Building 3 (Figure 3c) has an area of about 5800 m², of which 23% is offices, 13% is laboratories, 10% is classrooms, 38% is public spaces, and the remaining is included within other spaces. Building 4 (Figure 3d) has an area about 7086 m², of which 46% is classrooms, 5% is laboratories, 7% is offices, 36% is public spaces, and the rest corresponds to other spaces. Besides, the buildings under study presented similar building materials based on masonry construction. In this regard, the external walls are formed by 10 cm thick concrete blocks. Exterior windows are composed of single-pane glasses and metal frames, with a typical U-value of 5.8 W/m²-K. Roofs are from reinforced concrete in all cases. Building 1 has a window-to-wall ratio (WWR) of 17%, Building 2 has a ratio of 5%, Building 3 has a ratio of 5%, and Building 4 has a ratio of 11%.

Internal heat gains of the buildings—i.e., lighting, plug loads, and people—were defined based on observations made during on-site inspections. Figure 4 summarizes the total installed power by story of each building. As observed, the lighting load is composed of LED and fluorescent technologies in all cases. Similarly, plug loads result from the sum of the office equipment, appliances, and miscellaneous loads. Office equipment refers to computers, printing machines, and other related devices. Appliances encompass devices such as microwaves, coffee makers, or similar. Miscellaneous loads include Wi-Fi devices, audio equipment, and others. All loads in Figure 4 were collected during on-site inspections. Table 4 summarizes the total installed power in terms of lighting and plug loads in each building. Moreover, the maximum occupancy for each case is also included in the table and was estimated assuming that classrooms and laboratories are 80% occupied. Regarding office spaces, the number of people is fixed and corresponds to the workers that occupy these spaces.

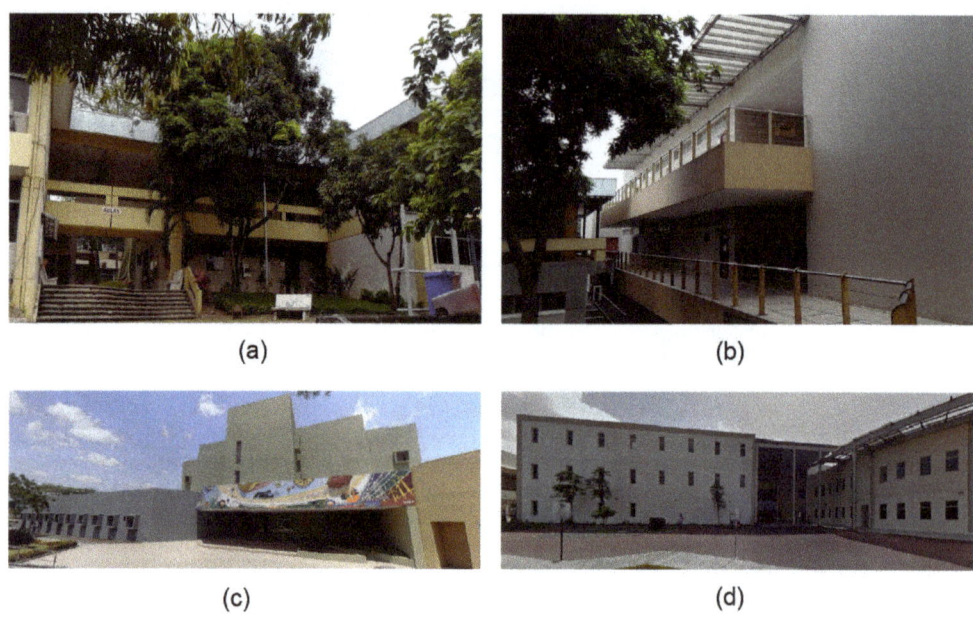

Figure 3. Classrooms under study (**a**) Building 1, (**b**) Building 2, (**c**) Building 3, and (**d**) Building 4.

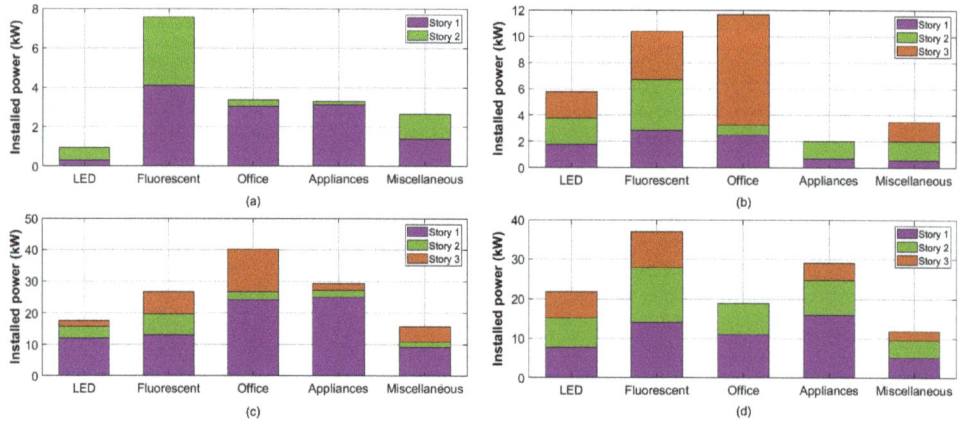

Figure 4. Lighting and plug loads. Installed power in (**a**) Building 1, (**b**) Building 2, (**c**) Building 3, and (**d**) Building 4.

Table 4. Building internal heat gains: total installed power and people.

Description	Building 1	Building 2	Building 3	Building 4
Lighting (kW)	8.5	16.2	44.2	58.6
Plug loads (kW)	9.3	17.2	85.1	59.9
People (#)	358	432	450	1132

Different schedules were set considering the space types and their activities. For instance, schedules in office spaces were fixed from 8:00 to 16:30 during the weekdays. On the other hand, schedules in classrooms and laboratories varied between 7:30 and 20:30 during the weekdays. Spaces classified as others did not present uniform schedules due to the fact that they are usually storage or mechanical/electrical rooms with low occupancy and use. Overall, lighting, plug loads, and occupancy schedules were built according to the indications of the users.

Due to its high temperatures and humidity, Guayaquil has a cooling-dominated climate. Therefore, HVAC systems for cooling are required during the whole year of operation of the buildings. The air-conditioners installed in the case studies are from the direct expansion (DX) technology, divided between splits and central units. During the inspections, it was observed that air-conditioners, both centralized and decentralized, are usually set by their users at 21 or 22 °C in all buildings. The air-conditioners' operating hours in office spaces extend to 8 h. In classrooms and laboratories, this depends on the lecture schedules, as mentioned above. Table 5 lists the total installed capacity in each building.

Table 5. Conditioned loads: total installed capacity by building.

Building	Installed Capacity (BTU/h)
1	516,000 (151.2 kW)
2	855,000 (250.6 kW)
3	3,324,000 (974.2 kW)
4	3,438,000 (1007.6 kW)

Considering the above, simulations were carried out in OpenStudio v. 2.7 and EnergyPlus v. 9.1 software using a TMY weather file. Spaces and schedules were created using custom inputs, depending on each area. The thermal properties of building materials were established based on OpenStudio libraries and following previous works [34,35]. Lighting and plug loads were defined according to Figure 4 and Table 4. The thermostat set-points were set between 21 and 22 °C, and the coefficient of performance (COP) of the air-conditioners was set to 3. The air infiltration rate was assumed to be 0.54 ACH as in previous works [34,35]. Since all users are assumed to be performing light office activities while remaining seated, their activity level was set to 100 W/person. The accuracy of the EnergyPlus software has been previously validated through the BESTEST procedure [36]. The expected results should be consistent with the energy consumption estimated in previous studies conducted on the Ecuadorian coast [19,34,35].

2.3. Benchmarking

This study proposes three methods for building energy benchmarking: the classic Energy Use Intensity (EUI), an EUI based on the final end-uses, and an EUI based on the number of people in each building. All of these approaches correspond to a simple normalization method [17]. To calculate the EUI, Equation (2) is used.

$$\text{EUI} = \frac{\sum_{i=1}^{12} E_i}{A_{building}} \qquad (2)$$

where E_i is the monthly total energy consumption, and $A_{building}$ is the total building floor area.

Similarly, the end-use-based EUIs are estimated via Equations (3) and (4) [2], while the people-based EUI is calculated via Equation (5).

For cooling,

$$EUI_{cooling} = \frac{\sum_{i=1}^{12} E_{c_i}}{A_{building}} \qquad (3)$$

where E_{c_i} is the monthly total energy consumption for cooling.

For base loads,

$$EUI_{base\ loads} = \frac{\sum_{i=1}^{12} (E_l + E_p)_i}{A_{building}} \qquad (4)$$

where $(E_l + E_p)_i$ is the monthly total energy consumption for lighting and plug loads.

For people,

$$EUI_{people} = \frac{\sum_{i=1}^{12} E_i}{n_{people}} \qquad (5)$$

where n_{people} is the total number of people in the building.

3. Results

In this section, the energy performance evaluation of the ESPOL campus and the results of the building simulation models are presented. Then, the building energy benchmarking analysis is conducted based on the three criteria proposed in the previous section.

3.1. Energy Performance of ESPOL Campus

The relationship between the annual energy consumption of the ESPOL campus buildings, the cooling requirements, and the number of occupants is shown in Figure 5 and expressed in Equation (6). As observed, the energy consumption is strongly dependent on these two parameters, similar to the results found in [37], which implies that every change in the number of occupants and climate can impact this. In this sense, the highest electricity consumption was reached in 2017 (13,143,660 kWh), when both parameters presented high values, and the sustainability program had not been launched. On the other hand, in 2018, energy consumption was lower, and savings of 5.4% were achieved (in reference to 2017), mainly due to the lower cooling requirements associated with a lower CDD. Based on the available data, the average EUI calculated for the ESPOL campus is 83 kWh/m²/year. In terms of the normalized consumption per user, the average EUI is about 1073 kWh/student, while the EUI of an average Ecuadorian citizen was estimated as 1517 kWh per capita in 2019 according to the National Energy Balance [38].

$$Y = 5536X_1 + 556X_2 - 10,599,324 \qquad (6)$$

where Y is the annual energy consumption in kWh, X_1 is the CDD (18.3 °C), and X_2 is the number of occupants.

3.2. Estimated Annual Energy Consumption of Modeled Buildings

The estimated annual energy consumption of the four case studies was obtained after running the building energy models in EnergyPlus. Table 6 summarizes the results obtained in each case. In this sense, Building 4 presented the highest energy consumption compared to the other buildings, accounting for about 752,994 kWh per year. On the other hand, Building 1 exhibited the lowest energy consumption, with 86,114 kWh per year. The obtained results from these models are within the expected energy consumption range determined for different building types located on the Ecuadorian coast; that is, on average, from 14 to 340 kWh/m²/year [19]. Similarly, these results agree with what was previously estimated for other buildings from the ESPOL campus, which is around 90 kWh/m²/year [34,35].

Table 6. Annual energy consumption of each building.

Building	Year of Construction	Conditioned Area (%)	Energy Consumption (kWh)
1	1986	70.4	86,114
2	2008	60.9	151,533
3	2006	73.7	384,106
4	2010	63.7	752,994

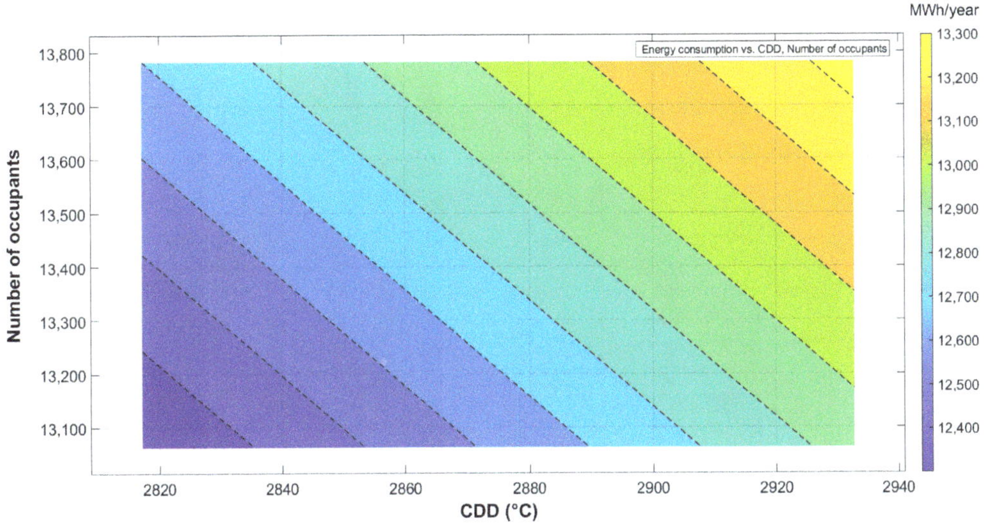

Figure 5. Contour plot of the relationship between the energy consumption, CDD, and number of occupants.

Figure 6 shows the results of the estimated annual energy consumption by end-use of each building. As can be observed from Figure 6a, air-conditioners were responsible for 72% of the total energy consumption in Building 1. Lighting and plug loads occupied 16% and 12% of the total, respectively. Likewise, in Building 2 (Figure 6b) more than 60% of its total energy was used to power air-conditioners, while plug loads consumed 20% and lighting 15%. The results for Building 3 can be observed in Figure 6c. In this case, 63% of the total energy consumption was used for cooling, 30% for plug loads, and the remaining for lighting. Finally, in Building 4, 62% of the energy was used for cooling, 23% for lighting, and 15% for plug loads, as depicted in Figure 6d. Overall, the energy consumption for air-conditioners corresponded to a share of more than 60% in all buildings. In contrast, the energy consumption for plug loads increased in buildings with a lower proportion of classrooms as observed in Buildings 2 to 4.

3.3. Energy Performance Indicators and Benchmarking

Figure 7 shows the EUI disaggregated by end-use for each analyzed building. As observed, the cooling EUI presented higher values during the months with higher temperatures in all cases. Additionally, Buildings 1 and 4 had similar cooling EUIs, and in both cases, lighting systems were more energy-consuming than plug loads. These buildings have in common the fact that the classroom area accounts for the largest floor area. On the other hand, Buildings 2 and 3 showed higher energy consumption for plug loads, which can be attributed to their larger office and laboratory spaces. Lighting and plug loads are considered to be base loads since they are almost constant throughout the year. In contrast, cooling loads vary according to the climate and exhibit higher values when the outside temperature and humidity are higher.

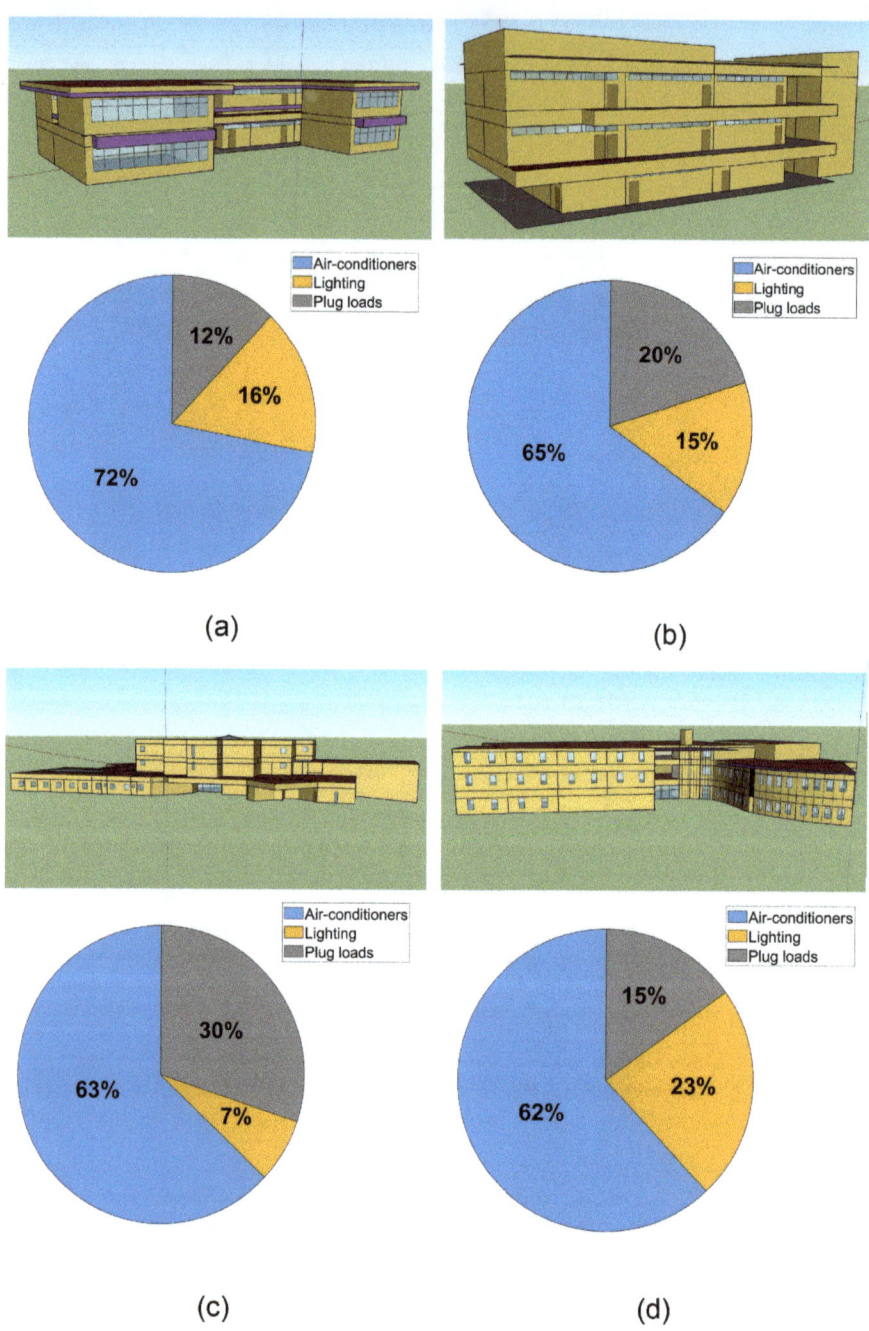

Figure 6. Energy models and estimated annual share of energy by end-use: (**a**) Building 1, (**b**) Building 2, (**c**) Building 3, and (**d**) Building 4.

Table 7 shows the results from the three proposed energy performance indicators. From the point of view of the classic approach, Buildings 1 and 4 exhibited higher values for the EUI. This could indicate that these buildings are the most energy-consuming. However, this indicator does not consider the building loads, use, or occupancy, and therefore it is not easy to draw a general conclusion. Thus, introducing other indicators would be relevant to the development of better-targeted energy efficiency strategies since it has been observed that lighting and plug loads consume 12–50% of the energy in buildings [39] while air-conditioners consume 40–60% [40,41]. On the other hand, the building defines the activities and needs of the building and its occupants, playing a critical factor in total energy consumption [42]. Considering the above, end-use-based and people-based indicators were estimated to examine the energy performance of the studied buildings in more detail. In the case of the cooling-based EUI, Buildings 1 and 4 presented the highest values. Regarding the base load, EUI values indicated that Buildings 2 and 4 were the most consuming in terms of lighting and plug loads. Finally, the people-based EUI showed a different pattern since, in this case, Buildings 3 and 4 were the most energy-consuming.

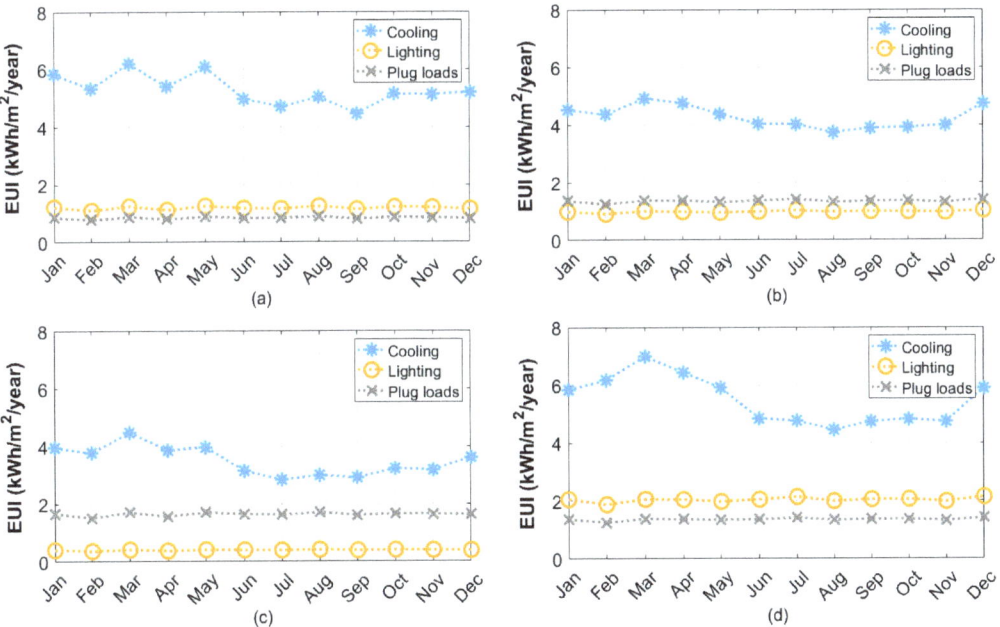

Figure 7. EUI disaggregated by end-use: (a) Building 1, (b) Building 2, (c) Building 3, and (d) Building 4.

Table 7. Results from the three proposed indicators.

	Classic	End-Use-Based		People-Based
Building	EUI (kWh/m^2/year)	EUI$_{base\ load}$ (kWh/m^2/year)	EUI$_{cooling}$ (kWh/m^2/year)	EUI$_{person}$ (kWh/person/year)
1	87.61	24.16	63.45	240.54
2	79.25	27.97	51.28	350.77
3	66.22	24.33	41.89	853.57
4	106.26	40.60	65.66	665.19

4. Discussion

Building energy benchmarking methods are widely used to evaluate the energy performance of related buildings from similar contexts. To be effective, the comparison must be between buildings of the same type due to their use and occupancy. Nevertheless, it is quite common to find mixed-use buildings in practice. If this is the case, those spaces could be analyzed independently according to their operation. An example of this is shown by Li and Chen [23]. Using regression analysis, the authors formulated an equation that allows the inclusion of the contribution of each space to the final EUI of the building. In this study, data limitations prevented a similar analysis. However, it was inferred that the space uses of the four buildings could explain the differences between their EUIs. The context in which the case study is located is also relevant and should be considered in the analysis. In fact, it has been observed that regional tariffs can influence the EUI, as reported by Chung and Yeung [43]. However, as stated by the authors, the results require careful analysis using more robust methods than simple normalization to draw final conclusions on this.

The potential of using EnergyPlus models for benchmarking has also been proved in the literature. As stated by Shabunko et al., these models have the advantage of generating time-series outputs of energy consumption, which are in agreement with observed data [44]. This could serve as an alternative when lacking actual disaggregated load data, and there is a need to start a pilot energy efficiency plan, as shown in this study. The resulting EUIs provide valuable information for the rapid identification of energy efficiency strategies that could be addressed in future retrofit projects or the modification of current operating settings in buildings. For instance, it is essential to change the temperature set-point of the air-conditioners in Buildings 1 and 4, since these presented higher cooling EUI values (see Table 7). Studies on this are available and should be taken into account to avoid compromising the thermal comfort of users when applying this measure [45]. Similarly, more efforts should be devoted to improving the occupants' behavior in buildings with larger floor areas (Buildings 3 and 4) by implementing energy policies. In general, the estimated EUIs reported in this work set an upper limit for new buildings on the ESPOL campus, as these indexes can be expressed within the terms of reference of new projects.

As observed in Figure 5, energy consumption in buildings depends on several parameters related to their design and systems, but also on the number of occupants and their behavior [46,47]. In this sense, universities account for a large number of users, mainly students, which vary from one year to another. The latter not only increases the latent loads of the buildings and therefore the energy for cooling, but also makes it difficult to predict and control the possible actions executed by the users that directly affect the energy consumption (i.e., opening windows, turning on/off lights or fans, and others). Particularly in classrooms, there may exist a hierarchical environment in which the teacher influences the general energy consumption of the space. This topic requires further exploration, as significant energy savings could be obtained by improving occupant behavior. Overall, it has been estimated that improving occupant behavior can result in energy savings of up to 20% [48].

Furthermore, this study supports evidence from earlier observations in hot and humid climates. When comparing these results with other studies, they were found to be in agreement. As observed in Table 8, most of the studies reported EUIs for different academic buildings and campuses. In general, we can note that the resulting ranges from our study are within the range of their studies: 49–637 kWh/m^2/year for university campuses and 47–628 kWh/m^2/year for academic buildings. Although the obtained results for ESPOL are close to the minimum in both ranges, this does not imply that this campus is more energy-efficient than the others, as each higher education institution differs in its administration and planning. Besides, studies in Table 8 rely only on the classic EUI, which complicated the comparison between indicators based on the number of people or energy end-uses.

Table 8. Summary of some EUIs reported in the literature for university campuses and academic buildings.

Reference (Year of Data Collection)	Location	Climate	University	Total Building Area (m^2)	EUI (kWh/m^2/year)	Academic Building	EUI (kWh/m^2/year)	Sustainability Program/Energy Saving Action Plan
This study (2019)	Guayaquil, Ecuador	Tropical	ESPOL	154,564	83	Classrooms (mixed)	66–106	Yes
[49] (2019)	Orlando, Florida, USA	Humid Subtropical	University of Central Florida	929,023	454	-	-	Yes
[50] (2019)	Singapore, Singapore	Tropical	National University of Singapore	1,459,900	≈190	-	-	Yes
[51] (2016)	Passo Fundo, Brazil	Humid, Subtropical	University of Passo Fundo (Campus I)	120,536	49	-	-	Yes
[25] (2017)	Taiwan	Subtropical	Public and Private Universities	321,868 (average)	56–93	-	-	Unknown
[52] (2019)	Singapore	Tropical	Universities Polytechnic/ITE campus	≥5000	121–367 (average)	-	-	Unknown
[13] (2016)	Queensland, Australia	Subtropical	Griffith University	-	-	Teaching buildings/Mixed buildings	145–148	Unknown
[53] (2017)	Singapore, Singapore	Tropical	Nanyang Technological University	-	-	Classrooms and laboratories	47–628	Unknown

5. Conclusions

In this study, the energy performance of a university campus and four existing university classrooms was evaluated. For this purpose, the available data of the campus electricity consumption were analyzed, and four classrooms were modeled in EnergyPlus, considering custom inputs. Subsequently, buildings were compared using three simple benchmarking approaches: the classic EUI, end-use based EUI, and people-based EUI. Through this comparison, a substantial difference was found in the energy performance of the studied buildings when considering different aspects.

Regarding the classic EUI, buildings with predominant classroom spaces (1 and 4) were found to be the most energy-consuming. Likewise, these buildings presented higher cooling and lighting EUIs than others, which can be attributed to their operating schedules. On the other hand, buildings with more extensive laboratory and office spaces (2 and 3) exhibited higher EUIs from plug loads compared to the others. The latter finding is due to the more extended use of equipment with higher loads in these areas. Finally, if we consider the people-based EUI results, Buildings 3 and 4 appeared less energy-efficient, having the characteristic in common that both have larger total floor areas than others. Overall, the results of this paper demonstrate that the energy efficiency of a building can be evaluated from different perspectives.

Identifying the most consuming space types within buildings could allow better-targeted energy efficiency measures to be proposed on a case-by-case basis and the elaboration of energy policies for buildings on campus. Similarly, the estimated EUIs can be set as an upper limit for new buildings, as these values can be expressed within the terms of reference of new projects. Moreover, when actual measurements are not available, these methods could be used as benchmarks to compare with related buildings located in similar contexts.

Author Contributions: Conceptualization, J.L.; methodology, J.L.; software, J.L.; validation, G.S.; formal analysis, J.L.; investigation, J.L. and R.H.-L.; resources, J.L., R.H.-L. and G.S.; writing—original draft preparation, J.L.; writing—review and editing, R.H.-L. and G.S.; visualization, J.L.; supervision, G.S. All authors have read and agreed to the published version of the manuscript.

Institutional Review Board Statement: Not applicable.

Informed Consent Statement: Not applicable.

Data Availability Statement: Data sharing not applicable.

Acknowledgments: The authors wish to thank ESPOL for providing the data to conduct this study.

Conflicts of Interest: The authors declare no conflict of interest.

Abbreviations

The following abbreviations are used in this manuscript:

CDD	Cooling degree days
COP	Coefficient of performance
DX	Direct expansion
EE	Energy efficiency
EUI	Energy Use Intensity
GHG	Global greenhouse gas
HVAC	Heating, Ventilation, and Air-Conditioning
IEA	International Energy Agency
LED	Light-emitting diode
SDG	Sustainable Development Goal
UN	United Nations
WWR	Window-to-wall ratio

Nomenclature

$A_{building}$	Total building floor area, m²
E_{c_i}	Monthly total energy consumption for cooling, kWh
E_i	Monthly total energy consumption, kWh
E_l	Monthly energy consumption for lighting, kWh
E_p	Monthly energy consumption for plug loads, kWh
N	Number of days in a year
n_{people}	Total number of people in the building
T_b	Reference or base temperature, °C
T_d	Daily mean temperature, °C
X_1	Annual CDD (T_b = 18.3 °C)
X_2	Number of occupants
Y	Annual energy consumption, kWh

References

1. United Nations Environment Programme. *2020 Global Status Report for Buildings and Construction: Towards a Zero-emission, Efficient and Resilient Buildings and Construction Sector*; Technical Report; United Nations Environment Programme: Nairobi, Kenya, 2020.
2. Kim, D.W.; Kim, Y.M.; Lee, S.E. Development of an energy benchmarking database based on cost-effective energy performance indicators: Case study on public buildings in South Korea. *Energy Build.* **2019**, *191*, 104–116. [CrossRef]
3. Ahn, Y.; Sohn, D.W. The effect of neighbourhood-level urban form on residential building energy use: A GIS-based model using building energy benchmarking data in Seattle. *Energy Build.* **2019**, *196*, 124–133. [CrossRef]
4. Tian, Z.; Si, B.; Shi, X.; Fang, Z. An application of Bayesian Network approach for selecting energy efficient HVAC systems. *J. Build. Eng.* **2019**, *25*, 100796. [CrossRef]
5. Jouhara, H.; Yang, J. Energy efficient HVAC systems. *Energy Build.* **2018**, *179*, 83–85. [CrossRef]
6. Baloch, A.A.; Shaikh, P.H.; Shaikh, F.; Leghari, Z.H.; Mirjat, N.H.; Uqaili, M.A. Simulation tools application for artificial lighting in buildings. *Renew. Sustain. Energy Rev.* **2018**, *82*, 3007–3026. [CrossRef]
7. Doulos, L.; Kontadakis, A.; Madias, E.; Sinou, M.; Tsangrassoulis, A. Minimizing energy consumption for artificial lighting in a typical classroom of a Hellenic public school aiming for near Zero Energy Building using LED DC luminaires and daylight harvesting systems. *Energy Build.* **2019**, *194*, 201–217. [CrossRef]
8. Han, H.J.; Mehmood, M.U.; Ahmed, R.; Kim, Y.; Dutton, S.; Lim, S.H.; Chun, W. An advanced lighting system combining solar and an artificial light source for constant illumination and energy saving in buildings. *Energy Build.* **2019**, *203*, 109404. [CrossRef]
9. Litardo, J.; Palme, M.; Borbor-Córdova, M.; Caiza, R.; Macías, J.; Hidalgo-León, R.; Soriano, G. Urban Heat Island intensity and buildings' energy needs in Duran, Ecuador: Simulation studies and proposal of mitigation strategies. *Sustain. Cities Soc.* **2020**, *62*, 102387. [CrossRef]
10. Sun, Y.; Luo, X.; Liu, X. Optimization of a university timetable considering building energy efficiency: An approach based on the building controls virtual test bed platform using a genetic algorithm. *J. Build. Eng.* **2021**, *35*, 102095. [CrossRef]
11. Liu, Q.; Ren, J. Research on the building energy efficiency design strategy of Chinese universities based on green performance analysis. *Energy Build.* **2020**, *224*, 110242. [CrossRef]
12. Yeo, J.; Wang, Y.; An, A.K.; Zhang, L. Estimation of energy efficiency for educational buildings in Hong Kong. *J. Clean. Prod.* **2019**, *235*, 453–460. [CrossRef]
13. Khoshbakht, M.; Gou, Z.; Dupre, K. Energy use characteristics and benchmarking for higher education buildings. *Energy Build.* **2018**, *164*, 61–76. [CrossRef]
14. Faghihi, V.; Hessami, A.R.; Ford, D.N. Sustainable campus improvement program design using energy efficiency and conservation. *J. Clean. Prod.* **2015**, *107*, 400–409. [CrossRef]
15. Gui, X.; Gou, Z.; Lu, Y. Reducing university energy use beyond energy retrofitting: The academic calendar impacts. *Energy Build.* **2021**, *231*, 110647. [CrossRef]
16. Arjunan, P.; Poolla, K.; Miller, C. EnergyStar++: Towards more accurate and explanatory building energy benchmarking. *Appl. Energy* **2020**, *276*, 115413. [CrossRef]
17. Chung, W. Review of building energy-use performance benchmarking methodologies. *Appl. Energy* **2011**, *88*, 1470–1479. [CrossRef]
18. Kim, S.C.; Shin, H.I.; Ahn, J. Energy performance analysis of airport terminal buildings by use of architectural, operational information and benchmark metrics. *J. Air Transp. Manag.* **2020**, *83*, 101762. [CrossRef]
19. Vallejo, C.; Villacreses, G.; Vásquez, F.; Godoy, F. *Evaluación Comparativa de los Consumos Energéticos de Edificaciones Públicas en la Región Costa y Galápagos*; Instituto Nacional de Eficiencia Energética y Energías Renovables (INER): Quito, Ecuador, 2018.
20. U.S. Department of Energy. Building Performance Database 2021. 2021. Available online: https://bpd.lbl.gov/ (accessed on 5 September 2021).
21. Economidou, M.; Atanasiu, B.; Despret, C.; Maio, J.; Nolte, I.; Rapf, O.; Laustsen, J.; Ruyssevelt, P.; Staniaszek, D.; Strong, D.; et al. Europe's Buildings under the Microscope. A Country-by-Country Review of the Energy Performance of Buildings. 2011.

Available online: http://dl.dropbox.com/u/4399528/BPIE/LR_%20CbC_study.pdforfromhttp://www.bpie.eu/eu_buildings_under_microscope.html (accessed on 5 September 2021)
22. Borgstein, E.H.; Lamberts, R. Developing energy consumption benchmarks for buildings: Bank branches in Brazil. *Energy Build.* **2014**, *82*, 82–91. [CrossRef]
23. Li, S.; Chen, Y. Internal benchmarking of higher education buildings using the floor-area percentages of different space usages. *Energy Build.* **2021**, *231*, 110574. [CrossRef]
24. Wang, J.C. A study on the energy performance of school buildings in Taiwan. *Energy Build.* **2016**, *133*, 810–822. [CrossRef]
25. Wang, J.C. Analysis of Energy Use Intensity and greenhouse gas emissions for universities in Taiwan. *J. Clean. Prod.* **2019**, *241*, 118363. [CrossRef]
26. ESPOL. Escuela Superior Politecnica del Litoral. Available online: https://www.espol.edu.ec/ (accessed on 5 September 2021).
27. Ministerio de Desarrollo Urbano y Vivienda. *NEC-HS-EE: Eficiencia Energetica*. Available online: https://www.habitatyvivienda.gob.ec/ (accessed on 5 September 2021).
28. Peel, M.C.; Finlayson, B.L.; McMahon, T.A. Updated world map of the Köppen-Geiger climate classification. *Hydrol. Earth Syst. Sci.* **2007**, *11*, 1633–1644. [CrossRef]
29. Bhatnagar, M.; Mathur, J.; Garg, V. Determining base temperature for heating and cooling degree-days for India. *J. Build. Eng.* **2018**, *18*, 270–280. [CrossRef]
30. Slade, M. The Future of Cooling: Opportunities for Energy-Efficient Air Conditioning. In *Report of the International Energy Agency*; International Energy Agency: Paris, France, 2018.
31. United Nations. THE 17 GOALS. Available online: https://sdgs.un.org/es/goals (accessed on 25 June 2021).
32. ESPOL. Sostenibilidad ESPOL. Available online: https://www.espol.edu.ec/es/la-espol/sostenibilidad/areas-de-trabajo/operacional (accessed on 5 September 2021).
33. Criollo, N.P.; Ramirez, A.D.; Salas, D.A.; Andrade, R. The Role of Higher Education Institutions Regarding Climate Change: The Case of Escuela Superior Politécnica del Litoral and its Carbon Footprint in Ecuador. In *ASME International Mechanical Engineering Congress and Exposition*; American Society of Mechanical Engineers: New York, NY, USA, 2019; Volume 59421, p. V005T07A025.
34. Litardo, J.; Hidalgo-León, R.; Macías, J.; Delgado, K.; Soriano, G. Estimating energy consumption and conservation measures for ESPOL Campus main building model using EnergyPlus. In Proceedings of the 2019 IEEE 39th Central America and Panama Convention (CONCAPAN XXXIX), Guatemala City, GT, USA, 20–22 November 2019; pp. 1–6.
35. Litardo, J.; Palme, M.; Hidalgo-León, R.; Amoroso, F.; Soriano, G. Energy Saving Strategies and On-Site Power Generation in a University Building from a Tropical Climate. *Appl. Sci.* **2021**, *11*, 542. [CrossRef]
36. ASHRAE. Energy estimating and modeling methods. In *ASHRAE Handbook-Fundamentals*; ASHRAE: Atlanta, GA, USA, 2017; Chapter 19.
37. Chihib, M.; Salmerón-Manzano, E.; Manzano-Agugliaro, F. Benchmarking energy use at University of Almeria (Spain). *Sustainability* **2020**, *12*, 1336. [CrossRef]
38. Ministerio de Energia y Recursos Naturales No Renovables. Balance Energetico Nacional 2020. Available online: www.recursosyenergia.gob.ec (accessed on 5 September 2021).
39. Anand, P.; Deb, C.; Ke, Y.; Yang, J.; Cheong, D.; Sekhar, C. Occupancy-based energy consumption modelling using machine learning algorithms for institutional buildings. *Energy Build.* **2021**, *252*, 111478. [CrossRef]
40. Shi, H.; Chen, Q. Building energy management decision-making in the real world: A comparative study of HVAC cooling strategies. *J. Build. Eng.* **2021**, *33*, 101869. [CrossRef]
41. Hidalgo-León, R.; Litardo, J.; Urquizo, J.; Moreira, D.; Singh, P.; Soriano, G. Some factors involved in the improvement of building energy consumption: A brief review. In Proceedings of the 2019 IEEE Fourth Ecuador Technical Chapters Meeting (ETCM), Guayaquil, Ecuador, 13–15 November 2019; pp. 1–6. [CrossRef]
42. Delzendeh, E.; Wu, S.; Lee, A.; Zhou, Y. The impact of occupants' behaviours on building energy analysis: A research review. *Renew. Sustain. Energy Rev.* **2017**, *80*, 1061–1071. [CrossRef]
43. Chung, W.; Yeung, I.M. A two-stage regression-based benchmarking approach to evaluate school's energy efficiency in different tariff regions. *Energy Sustain. Dev.* **2021**, *61*, 15–27. [CrossRef]
44. Shabunko, V.; Lim, C.; Mathew, S. EnergyPlus models for the benchmarking of residential buildings in Brunei Darussalam. *Energy Build.* **2018**, *169*, 507–516. [CrossRef]
45. Guevara, G.; Soriano, G.; Mino-Rodriguez, I. Thermal comfort in university classrooms: An experimental study in the tropics. *Build. Environ.* **2021**, *187*, 107430. [CrossRef]
46. Zhang, Y.; Bai, X.; Mills, F.P.; Pezzey, J.C. Rethinking the role of occupant behavior in building energy performance: A review. *Energy Build.* **2018**, *172*, 279–294. [CrossRef]
47. Laaroussi, Y.; Bahrar, M.; El Mankibi, M.; Draoui, A.; Si-Larbi, A. Occupant presence and behavior: A major issue for building energy performance simulation and assessment. *Sustain. Cities Soc.* **2020**, *63*, 102420. [CrossRef]
48. He, Z.; Hong, T.; Chou, S. A framework for estimating the energy-saving potential of occupant behaviour improvement. *Appl. Energy* **2021**, *287*, 116591. [CrossRef]
49. University of Central Florida. Open Energy Information System. 2021. Available online: https://oeis.ucf.edu/ (accessed on 5 September 2021).

50. National University of Singapore. NUS Sustainability Highlights 2019. Available online: https://uci.nus.edu.sg/oes/wp-content/uploads/sites/11/2020/10/NUS-Sustainability-Highlights-2019.pdf (accessed on 5 September 2021).
51. Salvia, A.L.; Reginatto, G.; Brandli, L.L.; da Rocha, V.T.; Daneli, R.C.; Frandoloso, M.A. Analysis of energy consumption and efficiency at University of Passo Fundo—Brazil. In *Towards Green Campus Operations*; Springer: Cham, Swizterland, 2018; pp. 519–533.
52. Building Construction Authority (BCA). *BCA Building Energy Benchmarking Report 2020*. 2020. Available online: https://www1.bca.gov.sg/docs/default-source/docs-corp-buildsg/sustainability/bca_bebr_abridged_fa_2020.pdf?sfvrsn=ea0f8\a99_8 (accessed on 5 September 2021).
53. Chang, C.C.; Shi, W.; Mehta, P.; Dauwels, J. Life cycle energy assessment of university buildings in tropical climate. *J. Clean. Prod.* **2019**, *239*, 117930. [CrossRef]

Article

Energy Performance of Buildings with Thermochromic Windows in Mediterranean Climates

Georgios E. Arnaoutakis * and Dimitris A. Katsaprakakis *

Department of Mechanical Engineering, Hellenic Mediterranean University, Estavromenos, 710 04 Heraklion, Crete, Greece
* Correspondence: arnaoutakis@hmu.gr (G.E.A.); dkatsap@hmu.gr (D.A.K.)

Abstract: This article presents comparative results on the energy performance of buildings in the Mediterranean. Many buildings in the Mediterranean exhibit low energy performance ranking. Thermochromic windows are able to improve the energy consumption by controlling the gains from sunlight. In this article, reference buildings in 15 cities around the Mediterranean are investigated. In this work, a dynamic building information modeling approach is utilized, relying on three-dimensional geometry of office buildings. Calculations of the energy demand based on computational simulations of each location were performed, for the estimation of heating and cooling loads. The presented study highlighted the need for high-resolution data for detailed simulation of thermochromic windows in buildings of Mediterranean cities. Temperature is one of the main climate parameters that affect the energy demand of buildings. However, the climate of Mediterranean cities nearby the sea may affect the energy demand. This was more pronounced in cities with arid Mediterranean climate with increased demand in air-conditioning during the summer months. On the other hand, cities with semi-arid Mediterranean climate exhibited relatively increased heating demand. With this parametric approach, the article indicates the energy saving potential of the proposed measures for each Mediterranean city. Finally, these measures can be complemented by overall building passive and active systems for higher energy reductions and increased comfort.

Keywords: thermochromic coatings; solar transmittance; solar reflectance

1. Introduction

The Mediterranean climate is characterized by mild winters and cool summers, especially in insular areas. The temperature in winter rarely falls below 0 °C, while the temperature during summer days varies around 30 °C, due to sea wind. These winds may, however, be absent in cities near the eastern Mediterranean basin and north Africa. These mild climate conditions result to moderate heating and cooling loads of buildings in Mediterranean cities. These loads are limited during the peak winter and summer seasons, reducing the conditioning requirements during the spring and autumn [1].

Nevertheless, buildings in Mediterranean cities exhibit considerable energy consumption. A significant effort is given to improve the energy performance of residential as well as commercial buildings with a number of innovative systems.

1.1. Existing Building Energy Consumption

An important building sector includes office and school buildings, as well as the evaluation and upgrade of their energy performance. School buildings represent a high percentage of public buildings with the number of school buildings in the Mediterranean estimated at 87,000 [2]. Many schools operate at low thermal comfort, while a heating strategy is absent [3].

The introduction of possible passive measures was examined by [4,5] for warm and cold Mediterranean climates in Greece and Spain, respectively. The insulation, as well as the replacement of openings and the use of shadings were evaluated. Both reports concluded

that the achieved saving can be up to 65%, while the reduction on annual CO_2 emissions can be higher than 70%. By applying passive measures in school buildings, an annual reduction of 17.7 and 15.9% in heating and cooling demand can be achieved, respectively [2].

A crucial parameter affecting the energy consumption for heating and cooling is air infiltration. The contribution of air infiltration on the total final energy specific consumption was reported between 2.43 and 16.44 kWh/m^2 for heating and between and 3.06 kWh/m^2 for cooling [6]. The consumption depends on three major parameters: (1) the location of the building whether this is built in urban, rural, high or low altitude environments, (2) the climate classification, and (3) the performance of the openings. Increasing the number of glazings can reduce the infiltration; however, this is a trade-off for the window–wall ratio, especially in the summer months in Mediterranean climates [7]. In addition, increasing the number of glazings reduces the environmental impact as reported in life-cycle assessment analyses [8]. There is, however, an optimum window–wall ratio for the best energy performance of buildings. Extensive studies limit the optimum window–wall ratio between 0.3 and 0.45 [9]. However, depending on the climate of the building's location, the optimum may vary to lower values down to 20% [10]. Changing either the number of glazings or the window–wall ratio is not always possible due to limitations related to cost or preservation of a building [11].

Reducing the window–wall ratio may increase the requirements for artificial lighting. Considering a lighting power density between 10 and 30 W/m^2, a commercial building may have annual electricity consumption between 20 and 25 kWh/m^2 [12]. This consumption can be reduced by optimizing the use of natural lighting. By adopting a building orientation that maximizes natural lighting during the day, electricity savings between 40 and 80% can be achieved in commercial buildings [13].

1.2. Reducing Energy Needs by Passive Measures

Green roofs constitute an attractive passive measure originating from traditional architecture approaches [14,15]. In Mediterranean climates, green roofs in office buildings in Cyprus [16] and residential buildings in Catania, Sicily [17] were investigated. Annual energy savings of 25% for heating and 20% for cooling were reported for the office building in Cyprus. In Sicily, the annual energy saving varied between 31 and 35% for cooling, while for heating was between 2 and 10%. Thermochromic coatings applied on building roofs were shown to decrease the energy consumption of the building by 7.7% [18], while annual energy savings up to 19% were reported for Mediterranean cities [19]. The solar reflectance of the reference thermochromic coating increased from $22 \pm 4\%$ to $47 \pm 12\%$ when titania was added. When idealized spectra are considered, a low transition temperature and a sharp hysteresis exhibit the highest energy savings [20]. The width of the hysteresis and the transmittance of thermochromic coatings can be modified by increased nitrogen flow rates during deposition [21], while the temperature of the window can be modified by controlling the incident irradiance by means of geometric concentration [22]. In addition to the thermochromic material, the matrix can be modified to obtain lower temperatures, as for example in nanocomposites and conjugated polymers that exhibit low transition temperatures down to 30 °C [23], making them promising matrices for application in windows. A temperature gradient of 6.9 °C resulting at 396 W/m^2 cooling power at 3 W/cm^2 was reported with hollow network nanoparticle nanocomposite [24], while the cooling power could be increased to 928 W/m^2 in a nanocomposite paint consisting of silica aerogel and titania nanoparticles [25].

1.3. Scope of the Study

Although thermochromic coatings exhibited considerable savings applied on roofs, their report and application as windows in buildings of Mediterranean climates is limited. Thermochromic coatings can be applied in new windows, but can be more important in existing building windows [26] with limitations in window adaptations. With this possibility in mind, we examine the performance of singly-glazed thermochromic windows in Mediterranean locations. A dynamic simulation is performed in this article to evaluate the performance of buildings with thermochromic windows. The evaluation is based on the heating and cooling energy consumption in heated and air-conditioned office buildings. The energy requirements after implementation of thermochromic windows in buildings of 15 Mediterranean cities with diverse climate conditions is evaluated and compared.

2. Methodology

2.1. Mediterranean Cities

The following cities in the Mediterranean were studied in this report. These are displayed in Figure 1. The main criteria for selection were their proximity to the Mediterranean sea and the availability of weather data. Cities as broadly as possible were chosen, across the Mediterranean coasts as well as further in continental areas.

Figure 1. Locations of the Mediterranean cities investigated in this study.

To aid the analysis of the results, the cities in this study can be grouped according to the future Köppen-Geiger classification to (a) cold semi-arid (BSk) in semi-arid cities, (b) hot-summer Mediterranean (Csa) in insular cities, and (c) dry subtropical desert (BWh) in arid cities. Accordingly:

a   Jaen, Marseille, Thessaloniki, Lisboa, and Izmir share semi-arid climate,
b   Cyprus, Palma, Heraklion, Cagliari, and Catania share insular climate,
c   Algiers, Alexandria, Beersheva, Tripoli, and Tunis share arid climate.

Figure 2 shows the annual statistics for the evolution of the air temperature in different Mediterranean locations. It can be seen that the temperature in all cities in this study varies between −10 and 40 °C during the coldest and warmest days of the year, respectively.

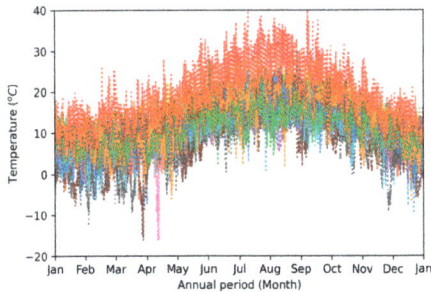

Figure 2. Annual temperature variation of the investigated Mediterranean cities.

2.2. Building Case Study

The simulations were performed on Energy Plus developed by the Lawrence Berkeley National Laboratory and the US Department of Energy. The simulation tool models the thermal and electrical loads and performs energy analysis of the entire building, enabling the user to use variable building geometries, envelope properties, lighting, heat, ventilation and air-conditioning (HVAC) systems, as well as set-point schedules. The American Society of Heating, Refrigerating and Air-Conditioning Engineers (ASHRAE) extreme dry- and wet-bulb temperature conditions were considered in simulations for heating and cooling design days, respectively.

In this study, an office building model with an area of 510 m² was considered. The window–wall ratio was 21.20% with windows distributed across all the walls of the building. This window–wall ratio is in good agreement with the 20–35% value for optimized energy performance of buildings in hot and dry regions [10,27]. Although this metric is adequate for one- and two- floor buildings, the window-to-floor-ratio should also be considered in multistory and high-rise buildings [28]. Tables 1 and 2 present the properties of the materials used in Energy Plus. Typical meteorological data were introduced to simulate the conditions of each location. To highlight the impact in energy consumption by the introduction of thermochromic windows, simulations in the absence of thermochromic coating were performed for the cities exhibiting the highest energy demand in each climate group, that is Larnaca, Tripoli, and Izmir. No dirt correction factors were applied and the windows were considered not diffusing over the solar spectrum.

Table 1. Construction materials and their properties as utilized in simulations.

Material	Thickness (m)	Conductivity (W/m·K)	Density (kg/m³)	Specific Heat (J/(kg·K)	Roughness	Thermal Absorptance	Solar Absorptance	Visible Absorptance
Wood Shingles	0.0178	0.115	513	1255	Very Rough	0.9	0.78	0.78
Wood Decking	0.0254	0.1211	593	2510	Medium Smooth	0.9	0.78	0.78
Roof Insulation	0.2216	0.049	265	836.8	Medium Rough	0.9	0.7	0.7
Gypsum	0.0127	0.16	784.9	830	Smooth	0.9	0.92	0.92
GP01 1/2 GYPSUM	0.0127	0.16	800	1090	Smooth	0.9	0.7	0.5
Stucco	0.0253	0.6918	1858	837	Smooth	0.9	0.92	0.92
Heavyweight Concrete	0.2033	1.7296	2243	837	Medium Rough	0.9	0.65	0.65
Wall Insulation	0.0339	0.0432	91	837	Medium Rough	0.9	0.5	0.5
Heavyweight Concrete	0.1016	1.311	2240	836.8	Rough	0.9	0.7	0.7
Floor Insulation	0.0464	0.045	265	836.8	Medium Rough	0.9	0.7	0.7
MAT-CC05 8 Heavyweight Concrete	0.2032	1.311	2240	836.8	Rough	0.9	0.7	0.7
Gypsum Top	0.0127	0.16	784.9	830	Smooth	0.9	0.92	0.92
Attic Floor Insulation	0.2379	0.049	265	836.8	Medium Rough	0.9	0.7	0.7
Gypsum Bottom	0.0127	0.16	784.9	830	Smooth	0.9	0.92	0.92
Roof Membrane	0.0095	0.16	1121.3	1460	Very Rough	0.9	0.7	0.7
Metal Decking	0.0015	45.006	7680	418.4	Medium Smooth	0.9	0.7	0.3
METAL Door Medium 18Ga_1	0.0013	45.3149	7833	502.08	Smooth	0.8	0.5	0.5
METAL Door Medium 18Ga_2	0.0013	45.3149	7833	502.08	Smooth	0.8	0.5	0.5
Std Wood 6inch	0.15	0.12	540	1210	Medium Smooth	0.9	0.7	0.7
Std 1.5 MW CONCRETE	0.038	0.858	1968	836.8	Rough	0.9	0.7	0.7
Std AC02	0.0127	0.0570	288	1339	Medium Smooth	0.9	0.7	0.2
Std MAT-CC05 4 MW CONCRETE	0.1	0.858	1968	836.8	Rough	0.9	0.7	0.2
Std Very High Reflectivity Surface	0.0005	237	2702	903	Smooth	0.9	0.05	0.05
Std PW05	1.91E-02	0.115	545	1213	Medium Smooth	0.9	0.78	0.78
Std Steel_Brown_Regular	1.50E-03	44.9696	7689	418	Smooth	0.9	0.92	0.92
Std Steel_Brown_Cool	1.50E-03	44.9696	7689	418	Smooth	0.9	0.73	0.73
Plywood3/4_in	0.0191	0.115	545	1213	Medium Smooth	0.9	0.7	0.78

Table 2. Properties of glazing layers used in simulations.

Magnitude	Glass	Thermochromic Layer	Clear Acrylic Plastic	Diffusing Acrylic Plastic
Thickness (m)	0.003	0.0075	0.003	0.0022
Solar transmittance	0.2442		0.92	0.9
Front solar reflectance	0.7058		0.05	0.08
Rear solar reflectance	0.7058		0.05	0.08
Visible transmittance	0.3192		0.92	0.9
Front visible reflectance	0.6308		0.05	0.08
Rear visible reflectance	0.6308		0.05	0.08
IR Transmittance at $0°$	0	0	0	0
Front IR Emissivity (2π)	0.9	0.84	0.9	0.9
Rear IR Emissivity (2π)	0.9	0.84	0.9	0.9
Conductivity (W/(m·K))	0.0199	0.6	0.9	0.9

3. Results

The monthly energy demand of the buildings studied are shown in Figures 3–5, alongside the respective temperatures and seasonal solar radiation. It is clear that the temperature profiles of each city affects the energy consumption of the building. It can be seen that the Mediterranean cities can be grouped in three categories according to their energy demand. That is (a) insular, (b) semi-arid, and (c) arid. It can be seen that the monthly energy demand in all cities during winter months varies from 4.5 to 5 MWh. There is a broader variation, however, during spring and autumn. There is increased energy demand of buildings in insular and arid climates, that can be higher by 500 W. In summer months, the energy demand of buildings in insular and arid climates can be up to 1000 W higher than in semi-arid climates. The annual energy demand of the building without thermochromic windows was increased by 104.83, 93.92, and 97.39 kWh for Larnaca, Tripoli, and Izmir, respectively.

Comparing the temperature in these cities, most of them have the warmest month in July and August. All of them have the coldest month in January. Among insular cities, it can be seen that Heraklion has the coldest winter season with the minimum temperature of −15.9 °C. This leads to an annual energy demand of 132.05 kWh/m², while the average annual energy demand of the locations under study was 130 kWh/m². Optimized thermochromic coatings applied in other Mediterranean cities yielded annual demand between 60 and 80 kWh/m² for south- and north-facing walls, respectively [29]. This difference in energy indicates (a) the requirement for optimization with regard to the orientation of the windows, becoming important when cost is introduced in the model, and (b) the requirement for thermochromic coatings with higher solar selectivity properties compared to the coatings utilized in this study.

Among semi-arid Mediterranean cities, in Marseille, the coldest winter with temperatures reaching minima of −15.4 °C. Tunis and Be'er Sheva have the hottest summer season with the maximum temperature of 40 °C. It can be seen that the energy demand of Tripoli peaks during June, July, and August, despite the fact that its temperature is the lowest among the investigated cities with arid climate. During these months, the energy demand increases to satisfy air-conditioning requirements and is presented in detail in the following section.

3.1. Arid Mediterranean Climates

The energy demand of the building is distributed across HVAC, lighting, and other needs including energy source equipment, that is fans and pumps.

As shown in Figure 6, the most energy demanding building in arid Mediterranean cities can be seen in Tripoli with 491.97 MJ/m², followed by Be'er Sheva and Alexandria with 485.32 and 483.21 MJ/m², respectively. Algiers and Tunis have the lowest demand among arid climate Mediterranean cities with 465.78 and 474.99 MJ/m², respectively. It can be seen that the use of thermochromic windows have a limited effect in lighting and other energy demands. There is, however, difference in HVAC demand of about 2% between the buildings in arid climate cities.

The heating consumption can be seen in Table 3. Algiers and Tunis have the highest consumption with 24 kW, followed by Be'er Sheva and Tripoli with 23 kW, while Alexandria has the lowest consumption with 20 kW.

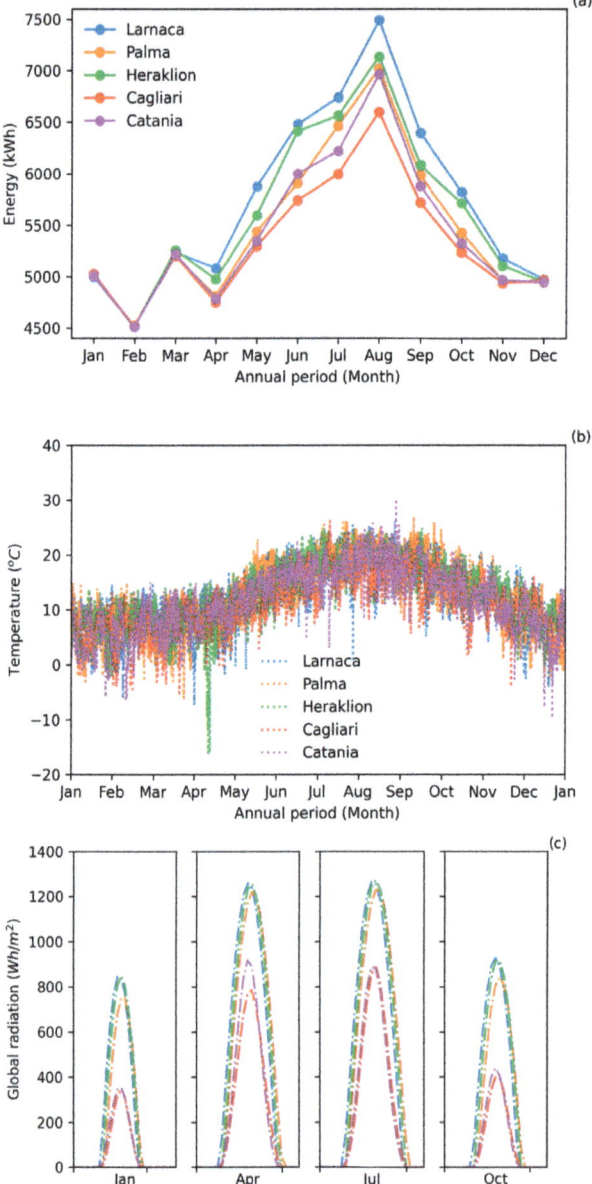

Figure 3. (a) Monthly energy demand of buildings, (b) temperature, and (c) seasonal global solar radiation in insular cities of the study.

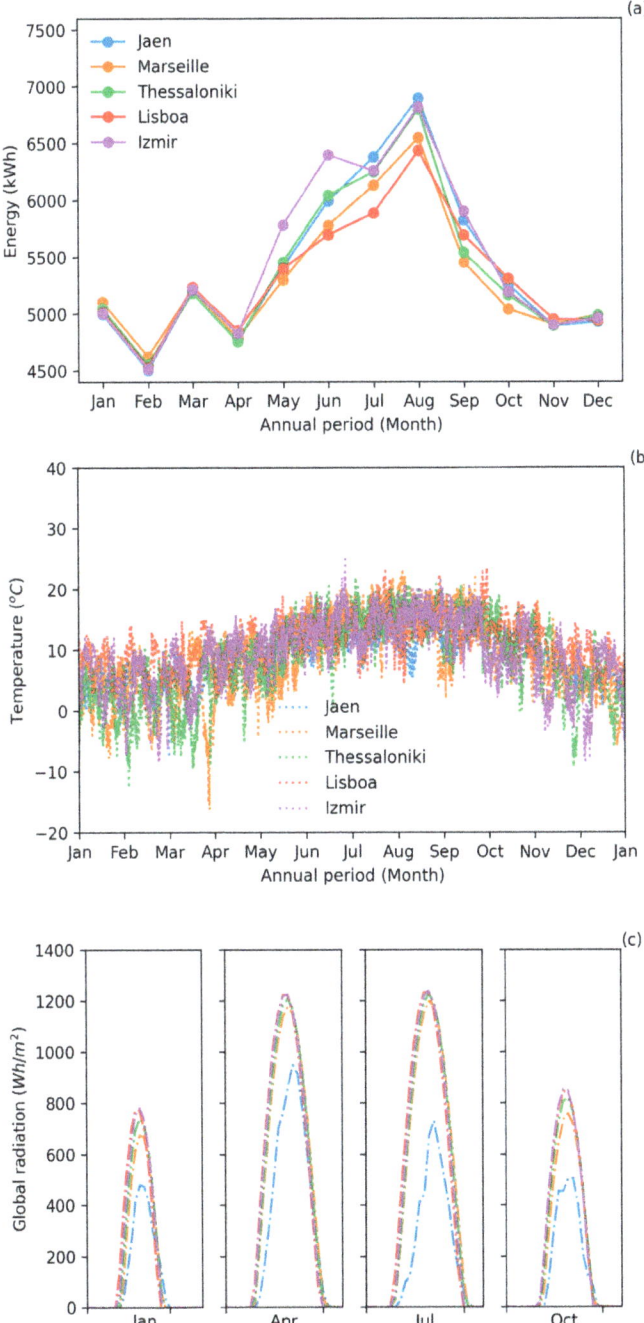

Figure 4. (**a**) Monthly energy demand of buildings, (**b**) temperature, and (**c**) seasonal global solar radiation in cities of the study with semi-arid climate.

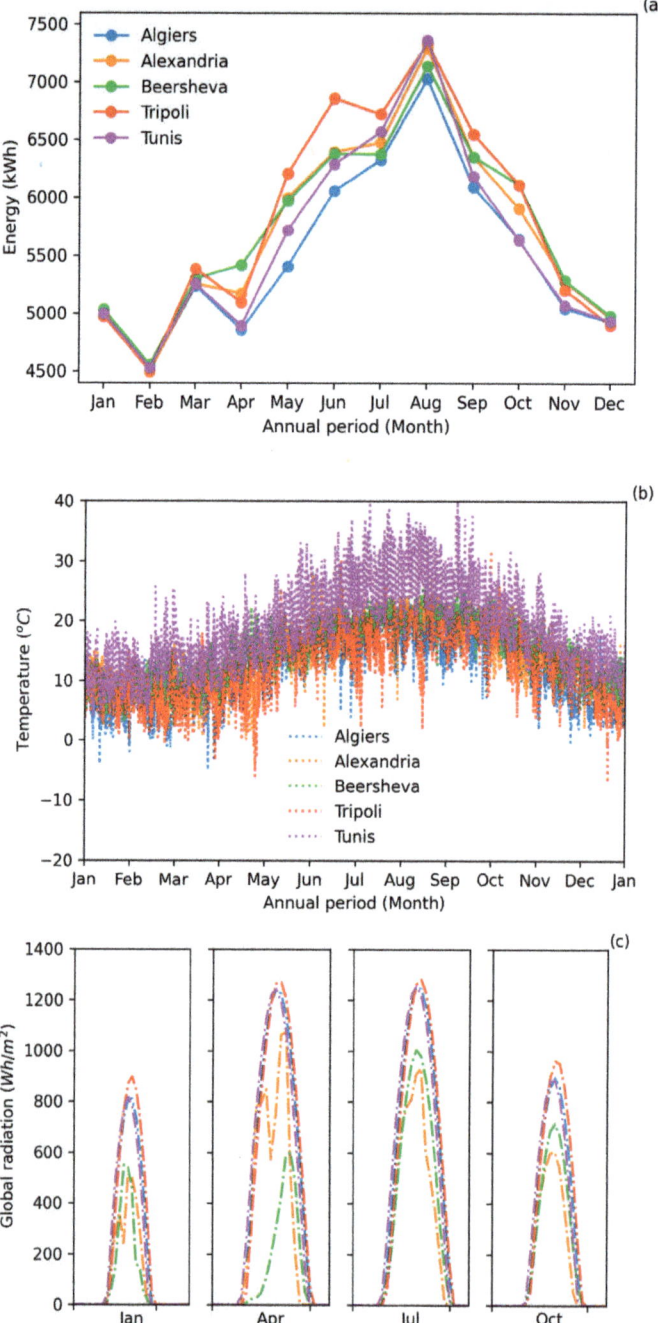

Figure 5. (**a**) Monthly energy demand of buildings, (**b**) temperature, and (**c**) seasonal global solar radiation in cities of the study with arid climate.

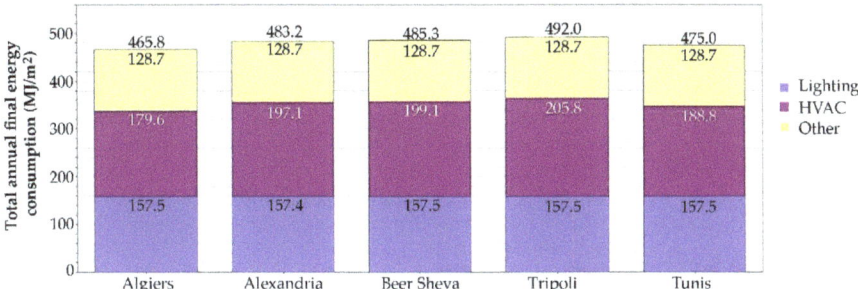

Figure 6. Annual building utility performance in arid Mediterranean climate.

Table 3. Heating consumption and hours of unmet comfort of buildings in arid climates.

City	Heating, Gas (W)	Unmet Comfort (h)
Algiers	24,854	1460
Alexandria	20,638	1085
Be'er Sheva	23,416	953
Tripoli	23,326	942
Tunis	24,067	1079

The remaining demand, shown in Figure 7, is distributed among fans (blue), cooling (purple), lighting (yellow), and other equipment (red). The demand of fans, lighting, and other equipment are the same regardless of the city. The cooling demand, however, varies considerably. Tripoli is the most demanding in cooling with 12.9 kW. Alexandria, Algiers, and Tunis require approximately 9 kW for cooling, while the least demanding of the cities is Be'er Sheva with 8.1 kW. This is in agreement with the highest energy demand depicted in Figure 5a. It can be seen that Tripoli exhibits the highest requirements among the studied cities in arid Mediterranean climate. This is largely due to the cooling requirements which are two times the cooling demand in Be'er Sheva and approximately higher by 1/3 than Algiers, Alexandria, and Tunis, see Figure 7. The extreme dry- and wet-bulb temperature conditions were utilized to simulate heating and cooling design days, respectively. This extreme scenario is commonly employed in HVAC sizing and accordingly, the demand may be overestimated due to fluctuations in local conditions. This suggests that in addition to temperature, climatic conditions such as relative humidity and wind speed may affect the energy demand of a building and should be further considered.

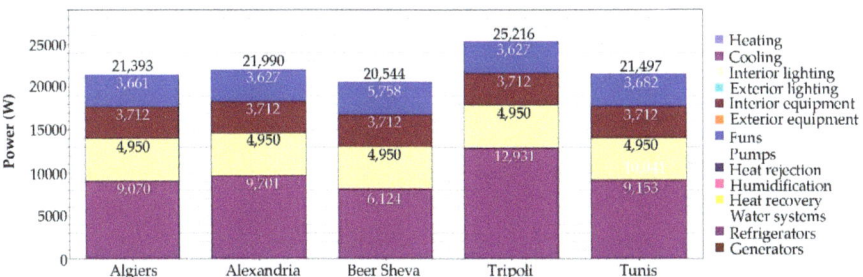

Figure 7. Demand end-use of components of buildings in arid Mediterranean climate. Components include fans (blue), cooling (purple), lighting (yellow), equipment (red).

These profiles result in several hours that comfort is not met. These hours are displayed in Table 3. In Algiers, 1460 h are not met, followed by Alexandria and Tunis with approximately 1000 h. Be'er Sheva and Tripoli share approximately 950 h of unmet comfort time.

3.2. Semi-Arid Mediterranean Climates

The most energy demanding building in semi-arid Mediterranean cities can be seen, in Figure 8, in Izmir with 219.45 MJ/m² followed by Jaen and Thessaloniki with 217.04 and 215.77 MJ/m², respectively. Marseille and Lisboa have the lowest demand among arid climate Mediterranean cities with 213 MJ/m². Again, lighting and other energy demand are similar; however, the demand of HVAC varies by 1% between the buildings in semi-arid Mediterranean climate cities.

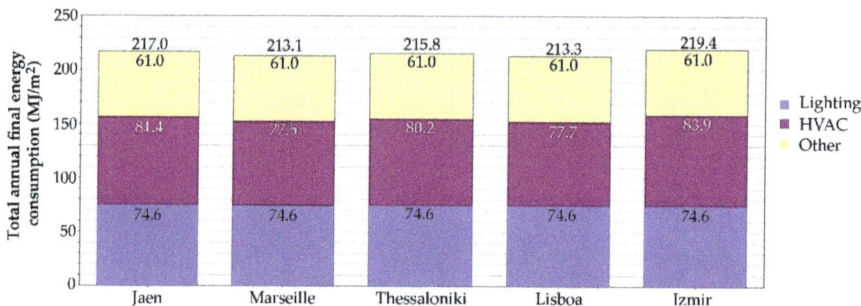

Figure 8. Annual building utility performance in semi-arid Mediterranean climate.

The heating consumption can be seen in Table 4. Jaen, Marseille, and Thessaloniki have the highest consumption with 26 kW, followed by Izmir with 24 kW and Lisboa with 23 kW.

Table 4. Heating consumption and hours of unmet comfort of buildings in semi-arid climates.

City	Heating, Gas (W)	Unmet Comfort (h)
Jaen	26,175	1185
Marseille	26,367	1930
Thessaloniki	26,320	1798
Lisboa	24,581	950
Izmir	25,331	1300

The remaining demand is shown in Figure 9. The cooling demand of Izmir is the highest with 10 kW. Jaen and Lisboa follow with 8.5 and 8.3 kW for cooling, respectively, while the least demanding cities are Marseille and Thessaloniki with 7.1 and 7.6 kW, respectively.

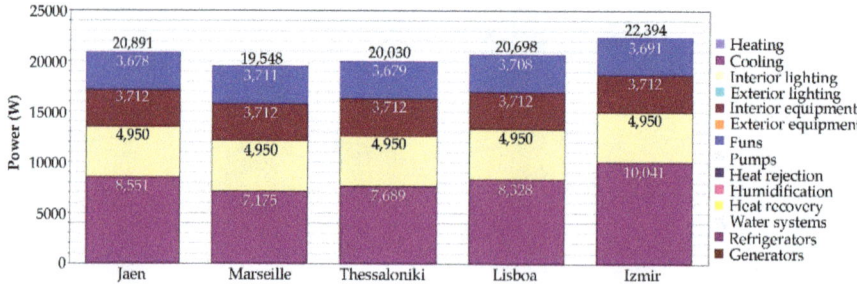

Figure 9. Demand end-use of components of buildings in semi-arid Mediterranean climate. Components include fans (blue), cooling (purple), lighting (yellow), equipment (red).

The unmet hours in this case are displayed in Table 4. In Marseille, 1930 h are not met, followed by Thessaloniki with almost 1800 h and Izmir with 1300 h. Jaen has 1185 h, while Lisboa has 950 h of unmet comfort time.

3.3. Insular Mediterranean Climates

The most energy demanding building in insular Mediterranean cities can be seen, Figure 10, in Larnaca with 484.56 MJ/m² followed by Heraklion and Palma with 474.28 and 462.83 MJ/m², respectively. Cagliari and Catania have the lowest demand among insular climate Mediterranean cities with 450 and 459 MJ/m². The demand of HVAC in this case varies by 3% between the buildings in semi-arid Mediterranean climate cities.

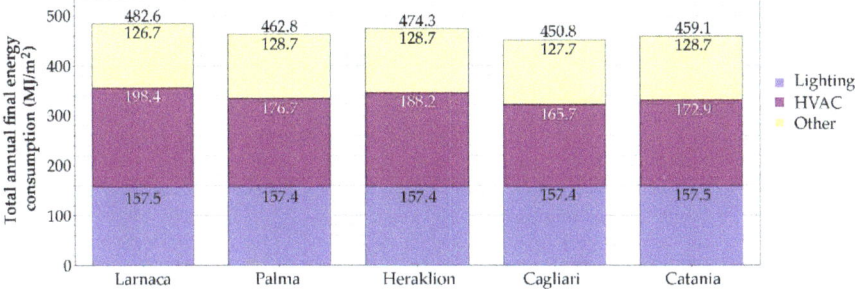

Figure 10. Annual building utility performance in insular Mediterranean climate.

The heating consumption can be seen in Table 5. Palma, Cagliari, and Catania have the highest consumption with 26 kW, followed by Heraklion and Larnaca with 25 kW.

Table 5. Heating consumption and hours of unmet comfort of buildings in insular climates.

City	Heating, Gas (W)	Unmet Comfort (h)
Larnaca	25,488	1206
Palma	26,342	2067
Heraklion	25,055	928
Cagliari	26,437	1407
Catania	26,411	1379

The remaining demand is shown in Figure 11. The cooling demand of Larnaca is the highest with nearly 10 kW. Catania has slightly lower cooling demand at 9.3 k. Palma and Heraklion follow with 8.1 kW for cooling, while the least demanding insular city of the study is Cagliari with 7 kW.

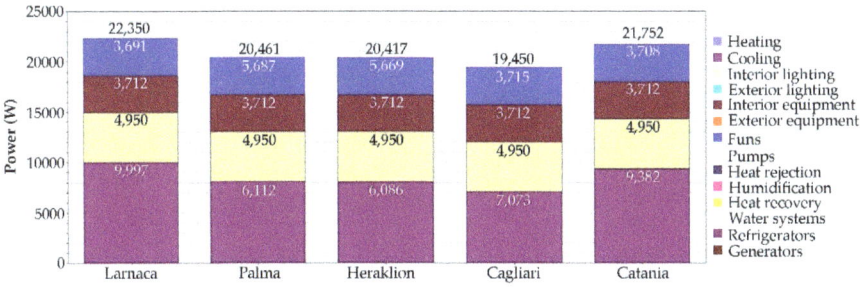

Figure 11. Demand end-use of components of buildings in insular Mediterranean climate. Components include fans (blue), cooling (purple), lighting (yellow), equipment (red).

The unmet hours in this case are displayed in Table 4. Palma has more than 2000 h unmet. Cagliari and Catania share approximately 1400 h of unmet comfort, while Larnaca follows closely with 1200 h. Finally, Heraklion has 928 h of unmet comfort time.

The HVAC equipment and the addition of thermochromic windows can optimize to a great extent the heating and cooling needs of buildings. However, in all cases there is a number of unmet hours of comfort. The number of hours was minimum in Be'er Sheva, Tripoli, Lisboa, and Heraklion. It is more likely that this is mainly due to the set of typical meteorological data used in this study and not an effect related to the local climate.

The study utilized the same building structure and construction materials. Depending on the micro-climate of each location, however, additional measures may be required to reduce the heating and cooling demand even further. Such measures can be shading as well as insulation with thermal capacity higher than what was utilized in this study. Furthermore, different thermochromic coatings with transition temperatures optimal to the peak temperatures of the location can lead to further reduction in energy consumption.

4. Conclusions

In this article, the potential for energy performance improvements in buildings was highlighted in cities sharing the Mediterranean climate. The operation of the building leads to considerable high energy consumption for space conditioning of buildings. Annual energy consumption can be reduced by introducing thermochromic windows, in particular for controlling the indoor space conditioning. The availability of high solar radiation in the Mediterranean region makes for a favorable technology for potential upgrades to zero energy buildings. The article also demonstrated the potential of energy saving in Mediterranean cities with significantly varying climate. This was highlighted by comparing the energy consumption of buildings in cities with insular, semi-arid, and arid climate.

The presented study highlighted the need for high resolution data for detailed simulation of thermochromic windows in buildings of Mediterranean cities. Temperature is one of the main climate parameters that affects the energy demand in buildings. The climate of Mediterranean cities nearby the sea may affect the energy demand. This was more pronounced in cities with arid Mediterranean climate with increased demand in air-conditioning during the summer months. On the other hand, cities with semi-arid Mediterranean climate exhibited relatively increased heating demand. It is expected that by optimally mixing and sizing energy saving systems, considerable energy savings can be achieved.

Author Contributions: Conceptualization, methodology, data curation, writing—original draft preparation, writing—review and editing, G.E.A.; funding acquisition, G.E.A. and D.A.K. All authors have read and agreed to the published version of the manuscript.

Funding: This work was supported by an academic fellowship of the HMU Postdoctoral Research Program, 2021.

Institutional Review Board Statement: Not applicable.

Informed Consent Statement: Not applicable.

Data Availability Statement: Data is available upon reasonable request.

Acknowledgments: This work was supported by an academic fellowship of the HMU Postdoctoral Research Program, 2021. We are grateful to the comments of the anonymous reviewers towards the improvement of the presented work.

Conflicts of Interest: The authors declare no conflict of interest.

References

1. Carnieletto, L.; Badenes, B.; Belliardi, M.; Bernardi, A.; Graci, S.; Emmi, G.; Urchueguía, J.F.; Zarrella, A.; Di Bella, A.; Dalla Santa, G.; et al. A European Database of Building Energy Profiles to Support the Design of Ground Source Heat Pumps. *Energies* **2019**, *12*, 2496. [CrossRef]
2. Gil-Baez, M.; Padura, Á.B.; Huelva, M.M. Passive Actions in the Building Envelope to Enhance Sustainability of Schools in a Mediterranean Climate. *Energy* **2019**, *167*, 144–158. [CrossRef]
3. Barbosa, F.C.; de Freitas, V.P.; Almeida, M. School Building Experimental Characterization in Mediterranean Climate Regarding Comfort, Indoor Air Quality and Energy Consumption. *Energy Build.* **2020**, *212*, 109782. [CrossRef]
4. Katsaprakakis, D.A.; Zidianakis, G.; Yiannakoudakis, Y.; Manioudakis, E.; Dakanali, I.; Kanouras, S. Working on Buildings' Energy Performance Upgrade in Mediterranean Climate. *Energies* **2020**, *13*, 2159. [CrossRef]
5. López-Ochoa, L.M.; Bobadilla-Martínez, D.; Las-Heras-Casas, J.; López-González, L.M. Towards Nearly Zero-Energy Educational Buildings with the Implementation of the Energy Performance of Buildings Directive via Energy Rehabilitation in Cold Mediterranean Zones: The Case of Spain. *Energy Rep.* **2019**, *5*, 1488–1508. [CrossRef]
6. Feijó-Muñoz, J.; Pardal, C.; Echarri, V.; Fernández-Agüera, J.; de Larriva, R.A.; Calderín, M.M.; Poza-Casado, I.; Padilla-Marcos, M.Á.; Meiss, A. Energy Impact of the Air Infiltration in Residential Buildings in the Mediterranean Area of Spain and the Canary Islands. *Energy Build.* **2019**, *188–189*, 226–238. [CrossRef]
7. Gasparella, A.; Pernigotto, G.; Cappelletti, F.; Romagnoni, P.; Baggio, P. Analysis and Modelling of Window and Glazing Systems Energy Performance for a Well Insulated Residential Building. *Energy Build.* **2011**, *43*, 1030–1037. [CrossRef]
8. Su, X.; Zhang, X. Environmental Performance Optimization of Window–Wall Ratio for Different Window Type in Hot Summer and Cold Winter Zone in China Based on Life Cycle Assessment. *Energy Build.* **2010**, *42*, 198–202. [CrossRef]
9. Goia, F. Search for the Optimal Window-to-Wall Ratio in Office Buildings in Different European Climates and the Implications on Total Energy Saving Potential. *Sol. Energy* **2016**, *132*, 467–492. [CrossRef]
10. Alwetaishi, M.; Benjeddou, O. Impact of Window to Wall Ratio on Energy Loads in Hot Regions: A Study of Building Energy Performance. *Energies* **2021**, *14*, 1080. [CrossRef]
11. Webb, A.L. Energy Retrofits in Historic and Traditional Buildings: A Review of Problems and Methods. *Renew. Sustain. Energy Rev.* **2017**, *77*, 748–759. [CrossRef]
12. Knoop, M.; Aktuna, B.; Bueno, B.; Darula, S.; Deneyer, A.; Diakite, A.; Fuhrmann, P.; Geisler-Moroder, D.; Hubschneider, C.; Johnsen, K.; et al. *Daylighting and Electric Lighting Retrofit Solutions*; Universitätsverlag der TU Berlin: Berlin, Germany, 2016.
13. Carletti, C.; Cellai, G.; Pierangioli, L.; Sciurpi, F.; Secchi, S. The Influence of Daylighting in Buildings with Parameters NZEB: Application to the Case Study for an Office in Tuscany Mediterranean Area. *Energy Procedia* **2017**, *140*, 339–350. [CrossRef]
14. Maiolo, M.; Pirouz, B.; Bruno, R.; Palermo, S.A.; Arcuri, N.; Piro, P. The Role of the Extensive Green Roofs on Decreasing Building Energy Consumption in the Mediterranean Climate. *Sustainability* **2020**, *12*, 359. [CrossRef]
15. Katsaprakakis, D.A.; Georgila, K.; Zidianakis, G.; Michopoulos, A.; Psarras, N.; Christakis, D.G.; Condaxakis, C.; Kanouras, S. Energy Upgrading of Buildings. A Holistic Approach for the Natural History Museum of Crete, Greece. *Renew. Energy* **2017**, *114*, 1306–1332. [CrossRef]
16. Ziogou, I.; Michopoulos, A.; Voulgari, V.; Zachariadis, T. Implementation of Green Roof Technology in Residential Buildings and Neighborhoods of Cyprus. *Sustain. Cities Soc.* **2018**, *40*, 233–243. [CrossRef]
17. Cascone, S.; Catania, F.; Gagliano, A.; Sciuto, G. A Comprehensive Study on Green Roof Performance for Retrofitting Existing Buildings. *Build. Environ.* **2018**, *136*, 227–239. [CrossRef]
18. Hu, J.; Yu, X.B. Adaptive Thermochromic Roof System: Assessment of Performance under Different Climates. *Energy Build.* **2019**, *192*, 1–14. [CrossRef]
19. Zinzi, M.; Agnoli, S.; Ulpiani, G.; Mattoni, B. On the Potential of Switching Cool Roofs to Optimize the Thermal Response of Residential Buildings in the Mediterranean Region. *Energy Build.* **2021**, *233*, 110698. [CrossRef]
20. Warwick, M.E.A.; Ridley, I.; Binions, R. Variation of Thermochromic Glazing Systems Transition Temperature, Hysteresis Gradient and Width Effect on Energy Efficiency. *Buildings* **2016**, *6*, 22. [CrossRef]
21. Vernardou, D.; Louloudakis, D.; Spanakis, E.; Katsarakis, N.; Koudoumas, E. Thermochromic Amorphous VO2 Coatings Grown by APCVD Using a Single-Precursor. *Sol. Energy Mater. Sol. Cells* **2014**, *128*, 36–40. [CrossRef]
22. Arnaoutakis, G.E.; Richards, B.S. Geometrical Concentration for Enhanced Up-Conversion: A Review of Recent Results in Energy and Biomedical Applications. *Opt. Mater.* **2018**, *83*, 47–54. [CrossRef]
23. Garshasbi, S.; Santamouris, M. Using Advanced Thermochromic Technologies in the Built Environment: Recent Development and Potential to Decrease the Energy Consumption and Fight Urban Overheating. *Sol. Energy Mater. Sol. Cells* **2019**, *191*, 21–32. [CrossRef]
24. Hu, F.; An, L.; Li, C.; Liu, J.; Ma, G.; Hu, Y.; Huang, Y.; Liu, Y.; Thundat, T.; Ren, S. Transparent and Flexible Thermal Insulation Window Material. *Cell Rep. Phys. Sci.* **2020**, *1*, 100140. [CrossRef]
25. An, L.; Petit, D.; Di Luigi, M.; Sheng, A.; Huang, Y.; Hu, Y.; Li, Z.; Ren, S. Reflective Paint Consisting of Mesoporous Silica Aerogel and Titania Nanoparticles for Thermal Management. *ACS Appl. Nano Mater.* **2021**, *4*, 6357–6363. [CrossRef]
26. Teixeira, H.; da Gomes, M.G.; Rodrigues, A.M.; Pereira, J. In-Service Thermal and Luminous Performance Monitoring of a Refurbished Building with Solar Control Films on the Glazing System. *Energies* **2021**, *14*, 1388. [CrossRef]

27. Alwetaishi, M.; Taki, A. Investigation into Energy Performance of a School Building in a Hot Climate: Optimum of Window-to-Wall Ratio. *Indoor Built Environ.* **2020**, *29*, 24–39. [CrossRef]
28. Saroglou, T.; Theodosiou, T.; Givoni, B.; Meir, I.A. A Study of Different Envelope Scenarios towards Low Carbon High-Rise Buildings in the Mediterranean Climate-Can DSF Be Part of the Solution? *Renew. Sustain. Energy Rev.* **2019**, *113*, 109237. [CrossRef]
29. Arnesano, M.; Pandarese, G.; Martarelli, M.; Naspi, F.; Gurunatha, K.L.; Sol, C.; Portnoi, M.; Ramirez, F.V.; Parkin, I.P.; Papakonstantinou, I.; et al. Optimization of the Thermochromic Glazing Design for Curtain Wall Buildings Based on Experimental Measurements and Dynamic Simulation. *Sol. Energy* **2021**, *216*, 14–25. [CrossRef]

Article

Thermal Performance Improvement of Double-Pane Lightweight Steel Framed Walls Using Thermal Break Strips and Reflective Foils

Paulo Santos * and Telmo Ribeiro

ISISE, Department of Civil Engineering, University of Coimbra, 3030-788 Coimbra, Portugal; telmo.ribeiro@dec.uc.pt
* Correspondence: pfsantos@dec.uc.pt

Abstract: The reduction of unwanted heat losses across the buildings' envelope is very relevant to increase energy efficiency and achieve the decarbonization goals for the building stock. Two major heat transfer mechanisms across the building envelope are conduction and radiation, being this last one very important whenever there is an air cavity. In this work, the use of aerogel thermal break (TB) strips and aluminium reflective (AR) foils are experimentally assessed to evaluate the thermal performance improvement of double-pane lightweight steel-framed (LSF) walls. The face-to-face thermal resistances were measured under laboratory-controlled conditions for sixteen LSF wall configurations. The reliability of the measurements was double-checked making use of a homogeneous XPS single panel, as well as several non-homogeneous double-pane LSF walls. The measurements allowed us to conclude that the effectiveness of the AR foil is greater than the aerogel TB strips. In fact, using an AR foil inside the air cavity of double-pane LSF walls is much more effective than using aerogel TB strips along the steel flange, since only one AR foil (inner or outer) provides a similar thermal resistance increase than two aerogel TB strips, i.e., around +0.47 m^2·K/W (+19%). However, the use of two AR foils, instead of a single one, is not effective, since the relative thermal resistance increase is only about +0.04 m^2·K/W (+2%).

Keywords: thermal performance; experimental assessment; double-pane; lightweight steel frame (LSF); partition walls; aerogel thermal break strips; aluminium reflective foils

---

1. Introduction

The most relevant energy consumer in European Union (EU) is the buildings' sector [1]. In fact, during 2019, forty percent of the EU27 final energy consumption was spent in buildings [2]. Within this share, almost 50% are used for space heating and cooling [1]. These facts are related to reduced energy efficiency of buildings and consequent wasted energy. Indeed, in the EU three out of four buildings are classified as inefficient [2]. To make Europe's building sector compatible with the Paris Agreement, two goals need to be achieved: (1) reduce energy demand through energy efficiency measures, and; (2) increase the use of renewable energy sources [3].

As is well known [4], the reduction of undesirable heat losses is one of the possible strategies to improve energy efficiency. This heat loss reduction could be achieved by mitigating each heat transfer mechanism across the building envelope: radiation, convection, and conduction. The most forthright and simplest approach to increase the thermal resistance of building envelope components is the usage of thermal insulation, reducing significantly the heat transfer by conduction. Nevertheless, the effectiveness of the thermal insulation depends also on their position within the building element, as previously demonstrated by Roque and Santos [5]. Additionally, this insulation material also endorses sound insulation, principally when porous batt insulation materials are used inside the air cavities [6].

Currently, extremely efficient insulation materials (sometimes designated by SIMs—super insulating materials) are emerging in the market, having very small thermal conductivities [7]. Aerogels [8] and vacuum insulating panels (VIPs) [9] are nowadays two of the most common examples of SIMs. Notice that increasing discontinuous thermal insulation along building envelopes may rise the relevance of thermal bridges, being this effect even more significant in steel structures, given the huge thermal conductivity of steel [10]. In fact, as concluded by Erhorn-Klutting and Erhorn [11] up to near one-third of the heating energy needs could be originated by thermal bridges in traditional buildings (reinforced concrete and masonry).

During the last years, the lightweight steel frame (LSF) construction system is being more used, mainly for low-rise residential houses [12], due to their intrinsic benefits. Some of these advantages are: fast construction, high mechanical strength, and low weight, high potential for recycling and reuse, reduced on-site disruption, great suitability for retrofitting, high architectural flexibility, economical transportation and handling, easy prefabrication, precise tolerances, superior quality, insect damage resistance, and humidity stability shape [4].

Nowadays, several techniques could be used to mitigate thermal bridges in LSF buildings' elements, such as slotted thermal steel studs [13–15], thermal break (TB) strips [13,16,17], and continuous thermal insulation layers (e.g., ETICS—external thermal insulation composite system) [18–20]. Moreover, the steel frame is so important that even minor changes in the stud flanges shape and size could have a relevant effect on the thermal performance of LSF walls [21].

Moreover, when there is an air cavity inside the wall, one effective way to improve the thermal performance is by reducing the heat transfer by radiation. This could be achieved by using reflective low-emissivity paint or foil inside the air gaps of the building components [22,23]. This thermal performance improvement solution has supplementary benefits, such as easy installation and low cost.

As recently mentioned by Bruno et al. [23], there is a very small number of research works related to thermal resistance improvement due to low-emissivity materials placed inside air cavities. This fact is even more perceptible in LSF double-pane building elements.

Recently, Santos and Ribeiro [24] studied the thermal performance of double-pane lightweight steel-framed walls with and without a reflective foil. This assessment was mainly experimental under laboratory-controlled conditions, but the measurements were compared with 2D finite element numerical simulations. Several air cavity thicknesses (0 mm up to 50 mm, with an increment of 10 mm) were evaluated, but only one aluminium reflective (AR) foil was considered (on the outer surface of the air cavity). It was concluded that "the use of a reflective foil is a very effective way to increase the thermal resistance of double pane LSF walls, without increasing the wall thickness and weight". However, when using an AR foil, it is not worthy to have an air cavity higher than 30 mm. One research gap that was not investigated in this work was if it is worthy to use two low-emissivity aluminium foils (one in each air-gap surface). Moreover, another interesting question is if the performance of the AR foil is similar when used in the outer or inner surface of the air cavity.

Regarding the use of TB strips, there is also a lack of scientific research works available. Perhaps the most relevant is the experimental campaign completed by Santos and Mateus [17] for the assessment of thermal break strip performance in load-bearing and non-load-bearing LSF walls. They concluded that an outer or inner TB strip has very similar thermal performances, being the best performance achieved for two TB strips (one on each steel stud flange) and for the aerogel TB strip material. Notice that these walls were single-pane LSF walls, not found in the literature any research work related with the use of TB strips in double-pane LSF walls, or in combined use with AR foils.

In this work, the authors seek the answers for some of these research gaps and questions, by assessing the thermal performance improvement of double-pane LSF walls using thermal break strips and reflective foils. The strategy was to start with a reference double-pane LSF wall (30 mm air gap) and compare the thermal resistance increase only due to aerogel TB strips and only due to AR foils, by performing measurements under laboratory-controlled conditions. Moreover, on the next set of measurements, the combined effect of both aerogel TB strips, and AR foils was evaluated. Notice that the aerogel was selected as the material for the TB strips since it is one of the highest performant materials available in the market, having a very reduced thermal conductivity (in this case, 0.015 W/m K). Moreover, aluminium was selected as the material for the reflective foil since it is the most currently used for this purpose, having a very reduced emissivity value (below 0.05).

This article is structured as follows. After this small introduction and contextualization, it is presented a section with the materials and methods, where the LSF walls and used materials are characterized. Moreover, the experimental lab tests are described, including the experimental setup, as well as the set-points and test procedures. Additionally, to ensure the reliability of the measurements and to check the test procedures, two verifications were performed: (1) comparison between the measured thermal conductivity of a homogeneous XPS panel with the value provided by the manufacturer, and; (2) comparison between the measured thermal resistance of four double-pane LSF walls, namely: (1) reference; (2) with a single TB strip; (3) with a single AR foil, and; (4) with both two TB strips and two AR foils, with the predictions provided by numerical simulation models. The obtained results are presented and discussed next, being grouped into three sets: (1) only TB strips; (2) only AR foils; (3) combined TB strips and AR foils. Finally, the main conclusions of this research work are summarized. Notice, that the various scenarios evaluated in this research work were based on previous papers from authors, namely reference [24] for the aluminium reflective foil scenarios and reference [17] for the thermal break strips.

2. Materials and Methods

In this section, the tested double-pane LSF walls are described, and the respective materials are characterized. Furthermore, the explanation about the lab tests is performed, through the presentation of the experimental setup and the description of the test procedures. Finally, to ensure the accuracy and reliability of the achieved experimental results, the test procedures are verified using numerical simulation results for comparison.

2.1. Walls and Materials Characterization

A representative horizontal cross-section of the reference double-pane LSF wall used in the lab tests is displayed in Figure 1. The frame structure of each LSF wall pane is composed of steel studs (C48 × 37 × 4 × 0.6 mm) spaced 400 mm apart and filled with mineral wool (MW) panels with a thickness of 48 mm. The LSF wall panes are separated by a 30 mm thick air cavity and its most superficial layers are made up of two gypsum plasterboards (GPB) panels with a total thickness of 25 mm, on each side.

The thickness and the thermal conductivities of the materials that composed the reference double-pane LSF wall used in the experimental tests are displayed in Table 1.

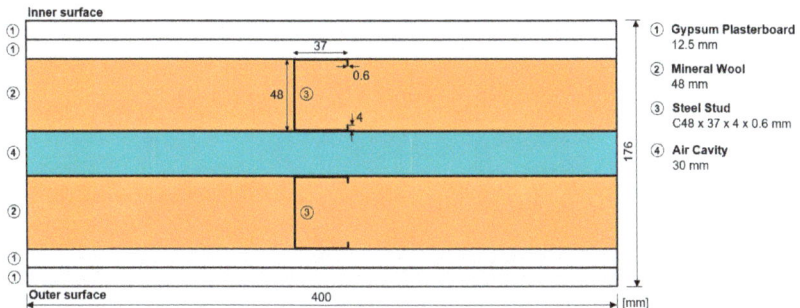

Figure 1. Geometry and materials of the reference double-pane LSF wall: horizontal cross-section.

Table 1. Reference double-pane LSF wall materials, thickness (d), and thermal conductivities (λ).

Material (Inner to Outer Layer)	d [mm]	λ [W/(m·K)]	ρ [kg/m³]	Reference
Gypsum plasterboard (2 × 12.5 mm)	25	0.175	600	[25]
Mineral wool	48	0.035	60	[26]
Steel stud (C48 × 37 × 4 × 0.6 mm)	-	50.000	7860	[27]
Air cavity	30	-	-	-
Mineral wool	48	0.035	60	[26]
Steel stud (C48 × 37 × 4 × 0.6 mm)	-	50.000	7860	[27]
Gypsum plasterboard (2 × 12.5 mm)	25	0.175	600	[25]
Total Thickness	176	-	-	-

In this research work, the thermal performance improvement due to thermal break (TB) strips and/or reflective foils was experimentally evaluated. The TB strips tested are 50 mm wide and 10 mm thick, and the material used was aerogel with thermal conductivity of 0.015 W/(m·K) [16]. Three configurations for the localization of the TB strips were considered (Figure 2): (i) along the inner steel stud flange (inner wall pane); (ii) along the outer steel stud flange (outer wall pane); (iii) on both steel stud flanges. Furthermore, the reflective foils tested are made of aluminium with an emissivity equal to 0.05 [23,24].

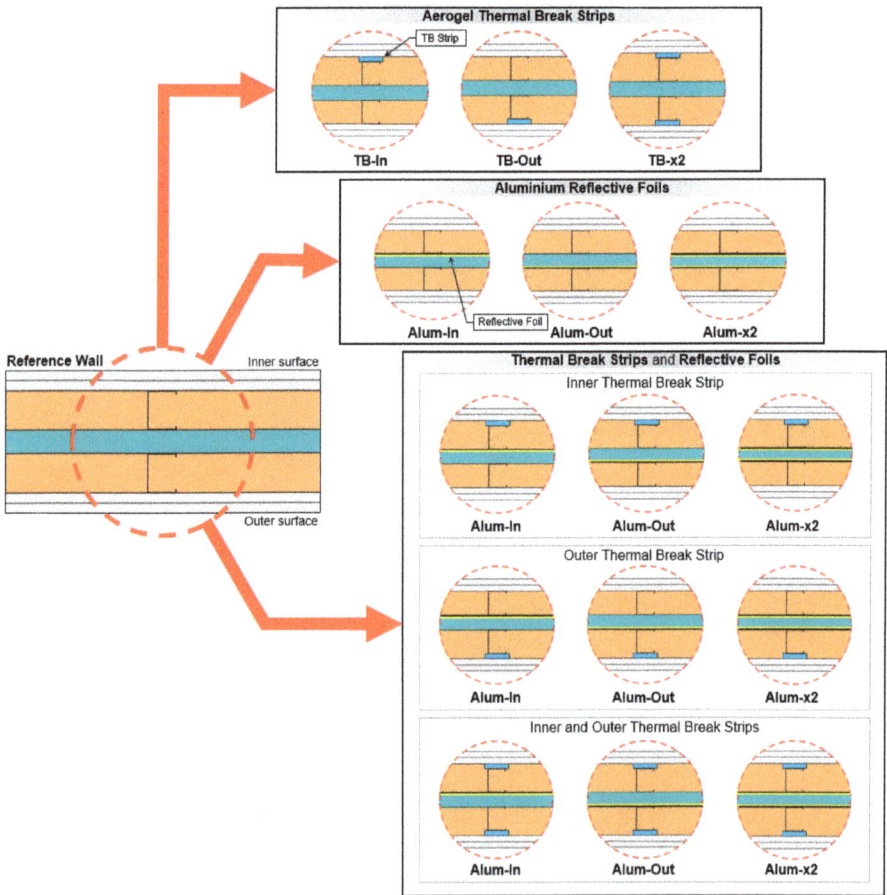

Figure 2. Wall configurations tested using aerogel thermal break (TB) strips and aluminium reflective foils.

Regarding the localization of the reflective foils, three configurations were considered (Figure 2): (i) along the inner side of the air cavity; (ii) along the outer side of the air cavity; (iii) along both sides of the air cavity. These two elements (aerogel TB strips and aluminium reflective foils) were tested separately, as well as combined, resulting in a total of fifteen wall configurations, as schematically illustrated in Figure 2.

2.2. Experimental Lab Tests

2.2.1. Experimental Setup

The lab measurements were performed using a mini hot box apparatus, as illustrated in Figure 3a. This equipment is composed of two climatic chambers: (i) a hot box heated by an electric resistance; (ii) a cold box cooled by a refrigerator. The double-pane LSF wall test sample was placed between these two chambers, as displayed in Figure 3b,c. To minimize the amount of heat that is lost through the lateral surfaces of the sample, its perimeter was covered with 80 mm thick polyurethane foam insulation (Figure 3a).

Figure 3. Mini hot box apparatus: (**a**) global view of the equipment; (**b**) LSF wall test sample; (**c**) LSF wall test sample with TB strips and a reflective foil.

The air cavity within the test frame was ensured by using a 30 mm thick EPS frame, separating the two LSF wall panes (Figure 3b,c), whose edges match the thickness of the thermal insulation of the climatic chambers' envelope (Figure 4b). The frontal view of the LSF wall pane (hot side) is displayed in Figure 4a, where it is possible to visualize the steel structure constituted by three vertical studs and filled with mineral wool. Despite the three vertical studs, when the sample is placed in the mini hot box apparatus, only the central stud is exposed to the temperature gradient provided by the climatic chambers. Figure 4b illustrates the EPS frame and the respective dimensions of its edges, as well as the aluminium reflective foil localization (in this case, on the hot side). The aerogel TB strips were applied covering all the steel profiles of the test sample (three vertical studs and two horizontal tracks), as displayed in Figure 4c. Inside each climatic chamber, small interior fans were used to promote internal air circulation and reduce the probability of air temperature stratification. Furthermore, on each side of the LSF wall sample (hot and cold), a black radiation shield was placed 10 cm apart from the wall surface.

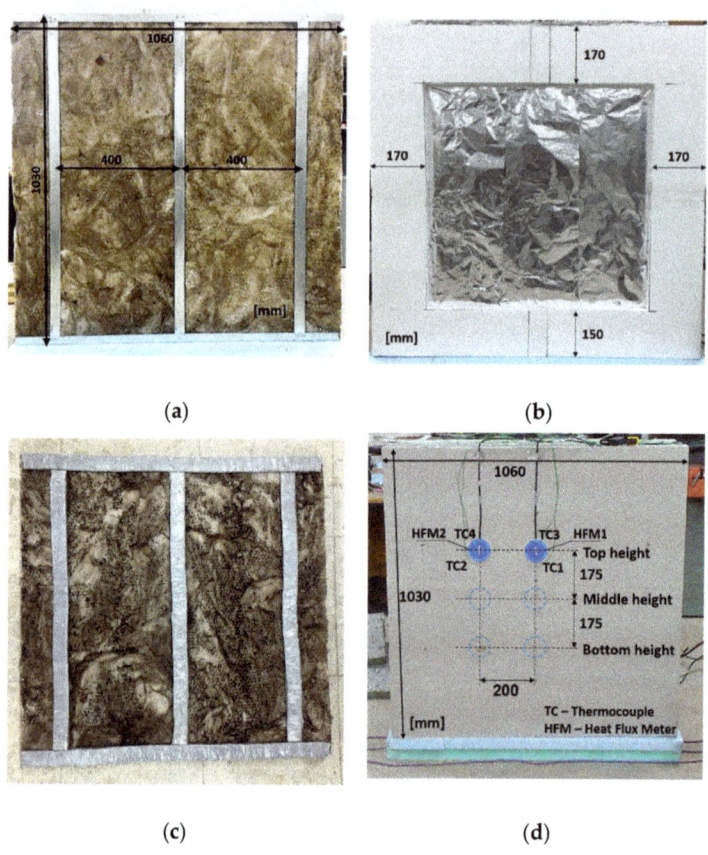

Figure 4. Double-pane LSF wall test sample: (**a**) LSF wall pane (hot side); (**b**) EPS frame and reflective foil; (**c**) aerogel TB strip; (**d**) arrangement of the monitoring system (cold side).

The arrangement of the monitoring system, constituted by heat flux meters (HFM) and thermocouples (TCs), is displayed in Figure 4d. To measure the heat flux through the test sample, four heat flux meters (Hukseflux model HFP01, precision: ±3%) were used, being two of them applied on the hot wall surface and another two on the cold wall surface. On each wall surface, since the LSF wall sample presents two distinct thermal behaviour zones, two locations for the heat flux meters were considered: (i) nearby the central vertical steel stud (HFM1); (ii) in the middle of the insulation cavity (HFM2).

Temperature measurements were performed using twelve type K (1/0.315) PFA insulated thermocouples (TCs), certified with class 1 precision. These TCs were calibrated within the temperature range [5 °C; 45 °C], with a 5 °C increment, by immersing them in a thermostatic stirring water bath (Heto CB 208). The measurements were performed using six TCs on each side of the wall sample, considering the following configuration: two of them measured the wall surface temperature (TC1 and TC2), another two measured the air temperature between the radiation shield and the wall surface (TC3 and TC4), and the remaining two measured the environment air temperature inside the climatic chamber (TC5 and TC6).

The temperature and heat flux data measured during each test were recorded using one PICO TC-08 data logger (precision ±0.5 °C), on each side of the wall test sample (hot and cold). The management of the data recorded was performed by connecting the two data loggers to a laptop and using the PicoLog version 6.1.10 software. In Table 2, the equipment used in the lab measurements is characterized in terms of brand, model, measurement range, and precision.

Table 2. Characteristics of the measurement equipment used in the lab experiments.

Equipment	Brand	Model	Measurement Range	Precision
Thermocouple	LabFacility	Type K * (1/0.315)	−75 to +260 °C	±1.5 °C
Heat flux meter	Hukseflux	HFP01	−2 to +2 kW/m^2	±3%
Data-logger	PICO	TC-08	−270 to +1820 °C	±0.5 °C

* Tolerance class 1 certified.

2.2.2. Set-Points and Test Procedures

The thermal performance of the double-pane LSF test samples was evaluated using the heat flux meter (HFM) method, adapted to measure, simultaneously, at both wall surfaces (hot and cold), as suggested by Rasooli and Itard [28]. Comparatively to the measurement on only one side (as prescribed by ISO 9869-1 [29]), the measurement at both wall surfaces allows to increase the precision and reduce the test duration. Regarding set-point temperatures, the values of 40 °C and 5 °C were programmed for the hot and cold boxes, respectively, being the measurements performed in a quasi-steady-state heat transfer condition.

The minimum duration of each experimental test performed was 24 h and the measurements were recorded at each 10 s. This recorded data was averaged to hourly values, allowing to obtain the hourly thermal resistance (R-value). Following the convergence criteria prescribed in ASTM C1155-95 [30], only the estimated hourly R-values with an absolute difference, in relation to the previous value, lower than 10% were considered in the measurements.

The application of the HFMs in two locations on the test sample with different thermal behaviour, allows the determination of two distinct conductive local R-values: (1) a lower value in the central vertical steel stud zone (R_{stud}); (2) a higher value in the middle of the insulation cavity zone (R_{cav}). The overall surface-to-surface value of the wall is calculated using an area-weighted average of these two conductive local R-values. Furthermore, for each wall sample, three tests were performed to ensure the repeatability of the measurements. The three tests correspond to three high locations (top, middle, and bottom), as illustrated in Figure 4d, and the measured overall conductive R-value of the LSF wall was considered equal to the average of these three tests.

2.3. Numerical Simulations

These numerical simulations were performed making use of bidimensional models built in the THERM [31] finite elements software. The main idea was to compare the measured conductive R-values with the results provided by these simulations, as presented later in Section 2.4.2. In these models, only a representative part of the double-pane LSF walls (400 mm wide) were simulated, as previously illustrated in Figure 1 for the reference LSF wall. The thermal conductivities of the materials used in these numerical simulations were previously displayed in Table 1. The finite element mesh was refined to achieve a maximum 3% error in these computations and the mesh void tolerance was 1 mm^2. The maximum number of iterations was 100 and the used quad-tree mesh parameter was set to 6, while the convergence tolerance was equal to 1×10^{-6}. Using this mesh configuration, the maximum number of finite elements was 15,429.

Regarding the boundary conditions, the air temperature was set to 5 °C and 40 °C, for the outer and inner environments, respectively. Notice that these values are equal to the set-points defined for cold and hot boxes, as previously described in Section 2.2.2.

Furthermore, the surface thermal resistances were modelled using the average values measured for each LSF wall surface and for each test, considering the air and surface temperature differences and the surface heat fluxes. The measured surface thermal resistances vary within the interval [0.06; 0.13] m²·K/W, thereby respecting the range defined by EN ISO 6946 [32] for horizontal heat flow, i.e., between 0.04 m²·K/W for external surface resistance (Rse) and 0.13 m²·K/W for internal surface resistance (Rsi).

In a previous work, a similar procedure was implemented for a double-pane LSF wall with and without one AR foil for different air cavity thicknesses [24]. Regarding the simulation of unventilated airspaces, it was concluded that both "CEN simplified" and "NFRC 100" models were able to reproduce with reasonable accuracy (around ±5%) the thermal behaviour of the air cavities. However, for larger thicknesses of the air cavities (greater than 20 mm) the "CEN simplified" model exhibited a better accuracy. Thus, in the present work, this calculation model was selected since the air cavity was 30-mm thick.

2.4. Verification of the Test Procedures and Measured Values

The authors already have a large experience in measuring the thermal performance of LSF walls under controlled laboratory conditions [17,19,21,24,33]. Nevertheless, some verifications were performed to check the test procedures and ensure the reliability of the measured values, as briefly explained in the next two subsections.

2.4.1. Homogeneous XPS Panel

The first verification was made using a homogenous XPS panel (Topox® Cuber SL), having a thermal conductivity of 0.034 W/(m·K) and a thickness of 60 mm, which were tested under the same conditions of the evaluated double-pane LSF walls. It was measured an XPS conductive thermal resistance equal to 1.748 m²·°C/W. Knowing the XPS panel thickness, easily it was computed the laboratory-measured thermal conductivity, which is equal to the value provided by the manufacturer, ensuring this way the good accuracy and working conditions of the data acquisition system and sensors.

2.4.2. Nonhomogeneous Double-Pane LSF Walls

Besides the previous verification for a homogeneous XPS panel with known thermal conductivity, it was performed another set of accuracy confirmation tests for some of the double-pane LSF walls assessed, since they are very heterogeneous regarding their thermal properties (e.g., mineral wool and steel).

Thus, besides the reference LSF wall without any thermal break (TB) strip nor aluminium reflective (AR) foil, additional selections included an LSF wall with one aerogel TB strip, a LSF wall with a single AR foil, and another LSF wall containing two TB strips (inner and outer) as well as double AR foils (inner and outer), embracing in this way the two thermal performance improvement strategies and a combined LSF wall configuration.

The measured thermal resistances of the four representative double-pane LSF walls are displayed in Table 3, as well as the R-values predicted by the THERM [31] bi-dimensional models (see previous Section 2.3). The differences between the measured and the predicted thermal resistances range from −0.059 m²·K/W (−2%) and +0.062 m²·K/W (+3%). Given all the simulation and experimental uncertainties involved, this comparison denotes a very good agreement between measurements and numerical simulations. Therefore, it was concluded that the implemented test procedures are adequate and the measured values are reliable.

Table 3. Predicted and measured conductive thermal resistance of double-pane LSF walls. Absolute and percentage differences.

Wall Description	R Value [m²·K/W] Predicted	R Value [m²·K/W] Measured	Differences [m²·K/W]	Differences [%]
Reference LSF wall	2.394	2.456	+0.062	+3%
Outer TB strip	2.744	2.749	+0.005	+0%
Outer AR foil	2.993	2.934	−0.059	−2%
Double TB strips and AR foils	3.821	3.808	−0.013	−0%

Besides the *R*-values comparison previously displayed in Table 3, an additional verification was performed making use of the finite elements THERM models, as illustrated in Figure 5. These plots exhibit the predicted heat flux distribution along the cross-section of four double-pane LSF walls formerly presented in Table 3.

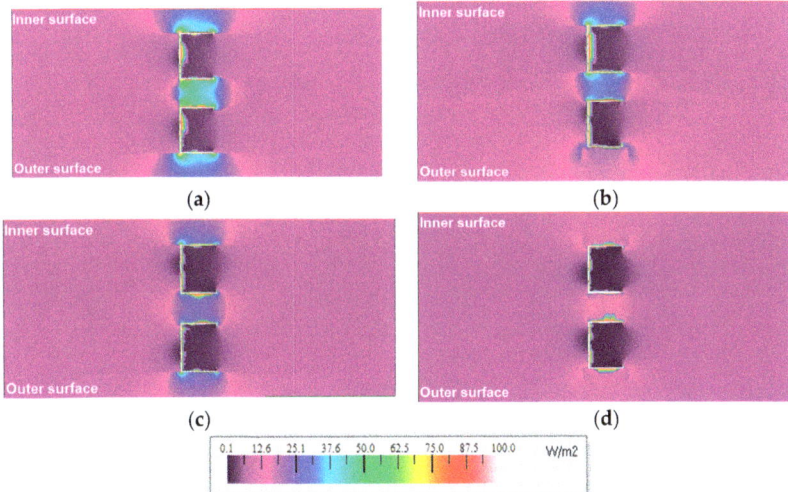

Figure 5. Predicted heat flux distribution within the LSF walls' cross-section: (**a**) reference LSF wall; (**b**) outer TB strip; (**c**) outer AR foil; (**d**) double TB strips and AR foils.

In Figure 5a, the higher heat flux is very well visible (thermal bridge effect) along the section containing the steel stud (nearly 50 mm wide). Within the air cavity, this increased heat flux is mainly related to convection and radiation mechanisms of heat transfer. Additionally, another mechanism of heat transfer (conduction) is well visible on both sides of the sheathing layers around the flanges of the steel studs, which diffuses heat to the surroundings. Moreover, within the two mineral wool insulation layers (one for each LSF wall pane) there is also an increased heat transfer (thermal bridge), also due to conduction heat transfer mechanism, which is concentrated within the steel web of both vertical studs.

When there is an aerogel TB strip placed in the outer steel flange (Figure 5b), there is a significant decrease of the heat conduction through the steel flange to the surrounding sheathing layers (gypsum plasterboard). However, the heat flux within the remaining parts of this steel stud's related thermal bridge, remained quite high, even with a small decrease in comparison with the reference LSF wall (Figure 5a).

As expected, when using an AR foil (Figure 5c) there is a significant decrease in the heat flux, mainly within the air cavity between the two steel studs, but also, as a consequence, within the gypsum plasterboard near the steel flanges. This feature illustrates quite well the importance of the radiative heat transfer across these double-pane LSF walls.

Finally, Figure 5d illustrates the heat flux distribution when both 2 TB strips and 2 AR foils are used in the double-pane LSF wall. As expected, this way the heat transfer by radiation is mitigated within the air cavity due to the existence of the AR foils, and the conductive heat transfer is also mitigated due to the existence of the TB strips (one near each steel flange). Consequently, the heat flux values are higher only within the 2 steel studs, being very reduced within the sheathing layers, the thermal insulation and the air-gap, confirming this way the best thermal performance previously mentioned in Table 3.

3. Results and Discussion

This section is divided into three parts. First, the measurement results related with the thermal resistance improvement of double-pane LSF walls, due to aerogel thermal break strips, are presented and discussed. After, a similar discussion is presented regarding the thermal performance enhancement due to aluminium reflective foils. Finally, the measured R-values for combined aerogel thermal break strips and aluminium reflective foils, in double-pane LSF walls, are displayed and discussed.

3.1. Thermal Performance Improvement Due to Aerogel Thermal Break Strips

Table 4 displays the measured thermal resistance values of the double-pane LSF walls, as well as the thermal performance improvement due to aerogel thermal break (TB) strips, providing the values graphically displayed in Figure 6.

Table 4. Measured thermal resistance of double-pane LSF walls and thermal performance improvement due to aerogel thermal break strips.

Wall Description	Abbreviation	R [m²·K/W]	ΔR [m²·K/W]	ΔR [%]
Reference wall (without thermal break strip)	NoTB	2.456	—	—
Inner thermal break strip	TB-In	2.746	+0.290	+12%
Outer thermal break strip	TB-Out	2.749	+0.293	+12%
Inner and outer thermal breaks	TB-x2	2.928	+0.472	+19%

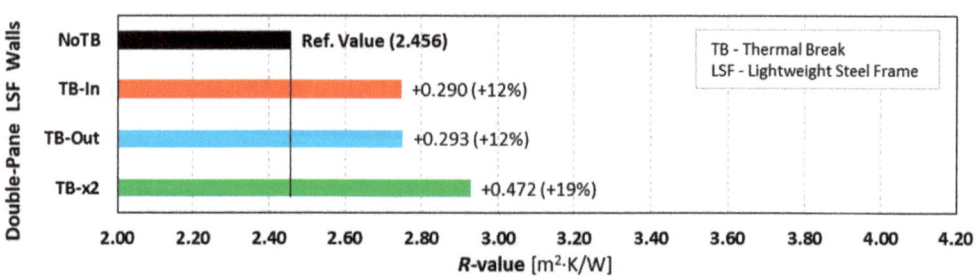

Figure 6. Measured thermal resistance improvement due to aerogel thermal break strips.

When there are no thermal break strips (reference LSF wall) the measured thermal resistance is 2.456 m²·°C/W. Adding an inner or an outer TB strip provides similar R-values (around 2.75 m²·°C/W), corresponding to a thermal resistance increase of about +0.29 m²·K/W (+12%). This R-value increase is similar to the one achieved by adding 10 mm of continuous mineral wool (MW). The higher measured R-value (2.928 m²·°C/W) is achieved when using two TB strips (inner and outer), allowing a thermal resistance improvement of +0.472 m²·K/W (+19%). Notice that the combined effect of these two TB strips (R-value increase of +19%), is smaller than the summation of two single TB strips (+24%), i.e., inner (+12%) and outer (+12%).

3.2. Thermal Performance Improvement Due to Aluminium Reflective Foils

The measured thermal resistance values and thermal performance enhancement due to aluminium reflective foils are displayed in Table 5, while Figure 7 exhibits a plot of these values.

Table 5. Measured thermal resistance of double-pane LSF walls and thermal performance improvement due to aluminium reflective foils.

Wall Description	Abbreviation	R-Value [m²·K/W]	ΔR [m²·K/W]	ΔR [%]
Reference wall (without aluminium foil)	NoAlum	2.456	—	—
Inner aluminium reflective foil	Alum-In	2.911	+0.455	+19%
Outer aluminium reflective foil	Alum-Out	2.934	+0.478	+19%
Inner and outer aluminium foils	Alum-x2	2.972	+0.516	+21%

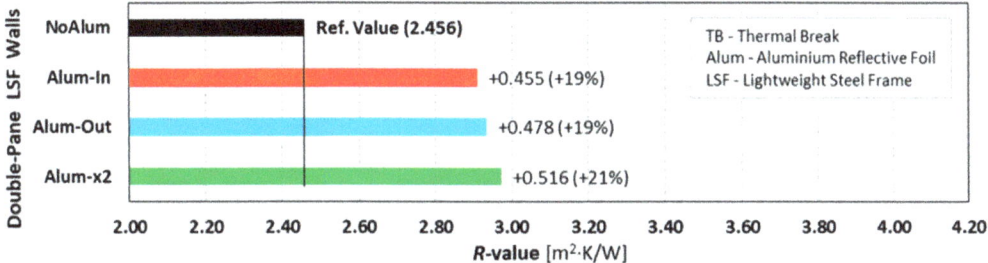

Figure 7. Measured thermal resistance improvement due to aluminium reflective foils.

Notice that the reference R-value (wall without aluminium reflective foil) remains the same, as in the previous subsection (2.456 m²·K/W). Adding an inner or an outer aluminium reflective foil provides similar R-values (around 2.92 m²·K/W), corresponding to a thermal resistance increase of about +0.46 m²·K/W (+19%). This R-value increase is similar to the one achieved by adding 16 mm of continuous mineral wool (MW). The higher measured R-value (2.972 m²·K/W) is achieved when using two reflective foils (inner and outer), allowing a thermal resistance improvement of +0.516 m²·K/W (+21%).

Comparing these values with the previous ones for the TB strips (Table 4), it can be concluded that the thermal performance improvement due to aluminium reflective (AR) foils is more effective, allowing to achieve higher thermal resistances of the double-pane LSF wall. In fact, the R-value improvement due to a single AR foil (inner or outer) is similar to the R-value improvement provided by two aerogel TB strips (+19%).

Another interesting conclusion is that the use of two reflective foils is not so effective, since the thermal performance improvement, in comparison with only one aluminium reflective foil, is very reduced (about +2%).

3.3. Thermal Performance Improvement Due to Thermal Break Strips and Reflective Foils

Table 6 exhibits the measured R-values for combined aerogel thermal break (TB) strips and aluminium reflective (AR) foils, as well as the corresponding thermal resistance increase. Figure 8 displays a graphical representation of these values, being grouped into three sets of R-values, depending on the number and location of TB strips (inner, outer, and both inner and outer). Notice that the reference R-value remains the same (2.456 m²·°C/W) for all three sets of measurements.

Table 6. Measured thermal resistance of double-pane LSF walls and thermal performance improvement due to aerogel thermal break strips and aluminium reflective foils.

Wall Description	Abbreviation	R-Value [m²·K/W]	ΔR [m²·K/W]	ΔR [%]
Reference wall (No TB Strip)	NoAlum	2.456	—	—
Inner TB Strip				
Inner aluminium reflective foil	Alum-In	3.334	+0.878	+36%
Outer aluminium reflective foil	Alum-Out	3.298	+0.842	+34%
Inner and outer aluminium foils	Alum-x2	3.476	+1.020	+42%
Outer TB Strip				
Inner aluminium reflective foil	Alum-In	3.308	+0.852	+35%
Outer aluminium reflective foil	Alum-Out	3.296	+0.840	+34%
Inner and outer aluminium foils	Alum-x2	3.458	+1.002	+41%
Inner and Outer TB Strips				
Inner aluminium reflective foil	Alum-In	3.667	+1.211	+49%
Outer aluminium reflective foil	Alum-Out	3.699	+1.243	+51%
Inner and outer aluminium foils	Alum-x2	3.808	+1.352	+55%

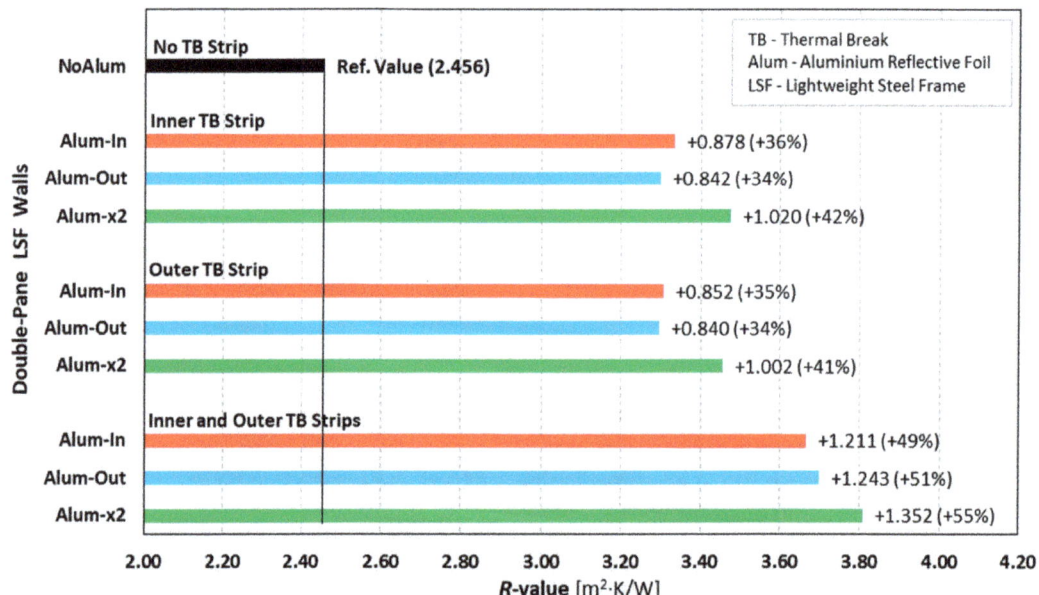

Figure 8. Measured thermal resistance improvement due to aerogel thermal break strips and aluminium reflective foils.

The combined thermal resistance improvement, due to a single TB strip (inner or outer) and a single AR foil (inner or outer), is around 0.85 m²·K/W (+35%), being this value slightly higher than the summation of individual R-values increase (+31%), obtained from Table 4 (+12%) and Table 5 (+19%).

Looking to the LSF walls with inner or outer TB strip, having two AR foils (inner and outer), the R-value increase is around +1.01 m²·K/W (+41%). Again, this thermal resistance improvement (+41%) is higher than the summation of individual R-values increase (+33%), obtained from Table 4 (+12%) and Table 5 (+21%), evidencing a bigger synergy outcome between the TB strips and the AR foils.

As expected, the higher measured thermal resistances are provided by the LSF walls with two TB strips. In this circumstance, when having an inner or an outer AR foil, the

R-value increase is about 1.23 m²·K/W (+50%). Once more, the summation of individual thermal performance improvement due to double TB strips (+19%) and due to one (inner or outer) AR foil (+19%) is smaller (only +38%) than their combined effect (about +50%).

The use of two AR foils, instead of a single one, only improved the thermal resistance by an additional 5%, i.e., +55% (+1.352 m²·°C/W) in relation to the reference LSF wall. The synergy effect remains, being even higher since this value (+55%) is considerably bigger than the summation (+40%) of individual improvement contribution from two TB strips (+19%, Table 4) and from two AR foils (+21%, Table 5). The R-value increase measured for double TB strips combined with double AR foils (about +1.35 m²·°C/W) is similar to the one achieved by adding 47 mm of continuous mineral wool (MW).

4. Conclusions

In this paper, the thermal performance improvement due to the use of aerogel thermal break (TB) strips and aluminium reflective (AR) foils in double-pane lightweight steel-framed walls were experimentally assessed. The lab measurements were performed using two mini climatic chambers (hot and cold) and the double-sided heat flow meter technic. Three sets of measurements were performed, considering the thermal performance improvement due to: (1) only aerogel TB strips; (2) only AR foils, and; (3) both TB strips and AR foils. Taking into account the three TB strips and AR foils locations, namely: (i) inner; (ii) outer; (iii) double (i.e., both inner and outer), as well as the reference LSF wall (without TB strips and AR foils), sixteen LSF walls' configurations were measured.

As expected, the key findings address the research questions of this research work, being the main conclusions summarized as follows:

- Both aerogel TB strips and AR foils allowed to improve the thermal performance of a reference double-pane LSF wall, which has a conductive thermal resistance of 2.456 m²·K/W.
- Placing the aerogel TB strip on the inner or outer steel stud flange provides similar conductive R-values, being the thermal resistance increment of about +0.29 m²·K/W (+12%).
- The use of inner or outer AR foils inside the air cavity also provides similar R-values, but the effectiveness of this improvement measure is higher, exhibiting a thermal resistance increment of about +0.47 m²·K/W (+19%).
- In fact, using an AR foil inside the air cavity of double-pane LSF walls is much more effective than using aerogel TB strips along the steel flange, since only one AR foil (inner or outer) provides similar thermal resistance increase than two aerogel TB strips, i.e., around +0.47 m²·K/W (+19%).
- However, the use of two AR foils, instead of a single one, is not effective, since the relative thermal resistance increase is only about +0.04 m²·K/W (+2%).
- The combined effect of both TB strips and AR foils allowed to achieve a maximum R-value increase of +1.35 m²·K/W (+55%).
- When combining these two thermal performance strategies, it was found a synergy effect between them, since the measured combined R-value increase, when using both TB strips and AR foils, is bigger than the summation of individual R-values increments.
- This synergy effect ranged from only +0.10 m²·K/W (+4%) for the thermal resistance increase due to a single TB strip and a single AR foil, up to +0.37 m²·K/W (+15%) when using two TB strips and two AR foils.

Notice that the higher effectiveness of the AR foils in comparison with the aerogel TB strips could be justified by the continuous air cavity (3 cm) and reflective foil, while the TB strips are restricted to the steel studs' flanges. Moreover, the steel studs in the double-pane LSF wall are separated in two different frames (inner and outer wall panes), not crossing the wall and, this way, the related steel thermal bridge effect is very reduced.

The main practical applications and implications of the research findings are related the design of double-pane LSF walls, or refurbishment, whenever their thermal performance is not satisfactory. As mentioned above, the use of AR foil is more efficient than TB strips, even when they are made with a super insulation material (aerogel). Moreover, adding this AR foil to the inner or outer side of the air cavity originates a similar thermal performance increase. Furthermore, the use of two AR foils, instead of a singleone, is not effective.

Cost assessment is not within the scope of this research, being an interesting research idea for future work. However, the acquisition cost of aerogel TB strips is much higher than the cost of an AR foil. For example, the cost of an aerogel Spacetherm® CBS (Cold Bridge Strip) is around 3.80 €/m (50 mm wide and 10 mm thick), while the cost of a Space-Reflex® AR foil (4 mm thick) is around 2.35 €/m². Assuming an LSF wall with a stud spacing equal to 400 mm and a high of 2.70 m, the unit consumption of a single TB strip is 3.30 m per wall square meter. Thus, the unit cost of this TB strip is around 12.54 €/m² for this LSF wall. Therefore, the unit cost of the AR foil is around five times lower than the aerogel TB strip. Consequently, it can be concluded that besides the higher thermal efficiency of the AR foil, this performance improvement strategy is also much cheaper.

The main limitations of this research are: only one air cavity was assessed (30 mm thick); only one TB strip material was assessed (aerogel); only one reflective foil was assessed (aluminium); the measurements were performed in a double-pane LSF wall test-sample (without any wall ties, connectors or other bridging elements inside the air cavity) under controlled lab conditions in a near steady-state regimen (not having into account any transient effects due to daily temperature variations).

Energy consumption is not within the scope of this research, being a good research suggestion for future work. Nevertheless, the achieved thermal resistance increments (maximum absolute increased value of +1.35 m²·°C/W, corresponding to +55%) will allow for sure to reduce the heat losses across the building opaque envelope during the winter season and, consequently, to reduce energy consumption for space heating. This energy consumption reduction will be larger for colder climates.

Author Contributions: Conceptualization, P.S.; methodology, P.S.; validation, P.S. and T.R.; formal analysis, P.S.; investigation, P.S. and T.R.; writing—original draft preparation P.S.; writing—review and editing, P.S. and T.R.; visualization, P.S.; supervision, P.S.; project administration, P.S.; funding acquisition, P.S. All authors have read and agreed to the published version of the manuscript.

Funding: This research was funded by FEDER funds through the Competitivity Factors Operational Programme—COMPETE and by national funds through FCT—Foundation for Science and Technology within the scope of the project POCI-01-0145-FEDER-032061.

Cofinanciado por: POCI-01-0145-FEDER-032061

Institutional Review Board Statement: Not applicable.

Informed Consent Statement: Not applicable.

Acknowledgments: The authors also want to thank the support provided by the following companies: Pertecno, Gyptec Ibéria, Volcalis, Sotinco, Kronospan, Hulkseflux, Hilti and Metabo.

Conflicts of Interest: The authors declare no conflict of interest.

References

1. European Union. Directive (EU) 2018/844 of the European Parliament and of the Council of 30 May 2018 Amending Directive 2010/31/EU on the energy performance of buildings and Directive 2012/27/EU on energy efficiency. *Off. J. Eur. Union* **2018**, *2018*, 75–91.
2. Climate Action Network (CAN) Europe. How to Roll out the Energy Transition in Buildings. *Factsheet* **2021**. Available online: https://caneurope.org/energy_transition_buildings_factsheet/ (accessed on 14 September 2021).

3. European Union. Directive (EU) 2018/2001 of the European Parliament and of the Council on the promotion of the use of energy from renewable sources. *Off. J. Eur. Union* **2018**, *328*, 82–209, European Parliament.
4. Santos, P.; Simões da Silva, L.; Ungureanu, V. *Energy Efficiency of Light-Weight Steel-Framed Buildings*, 1st ed.; European Convention for Constructional Steelwork (ECCS), Technical Committee 14—Sustainability & Eco-Efficiency of Steel Construction: Brussels, Belgium, 2012; ISBN 978-92-9147-105-8.
5. Roque, E.; Santos, P. The Effectiveness of Thermal Insulation in Lightweight Steel-Framed Walls with Respect to Its Position. *Buildings* **2017**, *7*, 13. [CrossRef]
6. Roque, E.; Santos, P.; Pereira, A.C. Thermal and sound insulation of lightweight steel-framed façade walls. *Sci. Technol. Built Environ.* **2019**, *25*, 156–176. [CrossRef]
7. Berardi, U.; Sprengard, C. An overview of and introduction to current researches on super insulating materials for high-performance buildings. *Energy Build.* **2020**, *214*, 109890. [CrossRef]
8. Lamy-Mendes, A.; Pontinha, A.D.R.; Alves, P.; Santos, P.; Durães, L. Progress in silica aerogel-containing materials for buildings' thermal insulation. *Constr. Build. Mater.* **2021**, *286*, 122815. [CrossRef]
9. Zach, J.; Peterková, J.; Dufek, Z.; Sekavčnik, T. Development of vacuum insulating panels (VIP) with non-traditional core materials. *Energy Build.* **2019**, *199*, 12–19. [CrossRef]
10. Santos, P.V.F.; Martins, C.; Simoesdasilva, L. Thermal performance of lightweight steel-framed construction systems. *Metall. Res. Technol.* **2014**, *111*, 329–338. [CrossRef]
11. Erhorn-Klutting, H.; Erhorn, H. ASIEPI P148—Impact of thermal bridges on the energy performance of buildings. In *Buildings Platform*; European Communities: Brussels, Belgium, 2009.
12. Soares, N.; Santos, P.; Gervásio, H.; Costa, J.J.; Da Silva, L.S. Energy efficiency and thermal performance of lightweight steel-framed (LSF) construction: A review. *Renew. Sustain. Energy Rev.* **2017**, *78*, 194–209. [CrossRef]
13. Martins, C.; Santos, P.; Simoesdasilva, L. Lightweight steel-framed thermal bridges mitigation strategies: A parametric study. *J. Build. Phys.* **2016**, *39*, 342–372. [CrossRef]
14. Váradi, J.; Toth, E. Thermal Improvement of Lightweight Façades containing Slotted Steel Girders. In *Twelfth International Conference on Civil, Structural and Environmental Engineering Computing*; Funchal: Madeira, Portugal, 2009; p. 107.
15. Lupan, L.M.; Manea, D.L.; Moga, L.M. Improving Thermal Performance of the Wall Panels Using Slotted Steel Stud Framing. *Procedia Technol.* **2016**, *22*, 351–357. [CrossRef]
16. Santos, P.; Lemes, G.; Mateus, D. Thermal Transmittance of Internal Partition and External Facade LSF Walls: A Parametric Study. *Energies* **2019**, *12*, 2671. [CrossRef]
17. Santos, P.; Mateus, D. Experimental assessment of thermal break strips performance in load-bearing and non-load-bearing LSF walls. *J. Build. Eng.* **2020**, *32*, 101693. [CrossRef]
18. Kempton, L.; Kokogiannakis, G.; Green, A.; Cooper, P. Evaluation of thermal bridging mitigation techniques and impact of calculation methods for lightweight steel frame external wall systems. *J. Build. Eng.* **2021**, *43*, 102893. [CrossRef]
19. Santos, P.; Gonçalves, M.; Martins, C.; Soares, N.; Costa, J.J. Thermal Transmittance of Lightweight Steel Framed Walls: Experimental Versus Numerical and Analytical Approaches. *J. Build. Eng.* **2019**, *25*, 100776. [CrossRef]
20. Kapoor, D.R.; Peterman, K.D. Quantification and prediction of the thermal performance of cold-formed steel wall assemblies. *Structures* **2021**, *30*, 305–315. [CrossRef]
21. Santos, P.; Poologanathan, K. The Importance of Stud Flanges Size and Shape on the Thermal Performance of Lightweight Steel Framed Walls. *Sustainability* **2021**, *13*, 3970. [CrossRef]
22. Jelle, B.P.; Kalnæs, S.E.; Gao, T. Low-emissivity materials for building applications: A state-of-the-art review and future research perspectives. *Energy Build.* **2015**, *96*, 329–356. [CrossRef]
23. Bruno, R.; Bevilacqua, P.; Ferraro, V.; Arcuri, N. Reflective thermal insulation in non-ventilated air-gaps: Experimental and theoretical evaluations on the global heat transfer coefficient. *Energy Build.* **2021**, *236*, 110769. [CrossRef]
24. Santos, P.; Ribeiro, T. Thermal Performance of Double-Pane Lightweight Steel Framed Walls with and without a Reflective Foil. *Buildings* **2021**, *11*, 301. [CrossRef]
25. Gyptec Ibérica. Technical Sheet: Standard Gypsum Plasterboard. 2021. Available online: https://www.gyptec.eu/documentos/Ficha_Tecnica_Gyptec_A.pdf (accessed on 19 April 2021). (In Portuguese).
26. Volcalis_MineralWool. Technical Sheet: Alpha Panel Mineral Wool. 2021. Available online: https://www.volcalis.pt/categoria_file/fichatecnica_volcalis_alphapainel-777.pdf (accessed on 19 April 2021). (In Portuguese).
27. Santos, C.; Matias, L. *ITE50—Coeficientes de Transmissão Térmica de Elementos da Envolvente dos Edifícios (in Portuguese)*; LNEC—Laboratório Nacional de Engenharia Civil: Lisboa, Portugal, 2006.
28. Rasooli, A.; Itard, L. In-situ characterization of walls' thermal resistance: An extension to the ISO 9869 standard method. *Energy Build.* **2018**, *179*, 374–383. [CrossRef]
29. ISO 9869-1. Thermal insulation—Building elements—In-situ measurement of thermal resistance and thermal transmittance. In *Part 1: Heat Flow Meter Method*; ISO—International Organization for Standardization: Geneva, Switzerland, 2014.
30. ASTM-C1155−95(Reapproved-2013). *Standard Practice for Determining Thermal Resistance of Building Envelope Components from the In-Situ Data*; ASTM—American Society for Testing and Materials: Philadelphia, PA, USA, 2013.

31. *THERM*; Software version 7.6.1; Lawrence Berkeley National Laboratory, United States Department of Energy: Berkeley, CA, USA, 2017. Available online: https://windows.lbl.gov/software/therm (accessed on 14 February 2019).
32. EN ISO 6946. *Building Components and Building Elements—Thermal Resistance and Thermal Transmittance—Calculation Methods*; CEN—European Committee for Standardization: Brussels, Belgium, 2017.
33. Santos, P.; Martins, C.; Simoesdasilva, L.; Bragança, L. Thermal performance of lightweight steel framed wall: The importance of flanking thermal losses. *J. Build. Phys.* **2013**, *38*, 81–98. [CrossRef]

Article

A Replicable Methodology to Evaluate Passive Façade Performance with SMA during the Architectural Design Process: A Case Study Application

Kristian Fabbri * and Jacopo Gaspari

Department of Architecture, University of Bologna, 40136 Bologna, Italy; jacopo.gaspari@unibo.it
* Correspondence: info@kristianfabbri.com or info@kristanfabbri.com

Abstract: Huge efforts have been made in recent decades to improve energy saving in the building sector, particularly focused on the role of façades. Among the explored viable solutions, climate-adaptive building shells [CABS] consider promising solutions to control solar radiation, both in terms of illuminance and heating levels, but are still piloting these solutions due to their complex designs and necessary costs. The present study aims to provide a speedy but reliable methodology to evaluate the potential impacts of adopting active/passive CABS systems during the preliminary design stage. The proposed methodology allows the evaluation and comparison, when multiple options are considered, of the effects of each solution in terms of the energy needs, thermal comfort and lighting, while reducing the required effort and time for an extensive analysis of the overall building level. This is based on the use of a "virtual test room" where different conditions and configurations can be explored. A case study in the city of Bologna is included for demonstration purposes. The achieved results support the decisions made regarding energy behavior (over/under heating), indoor comfort, lighting and energy at an early design stage.

Keywords: climate-adaptive building shells; sustainable design; energy efficiency; shape-memory alloy; climate change

Citation: Fabbri, K.; Gaspari, J. A Replicable Methodology to Evaluate Passive Façade Performance with SMA during the Architectural Design Process: A Case Study Application. *Energies* **2021**, *14*, 6231. https://doi.org/10.3390/en14196231

Academic Editor: Paulo Santos

Received: 14 September 2021
Accepted: 27 September 2021
Published: 30 September 2021

Publisher's Note: MDPI stays neutral with regard to jurisdictional claims in published maps and institutional affiliations.

Copyright: © 2021 by the authors. Licensee MDPI, Basel, Switzerland. This article is an open access article distributed under the terms and conditions of the Creative Commons Attribution (CC BY) license (https://creativecommons.org/licenses/by/4.0/).

1. Introduction and Background

The ever-increasing evidence regarding the potential impacts of climate change on the building sector is strictly related to not only to the recurrence of extreme events, such as heatwaves, flooding, wildfires, but also to the structural alterations of local climate conditions, which affect the population in everyday life and remarkably influence the expected response of buildings and cities [1–3].

The 2018 IPCC report "Global warming of 1.5 °C" warned about the potential consequences of temperature increase, considering several interrelated and variable parameters on key sectors [4–6], especially those linked to United Nations Sustainable Development Goals [UN SDGs]. The data observations over the last decades showed a relevant increase in the average temperature in many cities in different regions and a related growth of energy demand for air conditioning (AC) can be easily detected both in terms of intensity and diffusion across the EU and other countries [7,8].

Additional studies have been provided by the scientific community to explore the sector-based analyses of energy demand and supply options to achieve a large reduction in GHG emissions: Creutzig [9], Jacobson [10], and Clack [11] analyze a 100% renewable energy transition by 2050. However, most of the investigated measures differ for several reasons, such as political and socio-economic implications, and they are often randomly adopted within the same country without having a clear perspective of the extent to which the conditions will evolve in the future. The broad scientific literature documents the achievements gained in meeting the highest expected standards in terms of energy savings; however, it also reveals the minor attention given to the interrelated phenomena and citizens' wellbeing; see Ferrao and Fernandex [10] and European Environment Agency [12]. Meanwhile, other studies specifically focused on the mitigation strategies to cope with more specific issues, such Urban Heat Island [UHI] [13–15].

Nonetheless, despite the huge efforts made in recent decades to develop effective measures to support the transition towards a low-carbon society, as described in European Commission Environment Directorate-General [15], in 2015, the level of greenhouse gases [GHG] in most of the projections will show a dramatic rise in the coming years [16].

Therefore, the progressive understanding and awareness raised about climate change and sustainable development are leading to calls for groundbreaking and innovative design approaches in architecture [17–20] to concretely contribute to facing this huge challenge.

A wide range of architectural solutions developed in recent years mainly focus on decreasing the demand for heating [21], and adopting more efficient insulation material and systems to comply with the Energy Performance of Buildings Directive [22], the EPBD Recast III [23] and the Directive 2012/27/EU [24]. However, the achievement of a building sector which is more resilient to future climate conditions [25,26] also requires the appropriate decisions about orientation, as well as effective mitigation solutions, exploiting both natural ventilation and shielding options against overheating.

Among the possible answers to these issues, the use of the so-called adaptive solutions [27] (able to change their configuration according to the variation of temperature or solar radiation) [28] can be considered very promising solutions to reduce the thermal load on the building envelope.

The scientific literature groups these solutions into the family of climate-adaptive building shells (CABS) which can be classified according to different criteria, from the climate response to the activation agent, and from their geometrical configuration to the adopted actuator. Their common element is that the system is essentially designed to dynamically vary the façade configuration as a response to environmental changes according to the detection of specific parameter variations [29].

1.1. CABS in the Scientific Literature

Designed to primarily meet a sustainable design in terms of provided performances, the very first generation of CABS was mainly based on electro-mechanical actuators driven by a centralized CPU, which processes the outdoor condition variations [30–33].

This enables a real-time optimized response to external variations and does not require any manual action to be performed by the end users, although this limits the end user's own preferences.

The development of the so-called smart materials are defined as: "A material which has built-in or intrinsic sensor(s), actuator(s) and control mechanism(s) whereby it is capable of sensing a stimulus, responding to it in a predetermined manner and extent, in a short appropriate time and reverting to its original state as soon as the stimulus is removed" [34]. The smart materials enlarged the opportunities of kinetic characterization [35].

Much more relevant are the deriving solutions, which are generally highly responsive and can be considered energetically autonomous, concretely contributing to the development of innovative low-energy design concepts with very limited mechanical devices. The more frequently used deriving solutions are: magnetorestrictive materials, piezo electric materials, shape-memory polymers, phase-change materials, photochromic glasses, and shape-memory alloys. These materials can be used as actuators to passively drive the shielding system due to their own capacity to autonomously and reversibly adapt to climatic variations without computerized remote control. This typology remarkably reduces several maintenance and economic problems related to electro-mechanical solutions.

Passive systems typically react to environmental parameters such as temperature, humidity, and light, etc., providing an instant response to their variations. Some studies [17] consider the use of smart materials as no longer separable from adaptive building envelope design, supported by advances in technologies, production methods and simulation capacities.

When the adaptive capacity is an intrinsic feature of the building envelope sub-system and does not require an external decision-making component, it is usually classified as an intrinsic control system, self-adjusting according to temperature, humidity, light, and stimuli, etc. [27,36], providing a classification via the related energy (radiant, thermal, potential, and kinetic). Depending on the properties of materials and on their responsiveness to the environmental stimuli, technological solutions can be classified with several nuances: unpowered kinetic [36], energy-free adaptability [37], zero-energy adaptability [28], and auto-reactive [36] or intrinsically controlled adaptability [29].

For the scope of this study, the more inclusive definition of the self-sufficient kinetic building envelope [38], referring to both the capacity to autonomously control the dynamism and the independence from supplied energy sources, is adopted.

The passive actuators, more frequently adopted in the current market, are those based on bi-metallic components, a change of state wax and shape-memory springs. This study investigates the use of shape-memory springs, defined with the acronym SMA: "shape-memory alloy", with a relation to two intrinsic advantages of these materials: the pseudo elasticity and the shape-memory effect. This type of spring is composed of two elements, nickel and titanium, commonly called nitinol, and responds to temperature variations through an elongation of 10%. The shape-memory effect (EMS) allows SMAs a dual process that combines a stored physical shape with a transition from a malleable to a rigid state.

At room temperature, SMAs, in the martensitic phase, are malleable and can be bent into various shapes; when heated at a certain temperature they reach the austenitic phase, becoming rigid and remembering their stored shape [39].

1.2. CABS Based on a Shape-Memory Alloy

Shape-memory alloys (SMAs) have a dual solid phase actuated by a specific environmental trigger. In the first phase the material is inert and easily deformed. In the second phase it recovers the initial molecular structure and the related shape that can be set "at will" according to a predefined layout. This is achieved through a "training" process of the phase transition that can also influence some mechanical characteristics of the material such as stiffness.

SMAs ensure the advantage of an on-off function, which behaves like a switch that allows a specific configuration to be obtained. However, the main problem often lies in creating the counteraction bias to bring the actuator back to its original shape [40,41].

The main limitation of such intrinsically controlled systems is that they cannot be directly controlled by the occupants, influencing the satisfaction level. A dedicated study [42] pointed out that this is not connected with thermal comfort (which is generally achieved) but mostly with visual requirements, as the optimization of the shading element positioning often reduces external views from the interior. The other relevant limitation deals with the system's responsiveness, which is "programmed" during the fabrication process which cannot be changed during the time. This capacity is tuned during the design process, making the system dependent on the precise environmental parameters fluctuation.

The actuator can be localized, meaning that the kinetic component has an actuator triggering a pre-defined movement, or can be distributed, meaning that the system has several actuators. A combination of several actuators, intended as the mean to move the system components, can of course increase the kinematic complexity.

2. Goals

The scope of the study is to develop and test a methodology to evaluate the effectiveness of adaptive façade solutions during the preliminary design stage, with relation to indoor microclimate conditions: temperature, thermal comfort, and daylight. The proposed methodology is then applied to a shielding solution based on SMA technology. In the following section the methodology, calculations and output are described, followed by a case study application. The same process can be replicated with other adaptive options, which compare the different performance levels and the possible effects on the façade.

3. Methodology

This study provides a rapid and comparative methodology to assess the impacts of shielding systems in terms of energy and lighting levels. The related workflow includes three different typologies of input data regarding: (a) *climate data*: to properly consider the context in which the system is expected to operate; (b) *building configuration*: to adequately consider the influence of geometry and architectural choices and possibly define a representative standard room to be adopted as test room; and (c) *building shell design*: to set the geometrical and technical features of the system. The use of a standard virtual test room allows for the investigation of several variations, limiting the complexity of interaction for replicating the process and extending the results to a full-scale building when one (or more) specific solution(s) is selected. Figure 1 provides a graphical representation of the workflow.

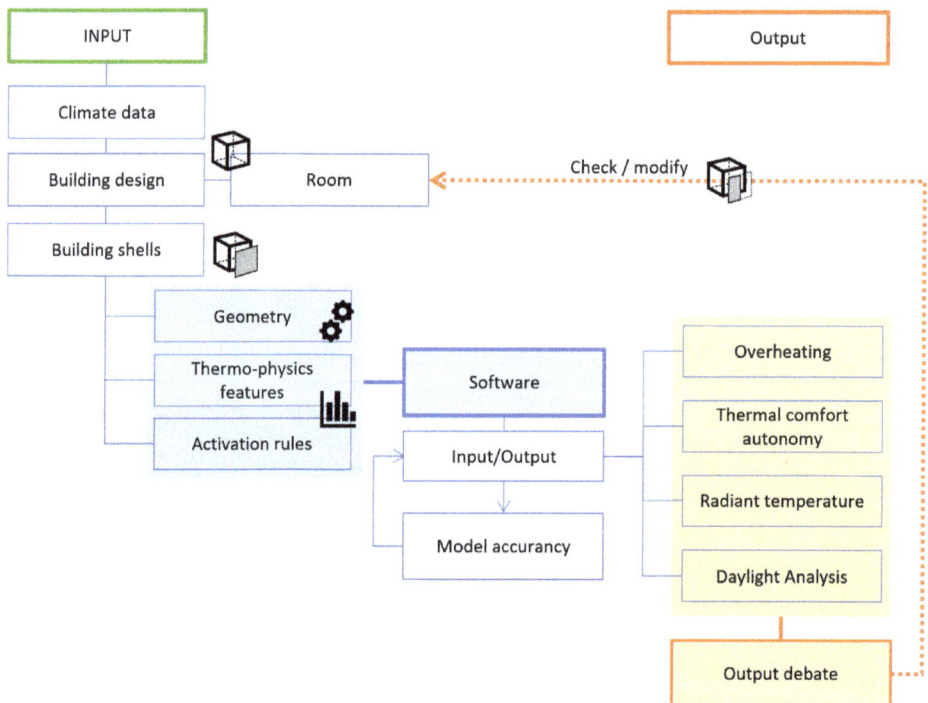

Figure 1. Methodology workflow.

With reference to the geometrical characteristics of the system, the following configuration options must be considered: totally shading the solution, partially shading the solution, simply filtering the solution. The rotation angle and the incident solar radiation must also be carefully taken into account. The standard virtual test room is modeled to be adiabatic, with no energy exchange with surrounding spaces except for the façade, n, for which the shielding system is installed. Internal walls have a reflection coefficient equal to 0.90 corresponding to a white plaster and a very similar floor finishing. The standard virtual test room is designed like a physical test laboratory.

The definition of the thermophysical properties of the shielding system is very dependent on its design and this is strictly connected with the architectural concept which underlines the design, as well as with the choice of the actuator, which can be based on an electromechanical device or on SMA component, as in the specific case study. Once these elements are preliminary decided, the related data are input for the model in Grasshopper, the software adopted for simulation purposes. As is the standard with most simulation processes, a check phase is required to assess the input/output data accuracy and reliability before the results can be successfully obtained.

The other relevant part of the process deals with the activation rules that influence the dynamism of the shielding system (if any) and may depend on the daytime, the lighting level, the indoor/outdoor temperature, or other factors. In the case of SMA, which is adopted for the demo case study, the activation criterion is the outdoor temperature, which determines the rotation angle based on the incident solar radiation. Depending on the rotation angle, the percentage of obstruction of the incoming solar and luminous radiation is defined for the shielding element and the indoor microclimate (temperature, illuminance, and radiation, etc.), for the standard virtual test room varies accordingly.

The final stage of the process deals with the simulation output which is intended to support a more detailed and in-depth design of the shielding system and actuator, once the preliminary choices are addressed. The simulation output regards:

(A) Overheating/underheating hours: the simulation allows the evaluation of the response with respect to comfort expressed as Predicted Mean Vote (PMV);
(B) Thermal Comfort Autonomy: measures the percentage of hours of the year that fall within the comfort parameters, without the aid of heating and cooling systems;
(C) Radiant temperature: measures the radiant temperature in °C at a specific time of the year, through points of a grid built inside the 3D room under study. The radiant temperature influences local thermal comfort and discomfort conditions;
(D) Parametric Daylight Analysis: it provides the illuminance level (lux) to evaluate the correct natural lighting of the interiors.

4. Case Study

In this section the proposed methodology is concretely applied to a case study for testing purposes on a building at the design stage, located in the city of Bologna (Figure 2). The building is under development and a specific study to achieve the highest efficiency and comfort level is required to support the design process during the different stages.

Figure 2. Italy and Bologna city.

Bologna is a city located in the mid-northern part of Italy, (lat. 44.31 N, long. 11.17 E), in the middle of the Po valley. Following the Koppen classification [43,44], the weather is a Humid Subtropical Climate(Cfa) and it is characterized by a range of latitudes from 20° to 35°. It is mildly cold in winter and hot and sunny in summer, with weak wind and focused rain occurring particularly during the autumn and spring seasons.

4.1. Building Design and Test Room

The case study is a free-standing residential building with a south-oriented main façade (Figure 3) where the living area of all the flat typologies is located. The overall shape is a quite compact and has a regular volume, with east-oriented bow windows on the north side and wide loggias along the entire length of the south side. The 1.8-m depth of the loggia partially shades the largely glazed façade during the summer period to avoid overheating, while ensuring a suitable solar gain during winter. However, without any shielding element, loggias cannot be used during the summer due to the exposure and high temperature of the location. Additionally, the south orientation may cause unsuitable glaring effects regardless of the loggia proportion. Therefore, during the preliminary design stage, the opportunity to add a shielding system was discussed and then which of the available options would best fit the design brief requirements was explored. During the preliminary discussion, some concern about the possible reduction in natural lighting and the possible user difficulties with managing the actuation mechanism emerged. This led to a chance to concretely apply the proposed methodology to select the most appropriate solution and optimize its configuration. After a preliminary evaluation, a self-sufficient solution was agreed upon (to avoid any energy dependency and to make the shielding system operating) and an SMA-based actuator was chosen, as described in the following paragraphs, to meet both the architectural design ambition and the expected indoor comfort levels.

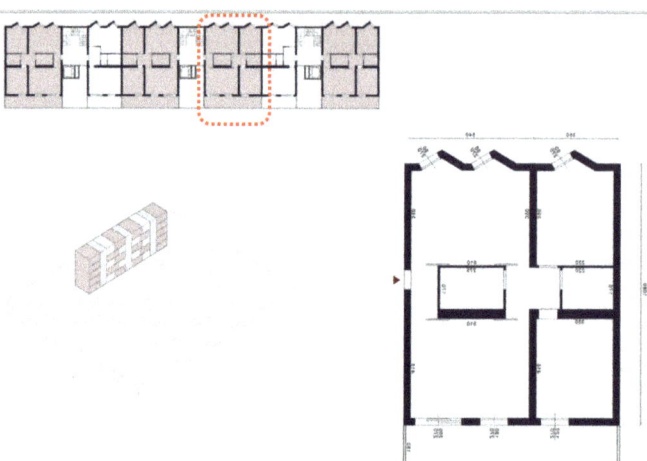

Figure 3. Building design: typical floor, axonometry of the building block and the typical apartment layout.

The standard virtual test room corresponded to the living room of a typical flat and was 5.10 m wide and 4.10 m long with an internal net height of 2.70 m. The loggia had a depth of 1.8 m (Figure 4).

4.2. Building Shell

The investigated adaptive shell solution is a frontal shielding screen of metal mesh aimed at filtering the incident solar radiation. Figure 5 shows the different shielding system configurations according to the different rotation angles of its vertical axis and the related effects on light inflow. The choice of using a metal mesh is mainly due to the façade orientation and the need to filter solar radiation, letting only part of it enter into the loggia. At the same time this material ensures an adequate resistance to the rotation impressed by the system's dynamism and offers less resistance to wind pressure on the façade surface (reducing the structural sections); despite this, the latter factor is not at the core of the present study. The use of rotating screens is to meet architectural preferences and does not influence the methodology testing process, which can be replicated with any other technical option. Specifically, in this case study, the torsion allows the element to shift from a flat configuration, parallel to the façade, to one in which its upper part is rotated 180°. This ensures a free visual perspective from the indoor spaces and the loggia. Furthermore, compared to other systems based on a rigid movement, the torsion of a light element requires less energy for system implementation and perfectly fits into passive actuation systems, which allow a change of configuration without the contribution of energy from the grid.

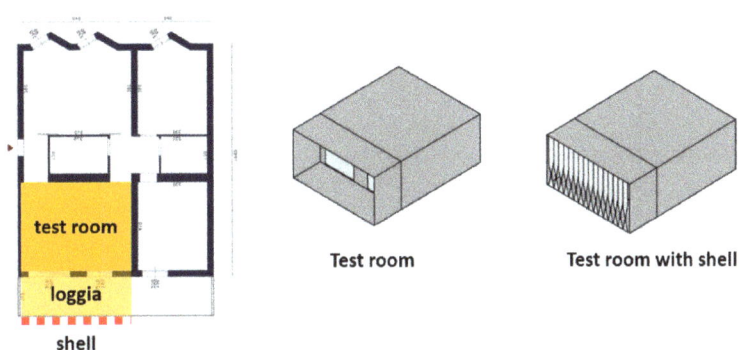

Figure 4. Typical flat layout with the test room evidenced.

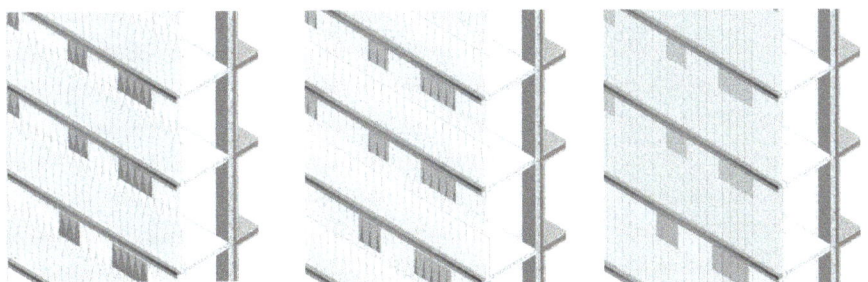

Figure 5. Building shell geometry and configuration of the different filtering conditions in relation to the angle of rotation.

Therefore, the lower weight, less material, and lower energy required to activate the torsion are the key design elements which emerged during the decision-making process and led to adopting a "passive" activation system.

4.3. Activation Rules

The chosen activation system is an actuator operating by means of two springs, one of which is a shape-memory alloy (SMA) positioned inside a "thermal box" (Milan Polytechnic patent) [45,46]. The upper end of the system is positioned on the actuator, while the lower part is fixed and keeps the shielding module under tension. When the shape-memory spring is elongated due to heating (at an air temperature of 25 °C and thermal chamber temperature of 50 °C), it is in its original form and causes the actuation system to move upwards. This upward movement is transformed into rotation by pins, which move a cylinder with a spiral hole to which the single façade module is connected. In this state, the shielding system assumes the maximum shielding configuration and covers 100% of the surface. When the temperatures decrease, even within the same day, the spring (in the martensitic phase) becomes malleable and is compressed downwards. The lower spring increases its elongation at the same time, causing a downward movement of the system, which means that the actuator triggers a rotation of the cylinder to which the shielding module is connected. This, albeit gradually, reaches a rotation of 180°. The rotation of the actuation system corresponds to a torsion of one end (in our case the upper end) of each shielding module.

The system is activated in relation to the external temperature and the solar radiation that comes into contact with the thermal box. The SMA spring stretches with a radiation of 300 W/m^2, which corresponds to a temperature of about 50 °C, and reaches inside the metal box with a dark glass in which the spring is placed. The box allows the spring to reach higher temperatures than those corresponding to current outdoor conditions.

With the solar radiation on a vertical plane higher than 300 W/m^2 (and beyond) the spring stiffens and overcomes the tension of the normal spring, recovering the original shape. The system transforms the compression generated by the spring into a rotational movement by means of a notched rod which transmits the axial force to a flange making it rotate. The ropes are attached to this latter element which rotates when the flange assumes a cylindrical configuration (maximum shielding in conditions of high incident radiation) or an hourglass configuration (opening the maximum possible area below the activation threshold). The system and possible shielding configurations are shown in Figure 6.

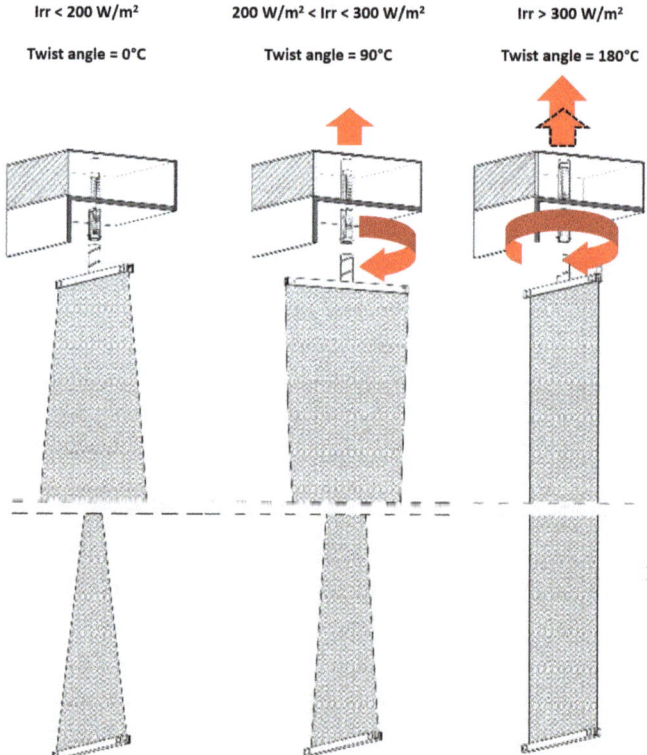

Figure 6. Configuration of the activation system (SMA) as a function of the solar radiation incident on the vertical plane (Irr, measured in W/m^2).

5. Outcomes

The following paragraphs summarize and comment on the outcomes of the application of the proposed methodology, as obtained by Grasshoper software according to the output listed (A-B-C-D) in Section 3.

5.1. Overheating

The simulation output shows the number of hours, expressed as a percentage, during which the indoor temperature is higher or lower than the standard value, conventionally fixed to 20 °C and the PMV comfort condition is equal to zero; therefore, representing a neutral comfort.

The results (Figure 7) show that the percentage of hours of overheating, in red, are about 30% (evidencing that the distribution is very close to the windows) with a significant positive reduction compared to the condition without shielding. For underheating, in blue, the lowest value is around 5%. In other words, in a building with an adaptive façade, the need for cooling energy decreases at approximately 30% less than the value during the summer and the heating energy requires a minor reduction of 5% during the winter season.

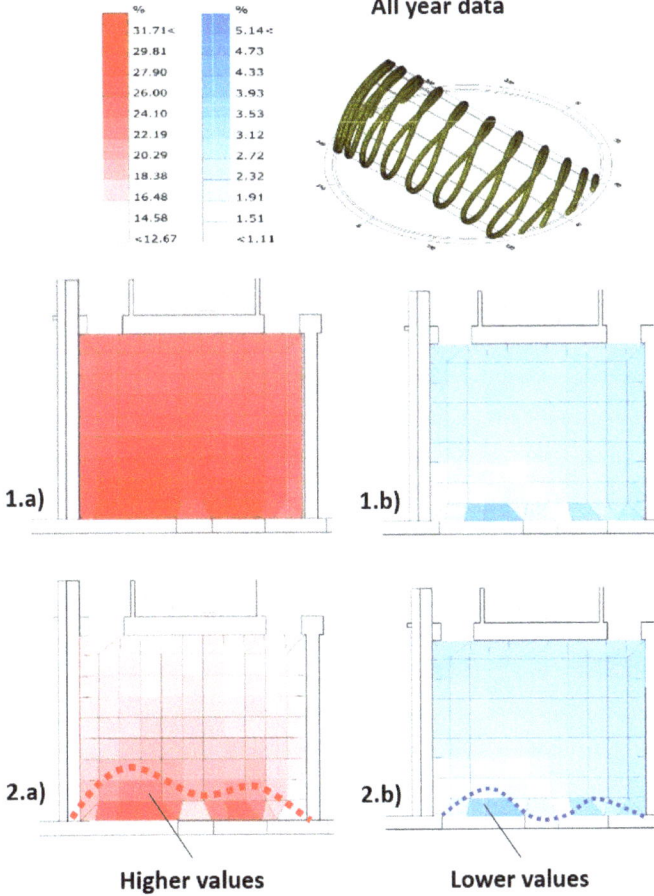

Figure 7. Overheating/underheating hours. Overheating (**1a**) summer season without shielding, (**1b**) summer season with adaptive shielding. Underheating (**2a**) winter season without shielding, (**2b**) winter season with shielding.

In terms of the activation of the shielding during the year, this controls and ensures solar gains during the winter regime (blue) and attenuates the solar radiation inflow during the summer regime (red).

5.2. Thermal Comfort Autonomy

The simulation output dealing with Thermal Comfort Autonomy (TCA) (Figure 8) allows an evaluation of how many hours, as a percentage, the environment is able to ensure the conditions of thermal comfort without activating any heating/cooling systems. Figure 8 shows that this percentage is around 25%, meaning that for approximately 25% of the hours during a year it is necessary to activate the heating or cooling system to keep the building air conditioned.

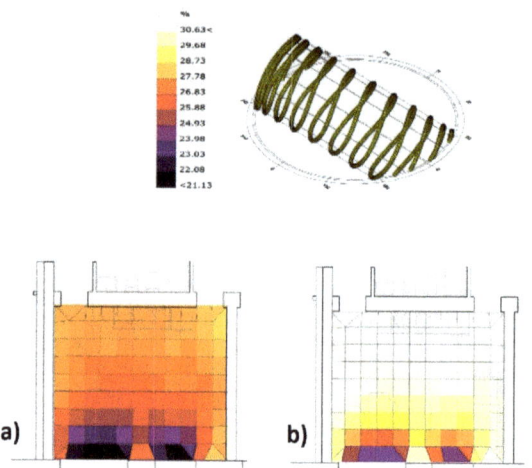

Figure 8. Thermal Comfort Autonomy, (**a**) without shielding, (**b**) with shielding.

5.3. Radiant Temperature

The output represents the distribution of the average radiant temperature value, expressed in degrees centigrade, within the room. Figure 9 reports the results for a representative day, in this specific case 5 August at 3 pm, which was selected as it was the peak day during the heat waves of summer 2018, and historically one of the hottest days of summer in Bologna.

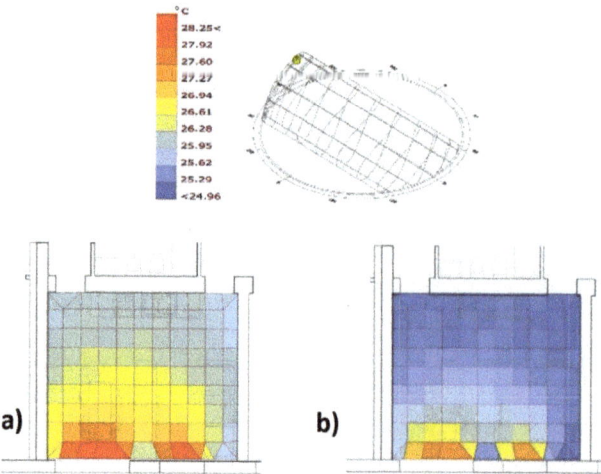

Figure 9. Radiant temperature: (**a**) without shielding, (**b**) with shielding.

The role of the façade in reducing the radiant temperature is evident, with a more marked approximate decrease of 3–4 °C near the window (Figure 9).

5.4. Daylight Analysis

The output displays the distribution of illuminance, measured in lux, within the building environment on 21 December and 21 July at 2 pm analyzing the configurations with and without shielding.

The results show that at 2 pm, both during the winter and summer solstices, the unshielded room shows an excessive level of lux, almost exceeding 1000 lux on the entire surface. By applying the shielding, this exceeding level of illumination is dimmed and generally remains at a satisfactory level. In summer conditions, the external cantilevered balcony largely intercepts the solar radiation that reaches the room's windows. However, the application of the shielding does not negatively affect the general conditions, having a value higher than 300 lux across almost the entire surface (Figure 10).

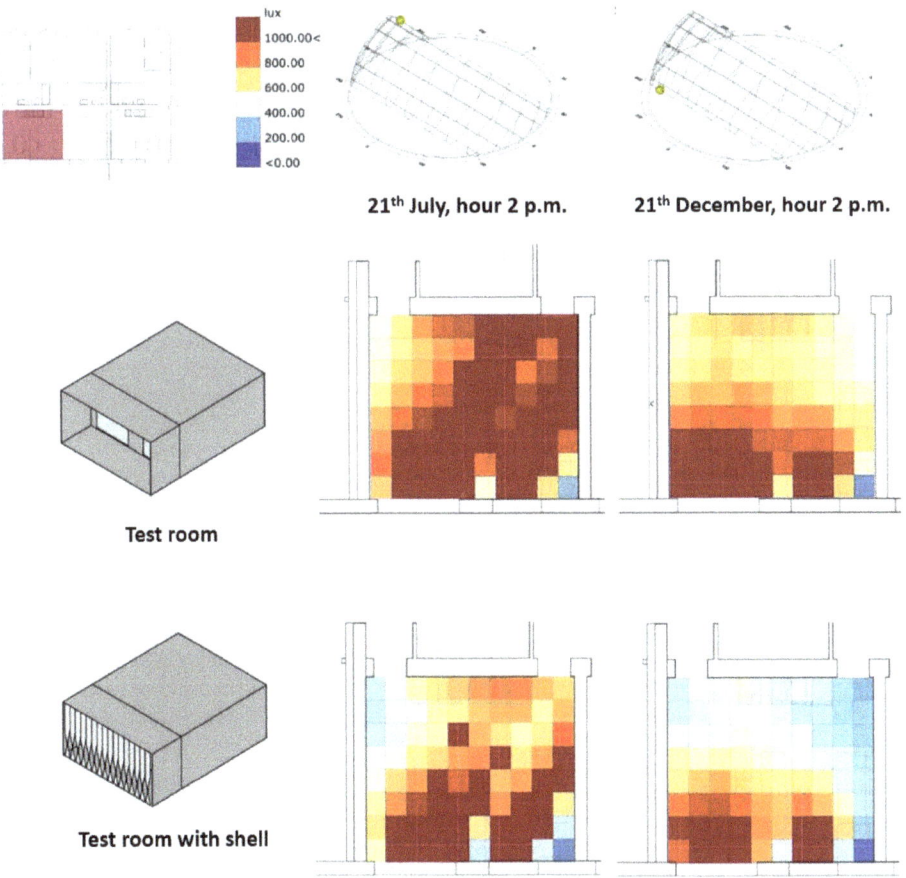

Figure 10. Daylighting Analysis. Above: without shields, below: with shields.

6. Discussion

The diffusion of adaptive façades in contemporary architecture is currently quite limited due to different factors. On the one hand, operating costs and maintenance costs are often generally considered unsustainable in the mid- to long-term; however, these factors can be greatly reduced by adopting self-sufficient systems such as SMA and avoiding the use of electro-mechanical systems. On the other hand, a common level of uncertainty is perceived around their real effectiveness, particularly during the early design stages.

Adaptive solutions are often driven by architectural and aesthetic choices, aiming to make the sustainable attributes of the building visible, rather than by their real performances and potential impacts on users' wellbeing. The very limited availability of methods or tools, in common design practice, to observe and evaluate the potential effects of adopting an adaptive solution, led to the exploration of a methodology, based on relatively simple steps, to support the fast preliminary evaluation of the adaptive façade performance. This led to the development of a feasible and reliable workflow whose results allowed for the assessment of the achievable performance level that, in the case of an SMA solution, based on a passive system, was adopted. However, this is not to be considered a foregone conclusion. The passive actuator significantly reduced the maintenance cost while optimizing the use of services and comfort levels with remarkable energy savings (which were not the specific focus of this article), but sometimes did not meet the users' preferences whose customization level was limited (override is often possible, but negatively affect the overall performance). Thus, having a clear projection of the potential effects, benefits and limitations during early design stages could be very helpful in addressing the architectural choices using an approach that best fits the investor and target users' expectations.

The advantages of the proposed method deal with a significant reduction in the time required to evaluate the potential effects and benefits of adopting the adaptive façade solution to the single investigated unit, according to a multidimensional approach regarding energy, comfort, and daylight. The conducted study demonstrated that the proposed methodology and the related outcomes could be successfully replicated in further research and design activities. Limiting the analysis to the single investigated unit could be considered a weakness of the study, but this approach was based on the same concept of a laboratory test: once the sample unit and the specific performance to be investigated were selected, the process allowed the selection of the most promising options. Yet, it was still necessary to deepen their behavior at the building scale in a further stage. This meant improvements and additional details could be explored without significantly changing the selected technologies.

According to the specific concept, CABS required a simulation phase, which was performed during the construction design phase and the proposed method avoided a time- and resource-consuming process which applied to multiple options. It was discovered only at the end of the flow that many of these options did not fit the scope properly.

The proposed methodology refers to the very preliminary design stage when there is a choice to apply a CABS, as well as a typology, and is still under discussion. It can be observed that the proposed methodology does not focus on the potential influence of the surrounding buildings, which is discussed in a second stage when the choice concerning the appropriateness of the solution is made. Furthermore, CABS are usually adopted to protect the façade in case of direct insulation, considering that the narrow construction of the urban fabric is not able to provide any kind of direct or indirect shading.

7. Conclusions

The proposed methodology can be easily applied to any kind of adaptive façade and the required resources and time can be considered as compatible with design activities and comparable to other consultancy typologies. This study required some additional time and resources due to the trial and error phase, but, once the workflow was fixed, the process could be replicated in a relatively smooth way. The adopted perspective was to reduce the effort required, focusing on a significant (recurrent) portion of the building to investigate the potential effect and scaling up of the outcomes, in order to obtain a reasonable projection of the effects on the overall building performance during the preliminary design, saving time and money. Once one or more options were explored and the suitable option was identified, the solution could be detailed and modelled to the entire building for construction purposes.

The study contributes to the building energy performance and building comfort with reference to adaptive solutions, which can provide a valuable response to the effects of climate change if applied to the retrofitting actions on existing buildings. These can also be included in policy recommendations for renovation actions, as well as new building initiatives. The current literature review, as well as the authors' previous studies in this filed [19], demonstrate that research and prototyping are particularly active in this sector. However, the concrete application of the tested pilots still remains episodic in the current conventional market due to two main reasons. On the one hand the technological solutions are often quite sophisticated and sometimes not cheap (this was the case during the 1960s with the first curtain wall applications). On the other hand, the effort, time and the costs of simulations during the early design stage still affects the decision-making process.

The adoption of rapid methods for achieving a comparative multidimensional approach to selecting the most appropriate options, can significantly reduce the required time and, consequently, can facilitate the diffusion of adaptive solutions with a potentially progressive cost reduction. Since this happened in the past with ventilated façades, it may also happen with adaptive façades considering that the frontiers of architectural design are very close to creating a building capable of sensing the surrounding environmental conditions and adapting to them, optimizing users' comfort and livability.

Author Contributions: Conceptualization, K.F. and J.G.; methodology, K.F. and J.G.; software, K.F.; validation, K.F. and J.G.; formal analysis, K.F.; investigation, K.F. and J.G.; resources, K.F. and J.G.; data curation, K.F. and J.G.; writing—original draft preparation, K.F. and J.G.; writing—review and editing, K.F. and J.G.; visualization, K.F.; supervision, K.F. and J.G.; project administration, K.F. and J.G.; funding acquisition, K.F. and J.G. All authors have read and agreed to the published version of the manuscript.

Funding: This research received no external funding.

Institutional Review Board Statement: Not applicable.

Informed Consent Statement: Not applicable.

Data Availability Statement: All datasets used and analyzed in this study are available from the corresponding author on reasonable request.

Acknowledgments: The authors are grateful to Xhorxhina Metohu and Francesco Polidori for data elaboration and for their contributions to simulations during the project development.

Conflicts of Interest: The authors declare no conflict of interest.

References

1. Ciscar, J.C.; Dowling, P. Integrated assessment of climate impacts and adaptation in the energy sector. *Energy Econ.* **2014**, *46*, 531–538. [CrossRef]
2. Tol, R.S.J. Population and trends in the global mean temperature. *Atmósfera* **2017**, *30*, 121–135. [CrossRef]
3. Schaeffer, R.; Salem, A.; Frossard, A.; De Lucena, P.; Soares, B.; Cesar, M.; Pinheiro, L.; Nogueira, P.; Pereira, F.; Troccoli, A.; et al. Energy sector vulnerability to climate change: A review. *Energy* **2012**, *38*, 1–12. [CrossRef]
4. Yamineva, Y. Lessons from the Intergovernmental Panel on Climate Change on inclusiveness across geographies and stakeholders. *Environ. Sci. Policy* **2017**, *77*, 244–251. [CrossRef]
5. Rogelj, J.; Shindell, D.; Jiang, K.; Fifita, S.; Forster, P.; Ginzburg, V.; Handa, C.; Kheshgi, H.; Kobayashi, S.; Kriegler, E.; et al. Mitigation Pathways Compatible with 1.5 °C in the Context of Sustainable Development. In Global Warming of 1.5 °C. An IPCC Special Report on the impacts of global warming of 1.5 °C above pre-industrial levels and related global greenhouse gas emission pathw. In *IPCC Special Report Global Warming of 1.5 °C*; IPCC: Geneva, Switzerland, 2018; 82p.
6. IPCC. Summary for Policymakers. In *Global Warming of 1.5 °C. An IPCC Special Report on the Impacts of Global Warming*; Intergovernmental Panel on Climate Change (IPCC): Geneva, Switzerland, 2018; ISBN 9789291691517.
7. Krinner, G.; Germany, F.; Shongwe, M.; Africa, S.; France, S.B.; Uk, B.B.B.B.; Germany, V.B.; Uk, O.B.; France, C.B.; Uk, R.C.; et al. Long-term climate change: Projections, commitments and irreversibilities. In *Climate Change 2013—The Physical Science Basis: Contribution of Working Group I to the Fifth Assessment Report of the Intergovernmental Panel on Climate Change*; Cambridge University Press: Cambridge, UK, 2013; Volume 9781107057, pp. 1029–1136. [CrossRef]

8. Arneth, A.; Denton, F.; Agus, F.; Elbehri, A.; Erb, K.; Elasha, B.O.; Rahimi, M.; Rounsevell, M.; Spence, A.; Valentini, R. Chapter 1: Framing and context. In *Climate Change and Land: An IPCC Special Report on Climate Change, Desertification, Land Degradation, Sustainable Land Management, Food Security, and Greenhouse Gas Fluxes in Terrestrial Ecosystems*; Shukla, P.R., Skea, J., Buendia, E.C., Masson-Delmot, V., Eds.; Intergovernmental Panel on Climate Change (IPCC): Geneva, Switzerland, 2019; pp. 77–129.
9. Creutzig, F.; Agoston, P.; Goldschmidt, J.C.; Luderer, G.; Nemet, G.; Pietzcker, R.C. The underestimated potential of solar energy to mitigate climate change. *Nat. Energy* **2017**, *2*, 2021. [CrossRef]
10. Jacobson, M.Z.; Delucchi, M.A.; Bauer, Z.A.F.; Goodman, S.C.; Chapman, W.E.; Cameron, M.A.; Bozonnat, C.; Chobadi, L.; Clonts, H.A.; Enevoldsen, P.; et al. 100% Clean and Renewable Wind, Water, and Sunlight All-Sector Energy Roadmaps for 139 Countries of the World. *Joule* **2017**, *1*, 108–121. [CrossRef]
11. Clack, C.T.M.; Qvist, S.A.; Apt, J.; Bazilian, M.; Brandt, A.R.; Caldeira, K.; Davis, S.J.; Diakov, V.; Handschy, M.A.; Hines, P.D.H.; et al. Evaluation of a proposal for reliable low-cost grid power with 100% wind, water, and solar. *Proc. Natl. Acad. Sci. USA* **2017**, *114*, 6722–6727. [CrossRef] [PubMed]
12. European Environment Agency EEA. *Agency European Environment Urban Adaptation to Climate Change in Europe*; European Environment Agency EEA: Copenhagen, Denmark, 2012.
13. Fabbri, K.; Costanzo, V. Drone-assisted infrared thermography for calibration of outdoor microclimate simulation models. *Sustain. Cities Soc.* **2020**, *52*, 101855. [CrossRef]
14. Gaspari, J.; Fabbri, K. A Study on the Use of Outdoor Microclimate Map to Address Design Solutions for Urban Regeneration. *Energy Procedia* **2017**, *111*, 500–509. [CrossRef]
15. Santucci, D.; Chokhachian, A.; Auer, T. Impact of environmental quality in outdoor spaces: Dependency study between outdoor comfort and people's presence. In Proceedings of the International Architecture Conference with AWARDs, Hong Kong, China, 7–9 June 2017; pp. 1–10.
16. Shen, P. Impacts of climate change on U.S. building energy use by using downscaled hourly future weather data. *Energy Build.* **2017**, *134*, 61–70. [CrossRef]
17. López, M.; Rubio, R.; Martín, S.; Croxford, B. How plants inspire façades. From plants to architecture: Biomimetic principles for the development of adaptive architectural envelopes. *Renew. Sustain. Energy Rev.* **2017**, *67*, 692–703. [CrossRef]
18. Aelenei, D.; Aelenei, L.; Vieira, C.P. Adaptive Façade: Concept, Applications, Research Questions. *Energy Procedia* **2016**, *91*, 269–275. [CrossRef]
19. Cattaruzzi, J.; Gaspari, J. Taxonomical investigation of Self-Sufficient Kinetic Building Envelopes. *J. Archit. Eng.* **2021**, *27*, 1–15. [CrossRef]
20. Gaspari, J. *Climate Responsive Building Envelopes. From Façade Shading Systems to Adaptive Shells*; Franco Angeli Editore: Milan, Italy, 2020; ISBN 9788835109419.
21. Campaniço, H.; Soares, P.M.M.; Cardoso, R.M.; Hollmuller, P. Impact of climate change on building cooling potential of direct ventilation and evaporative cooling: A high resolution view for the Iberian Peninsula. *Energy Build.* **2019**, *192*, 31–44. [CrossRef]
22. EU. European Parliament Directive 2010/31/EU of the European Parliament and of the Council of 19 May 2010 on the energy performance of buildings (recast). *Off. J. Eur. Union* **2010**, 13–35. [CrossRef]
23. EU. European Parliament Directive 2018/844 amending Directive 2010/31/UE on the energy performance of building and Directive 2012/27/UE on energy efficiency. *Off. J. Eur. Union* **2018**, *156*, 75–91.
24. EU. European Parliament Directive 2012/27/EU of the European Parliament and of the Council of 25 October 2012 on energy efficiency, amending Directives 2009/125/EC and 2010/30/EU and repealing Directives 2004/8/EC. *Off. J. Eur. Union* **2012**, *315*, 1–56.
25. Cellura, M., Guarino, F.; Longo, S.; Tumminia, G. Energy for Sustainable Development Climate change and the building sector: Modelling and energy implications to an of fi ce building in southern Europe Special Report on Emissions Scenarios. *Energy Sustain. Dev.* **2020**, *45*, 46–65. [CrossRef]
26. Andric, I.; Gomes, N.; Pina, A.; Ferrão, P.; Fournier, J.; Lacarrière, B.; Corre, O. Le Modeling the long-term effect of climate change on building heat demand: Case study on a district level. *Energy Build.* **2016**, *126*, 77–93. [CrossRef]
27. Loonen, R.; Rico-Martinez, J.M.; Favoino, F.; Marcin, B.; Ménézo, C.; La Ferla, G.; Aelenei, L. Design for façade adaptability: Towards a unified and systematic characterization. In Proceedings of the 10th Energy Forum-Advanced Building Skins, Bern, Switzerland, 3–4 November 2015; pp. 1284–1294.
28. Barozzi, M.; Lienhard, J.; Zanelli, A.; Monticelli, C. The Sustainability of Adaptive Envelopes: Developments of Kinetic Architecture. *Procedia Eng.* **2016**, *155*, 275–284. [CrossRef]
29. Loonen, R.C.G.M.; Trčka, M.; Cóstola, D.; Hensen, J.L.M. Climate adaptive building shells: State-of-the-art and future challenges. *Renew. Sustain. Energy Rev.* **2013**, *25*, 483–493. [CrossRef]
30. Doumpioti, C.; Greenberg, E.L.; Karatzas, K. Embedded intelligence: Material responsiveness in façade systems. In Proceedings of the Life Information: On Responsive Information and Variations in Architecture—Proceedings of the 30th Annual Conference of the Association for Computer Aided Design in Architecture, ACADIA 2010, New York, NY, USA, 21–24 October 2010.
31. Fox, M.A.; Yeh, B.P. Intelligent Kinetic Systems in Architecture. In *Managing Interactions in Smart Environments*; Springer: London, UK, 2000; pp. 91–103. [CrossRef]
32. Michael, W.; Jude, H. *Intelligent Skins*; Routledge: London, UK, 2002.

33. Ramzy, N.; Fayed, H. Kinetic systems in architecture: New approach for environmental control systems and context-sensitive buildings. *Sustain. Cities Soc.* **2011**, *1*, 170–177. [CrossRef]
34. Mohammadhosseini, Y. Smart Material Systems and Adaptiveness for Beauty of Modern Architecture. *Int. J. Archit. Energy Urban.* **2020**, *1*, 73–81.
35. Michelle, A.; Daniel, S. *Smart Materials and New Technologies*; Architectu: Oxford, UK, 2005; Volume 148.
36. Crespi, M.; Persiani, S.G.L. Rethinking Adaptive Building Skins from a Life Cycle Assessment perspective. *J. Facade Des. Eng.* **2019**, *7*, 21–43. [CrossRef]
37. Formentini, M.; Lenci, S. An innovative building envelope (kinetic façade) with Shape Memory Alloys used as actuators and sensors. *Autom. Constr.* **2018**, *85*, 220–231. [CrossRef]
38. Ricci, A.; Ponzio, C.; Fabbri, K.; Gaspari, J.; Naboni, E. Development of a self-sufficient dynamic façade within the context of climate change. *Archit. Sci. Rev.* **2021**, *64*, 87–97. [CrossRef]
39. Coelho, M.; Zigelbaum, J. Shape-changing interfaces. *Pers. Ubiquitous Comput.* **2011**, *15*, 161–173. [CrossRef]
40. Doumpioti, C. Responsive and autonomous material interfaces. In Proceedings of the Integration through Computation—Proceedings of the 31st Annual Conference of the Association for Computer Aided Design in Architecture, ACADIA 2011, Banff, AB, Canada, 13–16 October 2011.
41. Lignarolo, L.; Lelieveld, C.; Teuffel, P. Shape morphing wind-responsive facade systems realized with smart materials. In Proceedings of the Adaptive Architecture: An International Conference, London, UK, 3–5 March 2011.
42. Bakker, L.G.; Hoes-van Oeffelen, E.C.M.; Loonen, R.C.G.M.; Hensen, J.L.M. User satisfaction and interaction with automated dynamic facades: A pilot study. *Build. Environ.* **2014**, *78*, 44–52. [CrossRef]
43. Köppen, W.; Geiger, R. Das Geographische System der Klimate. *Handb. Klimatol.* **1936**, 7–30. [CrossRef]
44. Kottek, M.; Grieser, J.; Beck, C.; Rudolf, B.; Rubel, F. World map of the Koppen- Geiger climate classification updated. *Meteorol. Z.* **2006**, *15*, 259–263. [CrossRef]
45. Vercesi, L.; Speroni, A.; Mainini, A.G.; Poli, T. A novel approach to shape memory alloys applied to passive adaptive shading systems. *J. Facade Des. Eng.* **2020**, *8*, 43–64. [CrossRef]
46. Pesenti, M.; Masera, G.; Fiorito, F. Exploration of Adaptive Origami Shading Concepts through Integrated Dynamic Simulations. *J. Archit. Eng.* **2018**, *24*, 04018022. [CrossRef]

Article

Experimental and Theoretical Study on the Internal Convective and Radiative Heat Transfer Coefficients for a Vertical Wall in a Residential Building

Piotr Michalak

Department of Power Systems and Environmental Protection Facilities, Faculty of Mechanical Engineering and Robotics, AGH University of Science and Technology, Mickiewicza 30, 30-059 Kraków, Poland; pmichal@agh.edu.pl; Tel.: +48-126-173-579

Abstract: Experimental studies on internal convective (CHTC) and radiative (RHTC) heat transfer coefficients are very rarely conducted in real conditions during the normal use of buildings. This study presents the results of measurements of CHTC and RHTC for a vertical wall, taken in a selected room of a single-family building during its everyday use. Measurements were performed using HFP01 heat flux plates, Pt1000 sensors for internal air and wall surface temperatures and a globe thermometer for mean radiant temperature measured in 10 min intervals. Measured average CHTC and RHTC amounted to 1.15 W/m^{2}K and 5.45 W/m^{2}K, compared to the 2.50 W/m^{2}K and 5.42 W/m^{2}K recommended by the EN ISO 6946, respectively. To compare with calculated CHTC, 14 correlations based on the temperature difference were applied. Obtained values were from 1.31 W/m^{2}K (given by Min et al.) to 3.33 W/m^{2}K (Wilkes and Peterson), and in all cases were greater than the 1.15 W/m^{2}K from measurements. The average value from all models amounted to 2.02 W/m^{2}K, and was greater than measurements by 75.6%. The quality of models was also estimated using average absolute error (AAE), average biased error (ABE), mean absolute error (MAE) and mean bias error (MBE). Based on these techniques, the model of Fohanno and Polidori was identified as the best with AAE = 68%, ABE = 52%, MAE = 0.41 W/m^{2}K and MBE = 0.12 W/m^{2}K.

Keywords: convection; radiation; heat transfer coefficient; correlation

1. Introduction

Heat transfer by convection and radiation plays an important role in numerous applications from nanoscale [1–3] to large industrial installations [4]. Among these, buildings are of special interest because of their important contribution to global energy consumption [5]. Therefore, for economic and environmental reasons, it is common for the energy performance of buildings to be legally regulated in many countries [6].

Heat transfer by convection and radiation plays important role in the internal environment of a building, influencing thermal comfort, the operation of heating, ventilation and air conditioning (HVAC) systems, and energy consumption [7–10]. Heat transfer by convection and radiation can be derived both theoretically and experimentally, however the latter method prevails when considering non-typical and more complex cases, or for proving theoretical investigations.

Energy performance assessments for buildings are usually based on the energy usage for the heating and cooling of the space [11], calculated using various computer tools. These tools employ different models to compute radiative and convective heat transfer coefficients. The nature of the physical phenomenon for a given heat transport mechanism determines the parameters necessary to compute the appropriate coefficients. Heat transfer by radiation is simple to describe, as it depends on the temperature, emissivity and position of the considered body and its surroundings. Moreover, numerous studies have proven that the difference between the air and mean radiant temperatures inside the building zone is relatively small, typically below 2 °C [12], and in practice, constant values of radiative heat transfer coefficients (RHTCs) are commonly used [13,14] to simplify calculations.

In terms of heat transfer by convection, the aforementioned temperatures, flow conditions, and physical properties of the fluid (air) are needed. Hence, mathematical description is more complex than in the previous case, and analytical solutions are available mainly for relatively simple cases. For these reasons, over the past few decades, experimental correlation models of the convective heat transfer coefficients (CHTCs) have been developed. They significantly reduce computation effort by combining an unknown coefficient with the main driving physical factors in given conditions within the studied case. Usually, these models are subdivided into correlations of dimensionless numbers or temperature differences [15,16]. The first type links Nusselt with Grashof and Prandtl numbers in the form of the power equation $Nu = C\,(Gr\,Pr)^n$ with constant C and n. In the second case, an equation of the type $h_c = C\,(\Delta T)^n$ is used. The latter form is simpler to use because it is founded on an easily measurable quantity, which is temperature and, in certain cases, geometrical parameters (length, area or height of a surface) and air velocity.

Initially, a lack of detailed data forced the use of results from various physical experiments in buildings' thermal calculations. The similarity of buildings' partitions to typical geometric figures resulted in the use of both theoretical and experimental results, obtained mainly for horizontal, vertical, and inclined planes. Many such studies have been developed in recent decades, and different flow regimes have been studied under natural and forced convection.

Fishenden and Saunders [17–19] investigated convective heat transfer for metal (aluminium, copper, platinum, or steel) vertical plates in air, water and mercury. To study the impact of air pressure on that phenomenon, the measurements were conducted in a cylindrical steel test chamber. The temperature difference was up to 55 °C. Finally, the authors gave experimental correlations with dimensionless numbers, with $n = 1/4$ and $n = 1/3$ for laminar and turbulent flow, respectively.

McAdams [20] summarised the results of several previous works of various researchers on natural convection and presented correlations for vertical and horizontal plates and cylinders in the form of dimensionless equations with $n = 1/4$. In quoted experiments, metal plates were used (aluminium, copper or platinum). He also presented an alignment chart to derive a heat transfer coefficient for the turbulent flow of various gases on vertical surfaces. In the case of so-called ordinary temperatures and atmospheric pressure, simplified dimensional equations with $n = 1/3$ and $n = 1/4$ for turbulent and laminar flow, respectively, were given. The same exponent was used by Lewandowski [21,22] for copper plate in glycerine or for water [23].

Based on previous theoretical and experimental works, Churchill and Usagi [24] developed empirical dimensionless correlations with $n = 9/4$ and $n = 8/3$ under laminar free and forced convection, respectively, for an isothermal vertical plate. No information on the temperature difference range was given.

As the interest in building modelling has grown, experimental research on heat transfer phenomena in this group of objects has gained increased importance. Related studies commonly test chambers with dimensions similar to typical rooms in buildings.

Correlations for interior surfaces of buildings for vertical glazing (with no radiator under the window), vertical walls (non-heated or near the heat source) and horizontal surfaces (heated floor or cooled ceiling) were presented by Khalifa and Marshall in [25]. Using a two-zone test cell they performed nine experiments covering the most popular heating configurations in buildings under controlled steady-state conditions, as follows: forced convection on the interior surface using a fan, a cell heated by a 1 kW fan heater, a uniformly heated floor, a uniformly heated vertical wall and a uniformly heated edge using a metalized plastic foil, a cell heated by a 1.5 kW oil-filled radiator located opposite the test wall and adjacent to the test wall, and finally with a single glazed window in a test wall with a radiator located beneath and opposite the window.

The larger hot zone represented a large enclosure (such as a living room). It had one vertical wall common with a smaller cold zone built to control the outdoor temperature of the hot zone. A total of 142 tests were conducted. Vertical walls and the roof were covered with aluminium foil to minimise the impact of internal longwave radiative heat exchange on temperature and heat flux measurements, and to neglect it in heat transfer calculations. To reach the steady-state conditions between two consecutive tests with different configurations, a 24 h pre-run period was applied, and then 12 h data logging between 22:00 and 10:00 was undertaken. The CHTCs obtained for each of the different studied elements were presented as dimensional correlations against the internal air to wall surface temperature difference. For the heated vertical wall $n = 1/4$ was obtained.

Alamdari and Hammond [26] used the model of Churchill and Usagi to derive empirical correlations for buoyancy-driven convective heat transfer for the internal surfaces of naturally ventilated buildings, covering laminar, transitional, and turbulent airflows. They noticed that the experimental results related to buoyancy-driven convection given in the literature were often given for fluids, surface dimensions, and temperature differences not found in buildings. Based on Churchill and Usagi considerations and experiments presented by other researchers, they developed several correlations of different complexity and applicable for vertical and horizontal partitions.

Natural convection from heated surfaces in a room was examined by Awbi and Hatton [27]. Experiments were conducted in a test chamber and a small box. The first experiment was of a similar concept as presented in [25]. The internal layer of vertical walls, ceiling, and floor was of varnished plywood. Five heating plates (0.5 m per 2.3 m) covered with a 2 mm thick aluminium plate (to minimise radiative heat flux) were used to provide surface heating to each tested wall. To minimise the effect of internal longwave radiation, temperature sensors were shielded by small steel tubes. The small test box was a cube with sides of approximately 1 m made of the same materials. Each test was conducted for several hours to reach steady-state conditions. Obtained equations for CHTCs were in the form of power functions of temperature difference. Correlation coefficients were above 0.90, indicating strong dependences.

Delaforce et al. [28] investigated convective heat transfer at internal surfaces using a rectangular test cell with sides of approximately 2 m each. The cell was located on an exposed grassland site and experiments were conducted during winter conditions. External walls were made from bricks and an external layer of expanded polystyrene. Their internal surface had no additional covering layer, so indoor radiative heat transfer was monitored using a radiometer. Temperatures were measured at 5-min intervals. Heating was provided by a 1 kW fan heater. Continuous heating, intermittent heating, and unheated operation were examined separately for the west wall, floor, and ceiling. Before each experiment, authors used a two-day warm-up period with continuous heating. Finally, they did not provide any empirical correlations, but only averaged the numerical values of the calculated CHTCs and compared them with values recommended by the Chartered Institute of Building Service Engineers (CIBSE) Guide. The greatest discrepancies were noticed for a floor and vertical wall, where the measured CHTC was 0.5 and 1.6 W/m^{2}K compared to 1.5 and 3.0 W/m^{2}K from CIBSE, respectively.

Test chambers can be also used to analyse the performance of selected buildings components under various conditions. Sukamto et al. [29] used this method to investigate the thermal performance of the ventilated wall. Because of forced airflow, the obtained CHTC was about 20 W/m^{2}K.

Furthermore, several authors studied the impact of the heating, ventilation and air conditioning systems (HVAC) on internal heat transfer in buildings. A more detailed review of embedded radiant heating/cooling systems is given in [30]. Cholewa et al. [31,32] investigated heated/cooled radiant floor and radiant ceiling using a test chamber. A similar method was used by Acikgoz and Kincay [33] for radiant cooling walls. Hydronic radiant heated walls in a climatic test chamber with a window were investigated in [34]. Guo et al., investigated convective heat transfer for night cooling with diffuse ceiling [35] and mixing [36] ventilation in a guarded hot box used as a test chamber. To evaluate the dependence of CHTC on ventilation conditions, they used several correlations linking them with ventilation air changes per hour (ACH). Various cases of CHTCs for different building partitions in the form of empirical correlations are given in reviews [10,16,37–40]. For example, in the review by Khalifa, with CHTC correlations derived in 1, 2 and 3-dimensional enclosures [39,40], the author presented 13 studies performed for the latter case. The author concluded that large differences may arise between the presented models.

Recently, and in addition to the aforementioned methods, various computer-based techniques have been developed, and simulation tools used, in convective heat transfer analyses [41–45]. Despite their abilities to deal with complex problems that are not possible to solve analytically, they are used for proving experimental results rather than to formulate mathematical relationships to obtain convection heat transfer coefficients in building applications. For these reasons, in energy auditing, energy certification, or in other similar applications requiring relatively low computational effort and a limited number of input parameters, constant CHTC and RHTC values are commonly used [46,47]. These are given in various standards, such as EN 1264 [48], ISO 11855 [49], EN ISO 6946 [50] and guidebooks [51] providing relevant formulas to obtain CHTC or their constant values in different cases.

The presented methods and techniques more or less accurately reflect the nature of heat transfer phenomena in buildings. They allow the separation of noise, stabilisation, and control of thermal conditions, and so on [52]. On the other hand, the applicability of the theoretical models should be verified in real conditions. For this reason, the best research object seems to be a real building.

Cost, practical problems, and limitations for building users resulting from the presence of measuring equipment in the building during its everyday use are the probable reasons for so few tests being carried out in these conditions. Hence, there are few CHTCs from experiments in realistic situations [53]. Most measurements have been made in special test chambers where internal surfaces were metal coated. Therefore, it seems reasonable to address indoor convective and radiative heat transfer in buildings during their everyday use.

Wallentén [53] experimentally analysed internal convective heat transfer at an external wall with a window in a room with and without furniture, located in a single-family house built in a sandwich construction of light-weight concrete and polystyrene. The room had the dimensions of 3 × 3.6 × 2.4 m. Internal surfaces of walls and floor were painted with a mat white and light brown colour, respectively. Two electric radiators with 500 W and 1000 W heating power were used. T-type thermocouples were used to measure indoor air and partition temperatures. Longwave radiation was calculated from diffuse grey surface temperatures. A total of 14 different experiments (radiator placement, presence of furniture, wall and window analysis, ventilation schedule), each lasting approximately one week, were performed across four years. A sampling time of 1 min was assumed and then measurements were averaged in 4, 10, 30 and 60 min intervals. The author did not provide any conclusions on the impact of the averaging time on final results. Several graphs with CHTC depending on the surface (vertical wall or window) and indoor air temperature were presented, but no correlations were given. As far as the wall is concerned, the data from measurements were quite scattered and showed little correlation. The spatial distribution of internal air temperature was also presented. An impact from the radiator or ventilation was visible, but the maximum difference between the floor and ceiling did not exceed 3 °C. Only near the window, at a distance of several cm, it reached 5 °C.

The review presented here shows several important outcomes. Thermal conditions in climatic chambers during the aforementioned tests were controlled according to the requirements necessary to perform the assumed tests. Noises could be easily limited using selected construction materials or HVAC system components. On the other hand, heat transfer in buildings during everyday use is influenced by different environmental factors, such as occupation, solar radiation, air ventilation, and so on. The very low number of studies dedicated to this issue makes it worthy to compare the models developed theoretically or on a basis of experiments in test chambers, with those derived from real conditions in a building during its normal operation. The identification of the most significant limitations during such an experiment is also a possibility, and could indicate possible improvements to similar studies in the future.

The environmental conditions during measurements and experimental setup are given in the next section. Calculation procedures to obtain CHTC and RHTC and uncertainty analysis are then presented. For comparative purposes, 14 theoretical and empirical correlations to obtain CHTC for a vertical wall under free convection are used. The results are then presented and compared with relevant studies. Finally, concluding remarks are given.

2. Materials and Methods

2.1. Introduction

The research was conducted from 30 January to 10 February 2021 in a residential building located in southern Poland, equipped with a hydronic heating system.

During the studied period the outdoor air temperature (Figure 1) varied from $-15.1\ °C$ (at 3:50 on 1 February) to 10.1 °C (at 9:30 on 4 February). Global solar irradiance (Figure 2) incident on the external surface of the considered wall was from 0 to 113.3 W/m^2 (at 13:20 on 31 January).

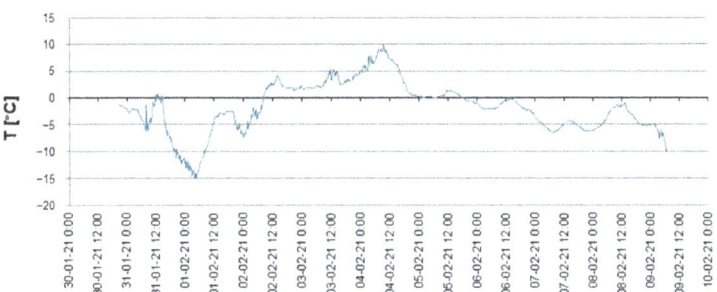

Figure 1. External air temperature during measurements.

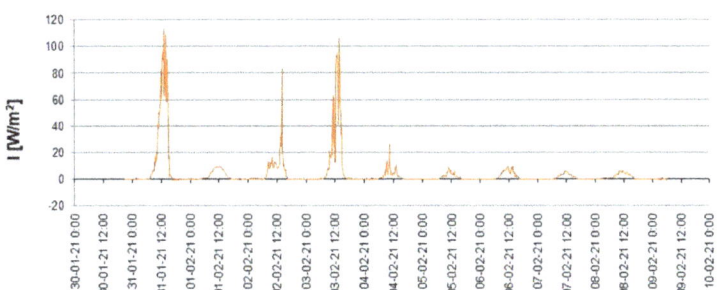

Figure 2. Global solar irradiance on the west-facing wall during measurements.

2.2. Experimental Setup

The west-oriented wall (strictly—its north part) in a living room was chosen for the measurements (Figure 3). Also located in this corner of the room are wooden stairs. The room is on a rectangular plane of 6 m × 4 m and has a height of 2.50 m.

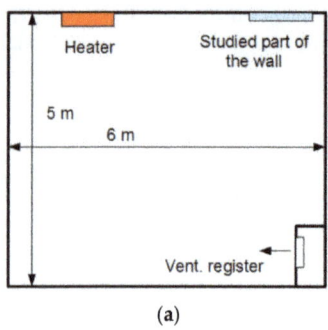 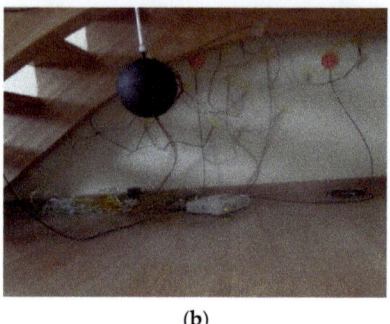

(a) (b)

Figure 3. (a) View of the room selected for experiment; (b) View of the measurement equipment during assembly.

The measured data was recorded in 10 min intervals with a MS6D Comet data logger. Periods from 4 to 60 min have typically been used in other studies [25,28,53].

In the experiment, two HFP01 heat flux sensors and three Pt1000 temperature sensors were mounted on the internal surface of the wall, while two Pt1000 internal air temperature sensors, one sensor of indoor radiant (ø 150 mm TP875.1 globe thermometer with the Pt100 sensor) and one Pt1000 of outdoor air temperature (mounted in a radiation shield) and a CMP11pyranometer for global solar irradiance incident were placed on the external wall. Their main parameters are given in Table 1.

Table 1. The main metrological parameters of the measuring sensors used.

Sensor	Measured Variable	Measurement Range	Accuracy
Pt1000 platinum resistance sensor	Air temperature	−50 °C ÷ +150 °C	Class A [1]
Pt1000 platinum resistance sensor	Wall surface temperature	−50°C ÷ +150 °C	Class A [1]
TP875.1 with the Pt100 sensor	Radiant temperature	−30 °C ÷ +120 °C	±0.2 °C
CMP11 Kipp&Zonen	Solar irradiance	0 ÷ 4000 W/m²	Spectrally Flat Class A [2]
HFP01 Hukseflux	Heat fluxdensity	−2000 ÷ 2000 W/m²	±3%

[1] According to EN 60751. [2] According to ISO 9060 and IEC 61724.

2.3. Calculation Procedure

The conduction heat flux flowing through the wall (q_w) and measured by the heat flux sensors (Figure 1) is the sum of radiative (q_r) and convective (q_c) fluxes entering an interior space of the room:

$$q_w = q_r + q_c. \tag{1}$$

The radiative heat flux is given by the relationship:

$$q_r = h_r(T_s - T_r), \tag{2}$$

Within the building enclosure, the considered wall at temperature T_s is exposed to remaining walls and heat transfer by radiation occurs between them. The mean radiant temperature of the given wall surroundings is measured by the globe thermometer [28–30]. Then, based on the Stefan–Boltzmann law and at the given emissivity of the wall, ε, the following equation can be written [31]:

$$q_r = \varepsilon\sigma\left(T_s^4 - T_r^4\right). \tag{3}$$

Comparing Equations (2) and (3) we obtain:

$$h_r = \varepsilon\sigma\frac{T_s^4 - T_r^4}{T_s - T_r} = \varepsilon\sigma(T_s + T_r)\left(T_s^2 + T_r^2\right). \tag{4}$$

Equation (4) shows that h_r can be obtained without direct measurement of radiative heat flux.

Convective heat flux is given by:

$$q_c = h_c(T_s - T_i). \tag{5}$$

With q_r known, q_c can be obtained from Equation (1), and then, from Equation (5) an unknown convective heat transfer coefficient can be computed.

2.4. Uncertainty Analysis

Experimental investigations inevitably have associated measurement errors. Both RHTC and CHTC are computed in the present study indirectly from formulas given by Equations (4) and (5), respectively.

In cases where the indirect measurement of a physical quantity, y, is a function of independent measurements $x_1, x_2, \ldots, x_n$:

$$y = f(x_1, x_2, \ldots, x_n) \tag{6}$$

the standard combined uncertainty u_c of y can be calculated applying the propagation model of uncertainty from the formula [54–59]:

$$u_c(y) = \sqrt{\left(\frac{\partial y}{\partial x_1} u(x_1)\right)^2 + \left(\frac{\partial y}{\partial x_2} u(x_2)\right)^2 + \ldots + \left(\frac{\partial y}{\partial x_n} u(x_n)\right)^2}. \tag{7}$$

In the next step, the expanded uncertainty of the measured quantity is calculated from the equation:

$$U = k \, u_c(y) \tag{8}$$

where k—coverage factor; k = 2 for 95% confidence level of the uncertainty.

Following the calculation procedure presented in the previous section, the uncertainty of RHTC is derived. From Equation (4) we obtain:

$$h_r = \varepsilon\sigma\left(T_s^3 + T_s T_r^2 + T_r T_s^2 + T_r^3\right). \tag{9}$$

Inserting Equation (9) into Equation (7) we get an expression to obtain the standard combined uncertainty of the measured RHTC. Three independent variables can be distinguished, namely: ε, T_r and T_s. Hence, Equation (7) can be written as:

$$u_c(h_r) = \sqrt{\left(\frac{\partial(h_r)}{\partial \varepsilon} u(\varepsilon)\right)^2 + \left(\frac{\partial(h_r)}{\partial T_r} u(T_r)\right)^2 + \left(\frac{\partial(h_r)}{\partial T_s} u(T_s)\right)^2}. \tag{10}$$

Then, partial derivatives in Equation (10) are computed, as follows:

$$\frac{\partial h_r}{\partial \varepsilon} = \sigma\left(T_s^3 + T_s T_r^2 + T_r T_s^2 + T_r^3\right), \tag{11}$$

$$\frac{\partial h_r}{\partial T_r} = \varepsilon\sigma\left(2T_s T_r + T_s^2 + 3T_r^2\right), \tag{12}$$

$$\frac{\partial h_r}{\partial T_s} = \varepsilon\sigma\left(3T_s^2 + T_r^2 + 2T_s T_r\right). \tag{13}$$

The uncertainty associated with the measurement of emissivity, mean radiant temperature and surface temperature can be derived based on sensors and measurement equipment data presented in Section 2.2.

The estimation of emissivity uncertainty is a very difficult task. Several authors have presented methods to address this issue, involving different approaches. Ficker [60] used two precise black body etalons of different emissivities heated to the same temperature to develop a comparative method to estimate virtual emissivity of infrared (IR) thermometers. Chen and Chen [61] determined the emissivity and temperature of construction materials by using an IR thermometer and two contact thermometers. For this purpose, they developed empirical regression equations. Höser et al. [62] found unknown emissivities of rock samples by comparing the temperature of the tested samples with a reference sample at a known temperature. They concluded that the relative uncertainty in rock emissivity is proportional to the relative uncertainty in the rock temperature (T) in the absolute sale. It can be written as:

$$u(\varepsilon) = \left| -4 \frac{u(T)}{T} \varepsilon \right|, \tag{14}$$

This method seems to be the most appropriate in the considered case and was applied here.

The remaining uncertainties of two temperatures, T_r and T_s, are both comprised of two elements. The first results from the accuracy of a temperature sensor. The latter is the effect of the data logging device accuracy [55,63]. Hence, we obtain:

$$u(T_r) = \sqrt{(u_{sens,r})^2 + (u_{dev,r})^2}, \tag{15}$$

and:

$$u(T_s) = \sqrt{(u_{sens,s})^2 + (u_{dev,s})^2}, \tag{16}$$

where "sens" and "dev" subscripts refer to sensor and measuring device, respectively. Subscripts after a comma, r and s, refer to radiant temperature and surface temperature, respectively.

Similar considerations should be made to obtain CHTC uncertainty. Firstly, Equation (5) is rearranged to the form:

$$h_c = \frac{q_c}{T_s - T_i}. \tag{17}$$

In the above formula, three independent variables can be distinguished that influence measured CHTC, i.e., q_c, T_i and T_s. Then, the uncertainty of CHTC is given by:

$$u_c(h_c) = \sqrt{\left(\frac{\partial(h_c)}{\partial q_c} u(q_c)\right)^2 + \left(\frac{\partial(h_c)}{\partial T_i} u(T_i)\right)^2 + \left(\frac{\partial(h_c)}{\partial T_s} u(T_s)\right)^2}. \tag{18}$$

Subsequent partial derivatives are as follows:

$$\frac{\partial h_c}{\partial q_c} = \frac{1}{T_s - T_i}, \tag{19}$$

$$\frac{\partial h_c}{\partial T_s} = \frac{-q_c}{(T_s - T_i)^2}, \tag{20}$$

$$\frac{\partial h_c}{\partial T_i} = \frac{q_c}{(T_s - T_i)^2}. \tag{21}$$

The uncertainty associated with the measurement of indoor air temperature $u(T_i)$ and surface temperature (T_s) can be derived similarly, as in the previous case on the basis of Equations (15) and (16). However, a measurement of heat flux density requires some additional explanation. In this case, according to the manufacturer's manual [64], measurement uncertainty consists of several components. These are calibration uncertainty, the difference between measurement and reference calibration conditions, the duration of sensor employment, and application errors resulting from working conditions. They can be included in the relationship:

$$u(q_c) = \sqrt{\sum_{i=1}^{n} u_i^2}, \tag{22}$$

where n is the number of uncertainty components and u_i is the i-th component.

2.5. Selected Correlations for CHTC Calculations

During the test, there was no measured air velocity near the wall. A gravity ventilation system is present in the building, but the ventilation register in the room, however, is located far from the studied wall and was closed, and it was assumed that it has no impact on the flow regime near the wall. Hence, for further consideration, 13 correlations were chosen to obtain convective heat transfer coefficients for vertical walls under natural convection [10,16,37,65,66]. In all following relationships, it is assumed that:

$$\Delta T = T_s - T_i. \tag{23}$$

The model of Wilkes and Peterson (quoted from [16]) is given by:

$$h_c = 3.05(\Delta T)^{0.12}. \tag{24}$$

In [19] it was stated that that model was developed on the basis of a test for temperature difference between 4.5 to 15.5 °C with two heated plates 2.4 × 0.8 m with 0.1 m air space. The resulting correlation was derived on the three data points only.

Khalifa in his review [39] quoted the relationship developed originally by Hottinger:

$$h_c = 2.50(\Delta T)^{1/4}. \tag{25}$$

Min et al. [40] determined natural convection coefficients by using three different sized rectangular test chambers. Hence, they are not applicable in the considered case. For the vertical wall, they obtained correlations in the case of the heated ceiling and heated floor. They also referred to other correlations from various other studies. The first of these, presented here, was developed using a 0.60 m square plate, the temperature difference up to 555 °C and the height H. For laminar flow:

$$h_c = 1.368 \left(\frac{\Delta T}{H}\right)^{1/4}. \tag{26}$$

The second correlation was developed during tests with a 1.2 m square plate. Apart from the temperature difference up to 100 °C, no additional information on the application range was given. The model was given by two equations, as follows:

$$h_c = 1.776(\Delta T)^{1/4} \tag{27}$$

and for the turbulent flow:

$$h_c = 1.973(\Delta T)^{1/4}. \tag{28}$$

The third is the correlation from King. Unfortunately, no data related to the test details or applicability ranges were provided. It is given by:

$$h_c = 1.517(\Delta T)^{1/3}. \tag{29}$$

Alamdari and Hammond [26] derived correlations for convective heat transfer from the internal surfaces in naturally ventilated buildings. Using data from other studies, and applying the mathematical model of Churchill and Usagi [24], they proposed the following relationship:

$$h_c = \left\{ \left[1.5\left(\frac{\Delta T}{L}\right)^{1/4}\right]^6 + \left[1.23(\Delta T)^{1/3}\right]^6 \right\}^{1/6}. \tag{30}$$

In addition, they provided a simplified formula valid within the limited range of temperatures and applicable to naturally ventilated buildings, given by the expression:

$$h_c = 0.134 L^{-1/2} + 1.11(\Delta T)^{1/6}, \qquad (31)$$

where L is a hydraulic diameter expressed as:

$$L = 4A/P \qquad (32)$$

Fohanno and Polidori [67] analysed laminar and turbulent heat transfer modelling at uniformly heated internal vertical building surfaces. The resulting correlations for CHTC were derived theoretically and compared with other models. Their model is given by the equation:

$$h_c = 1.332 \left(\frac{\Delta T}{H}\right)^{1/4}. \qquad (33)$$

Musy et al. [68], quoting the work of Allard, presented the following equation for walls with natural convection:

$$h_c = 1.5(\Delta T)^{1/3}. \qquad (34)$$

Churchill and Chu [65] investigated uniformly heated and cooled vertical plates. Their correlation is given by:

$$h_c = \frac{0.0257}{H} \left(0.825 + 7.01(\Delta T)^{1/6} H^{3/6}\right)^2. \qquad (35)$$

Khalifa and Marshall [25] investigated internal convection in a real-sized indoor test cell. Various combinations of relative positions of a heater, radiant panels and wall were studied. Two cases similar to that studied in the present paper were chosen. The first one is the case with a room heated by a radiator located adjacent to the test wall:

$$h_c = 2.20(\Delta T)^{0.21}. \qquad (36)$$

The second is a room heated by an oil-filled radiator located under a window:

$$h_c = 2.35(\Delta T)^{0.21}. \qquad (37)$$

Rogers and Mayhew [69] presented the correlation for vertical plates under laminar or transitional flow conditions:

$$h_c = 1.42 \left(\frac{\Delta T}{H}\right)^{1/4}. \qquad (38)$$

2.6. Statistical Analysis

Additionally, various statistical measures can be used to perform error analysis of the presented models against measured CHTC [70]. Among them, there is an average absolute error (AAE), average biased error (ABE), mean absolute error (MAE) and mean bias error (MBE) [71,72]. Assuming that $h_{m,i}$ is a i-th measured value of CHTC, $h_{p,i}$ is a i-th value of CHTC predicted by the model and m is a total number of conducted measurements, they are given by the following equations:

$$AAE = \frac{1}{m} \sum_{i=1}^{m} \frac{|h_{p,i} - h_{m,i}|}{h_{m,i}} \times 100\%, \qquad (39)$$

$$ABE = \frac{1}{m} \sum_{i=1}^{m} \frac{(h_{p,i} - h_{m,i})}{h_{m,i}} \times 100\%, \qquad (40)$$

$$\text{MAE} = \frac{1}{m}\sum_{i=1}^{m}\left|h_{p,i} - h_{m,i}\right|, \tag{41}$$

$$\text{MBE} = \frac{1}{m}\sum_{i=1}^{m}\left(h_{p,i} - h_{m,i}\right). \tag{42}$$

The average absolute error (AAE) is the average of all the absolute errors calculated for the consecutive measurements in a given dataset, and indicates the average error of a correlation. The average biased error (ABE) shows the degree of overestimation (ABE > 0) or underestimation (ABE < 0) of the considered correlation. The mean absolute error (MAE) shows the average magnitude of deviations of a modelled variable against the reference values. Low MAE indicates the high accuracy of a model. MBE is used to determine the overall bias of the correlation. Positive MBE means the overestimation of the model.

3. Results and Discussion
3.1. Radiative Heat Transfer Coefficient
3.1.1. Introduction

The radiative heat transfer coefficient was computed from Equation (4). During the whole period, the radiant temperature was higher than that of the wall surface (Figure 4). This difference varied from 1.54 °C (at 13:10 on 7 February) to 0.28 °C (at 6:50 on 6 February), with an average of 0.91 °C.

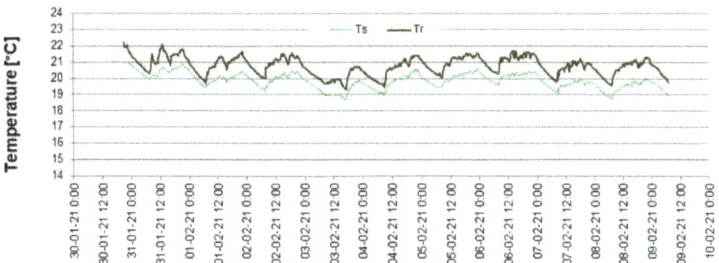

Figure 4. Average wall surface temperature (T_s) and radiative temperature (T_r).

RHTC, calculated under the aforementioned conditions, varied during the measurement period (Figure 5) from 5.373 W/m²K (at 16:40 on 3 February) to 5.516 W/m²K (at 21:00 on 31 January), with an average of 5.445 W/m²K. The EN ISO 6946 standard [50] gives the value h_r = 5.7 W/m²K for the blackbody at 20 °C. Consequently, for ε = 0.95 resulting h_r = 5.415 W/m²K is very close to that average value.

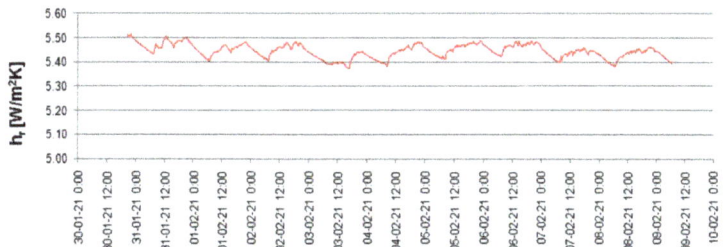

Figure 5. Radiative heat transfer coefficient from measurements.

Assuming that:
$$T_m = 0.5\,(T_r + T_s) \tag{43}$$

For cases with small T_m values, Equation (4) can be linearised [73,74] to the form:

$$h_r = 4\varepsilon\sigma T_m^3. \qquad (44)$$

Despite the nonlinear dependence of the RHTC on T_m, given by Equation (44), h_r almost linearly changes with T_s (Figure 6). This occurs primarily because of the relatively narrow range of T_s variation within the range below 2.5 °C. Similar observations were reported by Evangelisti et al. [65].

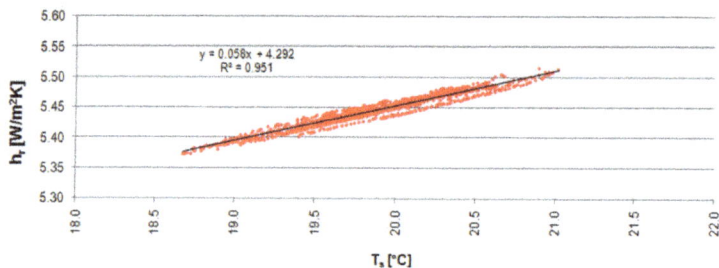

Figure 6. Radiative heat transfer coefficient from measurements.

3.1.2. Uncertainty Analysis

Before the calculations were made, the emissivity of the internal walls was assumed at $\varepsilon = 0.95$. This value was confirmed using a DIT-130 IR-thermometer (declared accuracy $\pm 1.5\%$ of a measured value + 2 °C) and the reference DFT-700 thermometer with a K-type thermocouple probe (accuracy $\pm 0.2\%$ of a full scale). Because of its low variation under the conditions considered here, it was reasonable to assume a constant emissivity value of the wall surfaces [75].

Based on the manufacturers' manuals, and assuming a temperature of 25 °C, relevant uncertainties were computed (Table 2). The resultant uncertainty of RHTC measurements is $u_c(h_r) = 0.18046$ W/m²K. Applying coverage factor k = 2 (see Equation (8)) we obtained the expanded uncertainty $U(h_r) = 0.36$ W/m²K. Assuming a temperature of 20 °C the value of $U(h_r) = 0.34$ W/m²K was obtained.

Table 2. Uncertainties in RHTC measurement.

Uncertainty	Value	Unit
$u(\varepsilon)$	0.030	—
$u(T_r)$	0.200	K
$u(T_s)$	0.200	K
$\frac{\partial(h_r)}{\partial \varepsilon}$	0.18033	W/m²K
$\frac{\partial(h_r)}{\partial T_r}$	0.004788	W/m²K²
$\frac{\partial(h_r)}{\partial T_s}$	0.004788	W/m²K²

3.2. Convective Heat Transfer Coefficient

3.2.1. Introduction

The measured internal air temperature varied from 20.44 °C (at 16:40 on 3 February) to 23.40 °C (at 22:10 on 30 January), with an average of 21.98 °C. It was higher than that of the wall surface (Figure 7). This difference varied from 1.54 °C (at 13:10 on 7 February) to 0.28 °C (at 6:50 on 6 February), with an average of 0.91 °C.

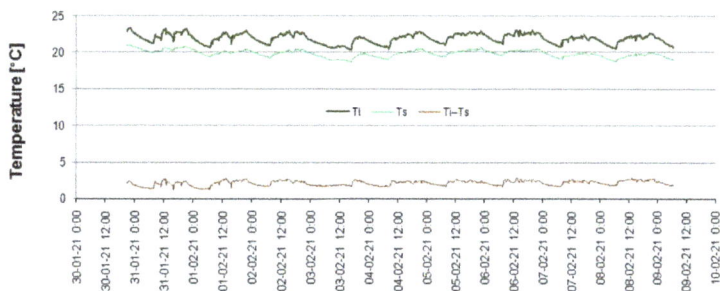

Figure 7. Wall surface temperature (T_s), internal air temperature (T_i) and their difference.

The calculated CHTC varied throughout the whole period (Figure 8) from 0.069 W/m²K (at 6:30 on 5 February) to 3.027 W/m²K (at 14:10 on 2 February), with an average of 1.153 W/m²K. The EN ISO 6946 standard gives the value h_c = 2.5 W/m²K for the internal surface of the vertical wall. The recommended internal surface resistance in such a case is R_{si} = 0.13 m²K/W. Taking the measured average h_r = 5.445 W/m²K we obtain R_{si} = 0.15 m²K/W.

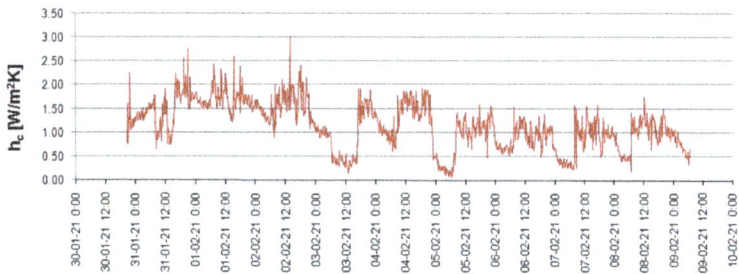

Figure 8. Convective heat transfer coefficient from measurements.

3.2.2. Uncertainty Analysis

At first, the uncertainty of the heat flux density measurement should be determined (Equation (22)). In the present study, two HFP sensors were used. Several authors have presented various experimental studies on the thermal performance of buildings using that sensor [76–78], but none of them discussed the problem of heat flux density measurement uncertainty. In some cases, the measurement accuracy of the HFP01 heat flux plate was assumed from 5% [79] to 20% [80]. For this reason, this issue is presented here in detail.

Following recommendations given by the manufacturer [64], the following factors were distinguished:

- calibration uncertainty given by the manufacturer: <3% (k = 2),
- non-stability uncertainty: <1% for every year of operation,
- correction of the resistance error,
- correction of the deflection error,
- error from the temperature dependence: <0.1% per 1 °C deviation from the 20 °C.

The calibration, non-stability and temperature errors are relatively easy to estimate because of clear criteria given by the manufacturer. The first is given as expanded uncertainty (see Equation (8)), and for uncertainty budget calculation it was assumed as 1.5%. The latter should be estimated for the worst conditions during measurement, i.e., the greatest temperature measured during the test: T_i = 23.40 °C.

The resistance error is related to the influence of the sensor thermal resistance added to the resistance of a given partition on the resultant heat flux density. While the manufacturer does not recommend the use of thermal paste because it tends to dry out, it was applied here because of the relatively short measurement period and moderate temperature. Assuming an average thickness of this paste of 0.1 mm and a thermal conductivity of 1.5 W/m·K it gives the additional thermal resistance of 6.67×10^{-6} m²K/W in comparison to 71×10^{-4} m²K/W of the sensor specified by the manufacturer. Hence, the resulting resistance is 0.93%.

Deflection error (called operational error in ISO9869) appears as a result of the difference between the thermal conductivity of the surrounding environment and that of a sensor. The sensor thermal conductivity is 0.76 W/m·K and is in the same order as values typically met in buildings materials: 0.3–0.4 W/m·K for ceramic blocks and 0.7–0.9 W/m·K for cement-lime plaster. Assuming a thermal resistance of the wall of approximately 4 m²K/W, we can see that the deflection error is at a negligible level. The average measured heat flux density of 7.42 W/m² calculated $u(q_c)$ = 0.1526 W/m².

The derivative components, given by Equations (19)–(21), were computed for the most unfavourable conditions during the experiment, i.e., the smallest T_s—T_i, which amounted to 1.23 °C for q_c = 2.54 W/m². Results are given in Table 3. Calculated $u_c(h_c)$ = 0.475218 W/m². Taking k = 2 (see Equation (8)) we get a resulting $U(h_c)$ = 0.95 W/m²K. On the other hand, for the maximum temperature difference of 2.77 °C and q_c = 2.11 W/m² the resulting uncertainty is $U(h_c)$ = 0.16 W/m²K. Other authors estimated the accuracy of experimentally measured CHTCs from approximately 20% [22,26] to 30% or even 35% at $\Delta T \approx 1$ °C [53].

Table 3. Uncertainties in CHTC measurement.

Uncertainty	Value	Unit
$u(q_c)$	0.05224	W/m²
$u(T_i)$	0.200	K
$u(T_s)$	0.200	K
$\frac{\partial(h_c)}{\partial q_c}$	0.81169	W/m²K
$\frac{\partial(h_c)}{\partial T_i}$	0.65884	W/m²K²
$\frac{\partial(h_c)}{\partial T_s}$	0.65884	W/m²K²

3.2.3. Comparison with Other Models

It can be noticed that fluctuations of the CHTC were connected with variations in the difference between internal air and wall surface temperature (Figure 9), but this relationship was rather weak (calculated coefficient of determination R^2 = 0.2498). Some authors recommended the use of nonlinear correlations [81,82], but they did not improve the quality of the fit. In case of the logarithmic, exponential, quadratic function and 4-th order polynomial, R^2 = 0.2252, 0.2459, 0.2308, and 0.2323, respectively.

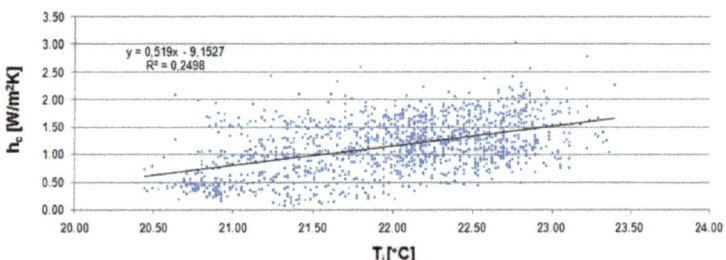

Figure 9. Convective heat transfer coefficient versus indoor air temperature.

There are several reasons for this. In the experimental studies performed in test chambers [7,25,27,39,40], thermal conditions during the measurements were stable. As noted in Section 2.5, the ventilation register in the room was closed, however, and located far from the studied wall, and it was assumed that it has no impact on the flow regime near the wall. The hydronic heater was located approximately 4 m from the measurement equipment. Solar irradiance incident on the external wall was very low (Figure 2), and that falling on internal surfaces could be neglected without any doubt, due to its reduction through windows and curtains. On the other hand, because of the everyday use of the building during the experiment, there could be disturbances in the airflow due to the movement of people in the building, however, indoor air temperature variation was below 3 °C. Hence, to find a solution to this problem, technical factors should be taken into account. Considerations presented in Section 2.4 indicate that an error of CHTC estimation results from uncertainty of heat flux density and air and surface temperature measurement. In the present case, the most important influence was that of the T_s—T_i temperature difference (Table 3). Both temperatures were measured by accurate sensors (Table 1). However, their difference is in the denominator of the fraction to calculate their uncertainty (Equations (19)–(21)). Hence, not only is the uncertainty of temperature measurement important, but the uncertainty of the temperature difference is as well. In the present study, ΔT varied from 1.23 °C to 2.77 °C. These were low values, difficult to measure with high accuracy and which, in certain cases, may lead to unacceptable uncertainties [83]. Numerous studies have pointed out this problem [84–86], and a common practice is to filter out the data when ΔT was lower by several Celsius degrees (typically 2–5 °C). This problem should be, however, investigated in more detail in the next experiment.

Several studies have presented the results of experiments on convection at vertical walls at temperature differences below 3 °C. In [26], the heat transfer coefficient for buoyancy-induced airflow near vertical surfaces for ΔT of 1 to 10 °C was between 1.5 to 4 W/m^{2}K. Khalifa [40] presented the results of numerous experimental and theoretical works on convection at various surfaces. For a vertical wall at ΔT = 2 °C, CHTC was between 1.2 to 3.0 W/m^{2}K. CHTC in a wall with an intermittently heated room [28] was between about 0.2 to 1.5 W/m^{2}K during the non-heating period. At continuous heating, the average CHTC was 1.6 W/m^{2}K, with a variation between maximum-minimum values of about 1.5 W/m^{2}K. CHTC for the temperature difference below 2 K was between 0.2 to 2.5 W/m^{2}K. The author did not provide any empirical correlation for h_c, and only compared averaged values with that recommended by the relevant standards. This significant dispersion of experimental results in a real building was also confirmed in [53]. CHTC was measured in that study at a wall with a radiator at the back wall and no ventilation, and varied from approximately 0.3 to 5.0 W/m^{2}K at ΔT < 2 °C. The dominant results for a wall temperature measured at half of its height ranged between 0.5 and 2.0 W/m^{2}K.

CHTC calculated from the 14 correlations presented in Section 2.5 varied from 1.305 W/m^{2}K (Equation (26)) to 3.328 W/m^{2}K (Equation (24)). In all cases, it was greater than measurements by 1.153 W/m^{2}K (Figure 10). The average value from all models was h_c = 2.024 W/m^{2}K and was greater than measurements by 75.6%. The results closest to the measured value were obtained for models given by the Equations (26), (31), (33) and (38) and they were greater than measurements by 13.4%, 14.9%, 10.2% and 17.5%, respectively.

Six correlations, given by Equations (26), (30), (31), (33), (35) and (38), included geometrical parameters of the considered wall (hydraulic diameter or height). The relevant columns in Figure 9 were filled with hatching. CHTC in this group of models varied from 1.271 W/m^{2}K (Equation (33)) to 1.829 W/m^{2}K (Equation (35)), with an average of 1.313 W/m^{2}K. The remaining eight models produced results from 1.912 W/m^{2}K (Equation (34)) to 3.328 W/m^{2}K (Equation (24)), with an average of 2.446 W/m^{2}K, i.e., more than twice the measured CHTC. The results presented here indicate that better compliance with measurements was obtained for models that take into account the geometry of the wall.

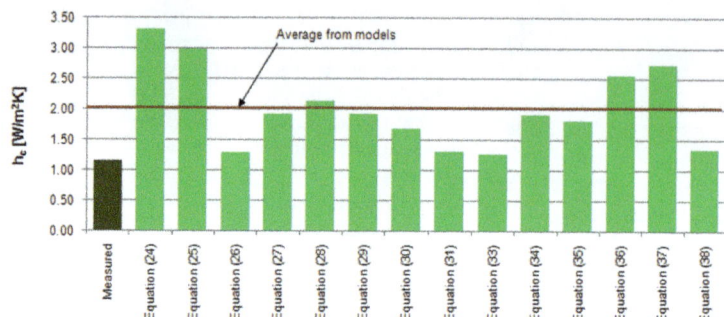

Figure 10. Average CHTC for different models for measurement conditions.

The theoretical calculation of CHTC revealed that various models provide similar results for ΔT up to 10 K (Figure 11). These can be gathered into several groups. The first consists of models given by Equations (26), (31), (33) and (38). All except for Equation (31) have an $(\Delta T/H)^{1/4}$ element. In Equation (31), $(\Delta T)^{1/6}$ was used along with $L^{-1/2}$ and resulted in a slightly lower rise in CHTC with ΔT. Similarities can also distinguished between models given by Equations (27) and (28), by (36) and (37), and by (29), (30), (34) and (35).

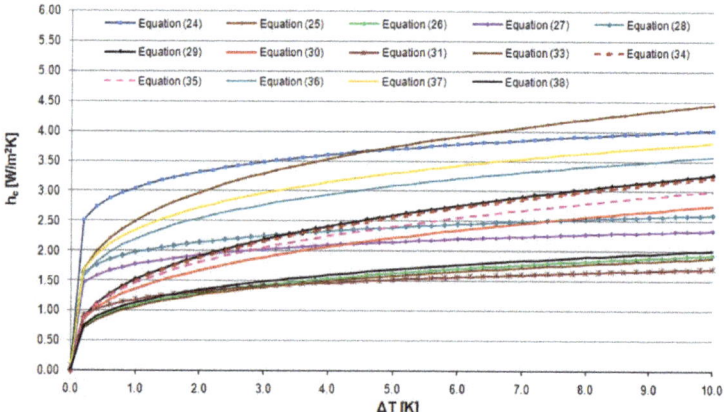

Figure 11. CHTC for different models.

All the models studied here had low coefficients of determination against measurements (Table 4). Their value in the case of the 10 min sampling time did not exceed $R^2 = 0.0437$ for the linear regression equation. In the next step, hourly averaged values of CHTCs (the typical length of the simulation time step in the building simulation tools) were computed from 10 min variables. The R^2 increase was negligible, in the third decimal place.

Table 4. Convective heat transfer coefficients from presented models and coefficients of the determination referred to measured values.

Model (Equation)	h_c [W/m²K]	R^2 [-]	AAE [%]	ABE [%]	MAE [W/m²K]	MBE [W/m²K]
24	3.328	0.0409	298	298	2.18	2.18
25	2.999	0.0426	258	258	1.86	1.86
26	1.305	0.0426	70	56	0.42	0.16
27	1.938	0.0409	132	132	0.80	0.79
28	2.153	0.0409	158	157	1.01	1.00
29	1.933	0.0437	131	131	0.81	0.79
30	1.692	0.0437	105	102	0.61	0.55
31	1.324	0.0415	72	58	0.42	0.18
33	1.271	0.0426	68	52	0.41	0.12
34	1.912	0.0437	129	128	0.79	0.77
35	1.829	0.0436	119	118	0.71	0.69
36	2.563	0.0421	206	206	1.42	1.42
37	2.738	0.0421	227	227	1.60	1.60
38	1.355	0.0426	74	62	0.43	0.21

The results presented here show significant differences between measurements and models. The best degree of convergence between predicted and measured results, given by AAE, was obtained in models given by Equations (26), (31), (33) and (38). MAE ranged from 0.41 W/m²K (Equation (33)) to 2.18 W/m²K (Equation (24)), showing relatively high deviations from the modelled CHTC versus measurements. The average bias in the models given by MBE varied from 0.12 W/m²K (Equation (33)) to 2.18 W/m²K (Equation (24)). From these results, it can be stated that the best match with measurements was obtained by the model of Fohanno and Polidori given by Equation (33).

Such discrepancies were also reported in other studies. For example, Kalema and Haapala [7] analysed the impact of interior heat transfer coefficient models on the thermal dynamics of a two-room test cell with radiator heating. Experimental results from the tests under steady-state and dynamic conditions were compared with that simulated in the thermal analysis program TASE. ASHRAE, Alamdari and Hammond, and Khalifa and Marshall correlation equations were used in the calculations of convective heat transfer coefficients. Measured air and surface temperatures were within the simulated minimum-maximum range of 4 °C. Only the window surface temperature was 0.7 °C greater than the calculated maximum temperature. All calculated heat fluxes were within the 10% error band of the measured values.

The discrepancies presented here may have an impact on simulation results when using computer tools for the dynamic simulation of buildings [10]. Commonly in such cases, constant values of internal CHTC and RHTC are assumed. If vertical walls are of interest, in the first case the values within the range from 2.0 W/m²K to 5.0 W/m²K [87–89] are met.

In the energy auditing of buildings, when monthly calculation methods are applied, constant values of total combined convective and radiative heat transfer coefficient (or internal surface resistance) are normally used [90–92].

In contrast to measured RHTC, it is not easy to present any recommendations on CHTC values to be used based on this study. Further detailed analyses are needed to find and minimise measurement uncertainties. The presented analysis shows that the most important contribution to CHTC uncertainty is the measurement of the temperature difference between internal air and wall surface.

4. Conclusions

In the present study, convective (CHTC) and radiative (RHTC) heat transfer coefficients for the internal surface of a vertical wall were calculated from measurements. For comparative purposes, in the first case, 14 correlations were used to obtain CHTC from temperature difference and, in several cases, the height of a wall.

The results showed significant differences between measurements and mathematical models. This was likely because in the presente study measurements were taken during the everyday use of a building. Hence, it was not possible to obtain the stable thermal conditions possible in the test chambers used in other experimental studies. Moreover, the internal surfaces of the test chambers were covered with materials different from those used in the building. For example, in the study of Khalifa and Marshall [10] authors used aluminium foil on the internal and external surfaces of the external partitions of their climatic chamber. This was employed to minimise the effect of longwave radiation exchange on the temperature and heat flux measurements, and meant that only convective heat transfer was studied. In the present study, combined convective and radiative heat transfer was considered.

It should also be noted that aluminium foil has different physical properties (especially roughness influencing the flow of air) than the cement-lime plaster used as the surface layers of ceramic walls. The flow of air could be also influenced because of stairs located near the sensors. The stairs are about 2 cm away from the wall and their construction allows air to flow, but its presence cannot be excluded from the measurement results.

Model quality was estimated using four statistical goodness-of-fit criteria: AAE, ABE, MAE and MBE. On this basis, the model of Fohanno and Polidori was chosen as the best, with AAE = 68%, ABE = 52%, MAE = 0.41 W/m^{2}K and MBE = 0.12 W/m^{2}K. The resulting h_c = 1.217 W/m^{2}K. The worst model was that of Wilkes and Peterson, for which AAE = 298%, ABE = 298%, MAE = 2.18 W/m^{2}K, MBE = 2.18 W/m^{2}K and h_c = 3.328 W/m^{2}K.

The experiment could be repeated over an extended period to avoid possible temporary disruptions and with the addition of airflow velocity measurement near the wall. Moreover, a shorter sampling time could be used and then time averaging in longer periods could be applied for comparison with hourly values calculated from other models.

The study showed, however, that calculated total surface resistance is close to that recommended by EN ISO 6946 standard, commonly used in energy auditing and certification of buildings. However, it is a rather crude estimation applied in less accurate, annual or monthly methods.

Because of the noticeable impact of a temperature difference between internal air and wall surface on CHTC uncertainty, in future studies, emphasis should be given to this issue.

Funding: This research received no external funding.

Conflicts of Interest: The authors declare no conflict of interest.

Symbols

h_c	convective heat transfer coefficient, W/m^{2}K
h_m	measured value of heat transfer coefficient, W/m^{2}K
h_p	predicted (by the certain correlation) value of heat transfer coefficient, W/m^{2}K
h_r	radiative heat transfer coefficient, W/m^{2}K
k	coverage factor, —
m	the total number of measurement samples, —
q_c	convective heat flux density, W/m^2
q_r	radiative heat flux density, W/m^2
q_w	wall conductive heat flux density, W/m^2
u_c	combined uncertainty, —
R^2	coefficient of determination, —
T_i	internal air temperature, °C
T_r	mean radiant temperature, °C
T_s	wall surface temperature, °C

U	expanded uncertainty, —
ε	surface emissivity, —
σ	Stefan–Boltzmann constant, $\sigma = 5.6697 \cdot 10^{-8}$ W/m^{2}K^4
AAE	average absolute error, %
ABE	average biased error, %
MAE	mean absolute error, W/m^{2}K
MBE	mean bias error, W/m^{2}K

References

1. Zaim, A.; Aissa, A.; Mebarek-Oudina, F.; Mahanthesh, B.; Lorenzini, G.; Sahnoun, M.; El Ganaoui, M. Galerkin finite element analysis of magneto-hydrodynamic natural convection of Cu-water nanoliquid in a baffled U-shaped enclosure. *Propuls. Power Res.* **2020**, *9*, 383–393. [CrossRef]
2. Abo-Dahab, S.M.; Abdelhafez, M.A.; Mebarek-Oudina, F.; Bilal, S.M. MHD Cassonnanofluid flow over nonlinearly heated porous medium in presence of extending surface effect with suction/injection. *Indian J. Phys.* **2021**, 1–15. [CrossRef]
3. Shafiq, A.; Mebarek-Oudina, F.; Sindhu, T.N.; Abidi, A. A study of dual stratification on stagnation point Walters'B nanofluid flow via radiative Riga plate: A statistical approach. *Eur. Phys. J. Plus* **2021**, *136*, 407. [CrossRef]
4. Palacio-Caro, I.D.; Alvarado-Torres, P.N.; Cardona-Sepúlveda, L.F. Numerical Simulation of the Flow and Heat Transfer in an Electric Steel Tempering Furnace. *Energies* **2020**, *13*, 3655. [CrossRef]
5. Markiewicz-Zahorski, P.; Rucińska, J.; Fedorczak-Cisak, M.; Zielina, M. Building Energy Performance Analysis after Changing Its Form of Use from an Office to a Residential Building. *Energies* **2021**, *14*, 564. [CrossRef]
6. Kwiatkowski, J.; Rucińska, J. Estimation of Energy Efficiency Class Limits for Multi-Family Residential Buildings in Poland. *Energies* **2020**, *13*, 6234. [CrossRef]
7. Kalema, T.; Haapala, T. Effect of interior heat transfer coefficients on thermal dynamics and energy consumption. *Energy Build.* **1995**, *22*, 101–113. [CrossRef]
8. Lomas, K.J. The U.K. applicability study: An evaluation of thermal simulation programs for passive solar house design. *Build. Environ.* **1996**, *31*, 197–206. [CrossRef]
9. Domínguez-Muñoz, F.; Cejudo-López, J.M.; Carrillo-Andrés, A. Uncertainty in peak cooling load calculations. *Energy Build.* **2010**, *42*, 1010–1018. [CrossRef]
10. Obyn, S.; van Moeseke, G. Variability and impact of internal surfaces convective heat transfer coefficients in the thermal evaluation of office buildings. *Appl. Therm. Eng.* **2015**, *87*, 258–272. [CrossRef]
11. Michalak, P.; Szczotka, K.; Szymiczek, J. Energy Effectiveness or Economic Profitability? A Case Study of Thermal Modernization of a School Building. *Energies* **2021**, *14*, 1973. [CrossRef]
12. Walikewitz, N.; Janicke, B.; Langner, M.; Meier, F.; Endlicher, W. The difference between the mean radiant temperature and the air temperature within indoor environments: A case study during summer conditions. *Build. Environ.* **2015**, *84*, 151–161. [CrossRef]
13. Filippín, C.; Beascochea, A. Performance assessment of low-energy buildings in central Argentina. *Energy Build.* **2007**, *39*, 546–557. [CrossRef]
14. Bueno, B.; Norford, L.; Pigeon, G.; Britter, R. A resistance-capacitance network model for the analysis of the interactions between the energy performance of buildings and the urban climate. *Build. Environ.* **2012**, *54*, 116–125. [CrossRef]
15. Herwig, H. What Exactly is the Nusselt Number in Convective Heat Transfer Problems and are There Alternatives? *Entropy* **2016**, *18*, 198. [CrossRef]
16. Bienvenido-Huertas, D.; Bermúdez, J.; Moyano, J.J.; Marín, D. Influence of ICHTC correlations on the thermal characterization of façades using the quantitative internal infrared thermography method. *Build. Environ.* **2019**, *149*, 515–525. [CrossRef]
17. Saunders, O.A. The effect of pressure upon natural convection in air. *Proc. R. Soc. Lond.* **1936**, *157*, 278–291. [CrossRef]
18. Saunders, O.A. Natural convection in liquids. *Proc. R. Soc. Lond.* **1939**, *172*, A17255–A17271. [CrossRef]
19. Fishenden, M.; Saunders, O.A. *An Introduction to Heat Transfer*; Clarendon Press: Oxford, UK, 1950.
20. McAdams, W.H. *Heat Transmission*; McGraw-Hill: New York, NY, USA, 1954.
21. Lewandowski, W.M.; Kubski, P. Effect of the use of the balance and gradient methods as a result of experimental investigations of natural convection action with regard to the conception and construction of measuring apparatus. *Wärme- Und Stoffübertragung* **1984**, *18*, 247–256. [CrossRef]
22. Lewandowski, W.M. Natural convection heat transfer from plates of finite dimensions. *Int. J. Heat Mass Transf.* **1991**, *34*, 875–885. [CrossRef]
23. Fujii, T.; Imura, H. Natural-convection heat transfer from a plate with arbitrary inclination. *Int. J. Heat Mass Transf.* **1972**, *15*, 755–767. [CrossRef]
24. Churchill, S.W.; Usagi, R. A general expression for the correlation of rates of transfer and other phenomena. *AIChE J.* **1972**, *18*, 1121–1128. [CrossRef]
25. Khalifa, A.J.N.; Marshall, R.H. Validation of heat transfer coefficients on interior building surfaces using a real-sized indoor test cell. *Int. J. Heat Mass Transf.* **1990**, *33*, 2219–2236. [CrossRef]
26. Alamdari, F.; Hammond, G.P. Improved data correlations for buoyancy-driven convection in rooms. *Build. Serv. Eng. Res. Technol.* **1983**, *4*, 106–112. [CrossRef]

27. Awbi, H.B.; Hatton, A. Natural convection from heated room surfaces. *Energy Build.* **1999**, *30*, 233–244. [CrossRef]
28. Delaforce, S.R.; Hitchin, E.R.; Watson, D.M.T. Convective heat transfer at internal surfaces. *Build. Environ.* **1993**, *28*, 211–220. [CrossRef]
29. Sukamto, D.; Siroux, M.; Gloriant, F. Hot Box Investigations of a Ventilated Bioclimatic Wall for NZEB Building Façade. *Energies* **2021**, *14*, 1327. [CrossRef]
30. Shinoda, J.; Kazanci, O.B.; Tanabe, S.; Olesen, B.W. A review of the surface heat transfer coefficients of radiant heating and cooling systems. *Build. Environ.* **2019**, *159*, 106156. [CrossRef]
31. Cholewa, T.; Rosiński, M.; Spik, Z.; Dudzińska, M.R.; Siuta-Olcha, A. On the heat transfer coefficients between heated/cooled radiant floor and room. *Energy Build.* **2013**, *66*, 599–606. [CrossRef]
32. Cholewa, T.; Anasiewicz, R.; Siuta-Olcha, A.; Skwarczynski, M.A. On the heat transfer coefficients between heated/cooled radiant ceiling and room. *Appl. Therm. Eng.* **2017**, *117*, 76–84. [CrossRef]
33. Acikgoz, O.; Kincay, O. Experimental and numerical investigation of the correlation between radiative and convective heat-transfer coefficients at the cooled wall of a real-sized room. *Energy Build.* **2015**, *108*, 257–266. [CrossRef]
34. Koca, A.; Gemici, Z.; Topacoglu, Y.; Cetin, G.; Acet, R.C.; Kanbur, B.B. Experimental investigation of heat transfer coefficients between hydronic radiant heated wall and room. *Energy Build.* **2014**, *82*, 211–221. [CrossRef]
35. Guo, R.; Heiselberg, P.; Hu, Y.; Johra, H.; Zhang, C.; Jensen, R.L.; Jønsson, K.T.; Peng, P. Experimental investigation of convective heat transfer for night cooling with diffuse ceiling ventilation. *Build. Environ.* **2021**, *193*, 107665. [CrossRef]
36. Guo, R.; Heiselberg, P.; Hu, Y.; Johra, H.; Zhang, C.; Jensen, R.L.; Jønsson, K.T.; Peng, P. Experimental investigation of convective heat transfer for night ventilation in case of mixing ventilation. *Build. Environ.* **2021**, *193*, 107670. [CrossRef]
37. Peeters, L.; Beausoleil-Morrison, I.; Novoselac, A. Internal convective heat transfer modeling: Critical review and discussion of experimentally derived correlations. *Energy Build.* **2011**, *43*, 2227–2239. [CrossRef]
38. Camci, M.; Karakoyun, Y.; Acikgoz, O.; Dalkilic, A.S. A comparative study on convective heat transfer in indoor applications. *Energy Build.* **2021**, *242*, 110985. [CrossRef]
39. Khalifa, A.J.N. Natural convective heat transfer coefficient–A review: I. Isolated vertical and horizontal surfaces. *Energy Convers. Manag.* **2001**, *42*, 491–504. [CrossRef]
40. Khalifa, A.J.N. Natural convective heat transfer coefficient–A review: II. Surfaces in two- and three-dimensional enclosures. *Energy Convers. Manag.* **2001**, *42*, 505–517. [CrossRef]
41. Dagtekin, I.; Oztop, H.F. Natural convection heat transfer by heated partitions within enclosure. *Int. Commun. Heat Mass Transf.* **2001**, *28*, 823–834. [CrossRef]
42. Vollaro, A.D.L.; Galli, G.; Vallati, A. CFD Analysis of Convective Heat Transfer Coefficient on External Surfaces of Buildings. *Sustainability* **2015**, *7*, 9088–9099. [CrossRef]
43. Zhong, W.; Zhang, T.; Tamura, T. CFD Simulation of Convective Heat Transfer on Vernacular Sustainable Architecture: Validation and Application of Methodology. *Sustainability* **2019**, *11*, 4231. [CrossRef]
44. Berrabah, S.; Moussa, M.O.; Bakhouya, M. 3D Modeling of the Thermal Transfer through Precast Buildings Envelopes. *Energies* **2021**, *14*, 3751. [CrossRef]
45. Maragkos, G.; Beji, T. Review of Convective Heat Transfer Modelling in CFD Simulations of Fire-Driven Flows. *Appl. Sci.* **2021**, *11*, 5240. [CrossRef]
46. Kaminska, A. Impact of Heating Control Strategy and Occupant Behavior on the Energy Consumption in a Building with Natural Ventilation in Poland. *Energies* **2019**, *12*, 4304. [CrossRef]
47. Lundström, L.; Akander, J.; Zambrano, J. Development of a Space Heating Model Suitable for the Automated Model Generation of Existing Multifamily Buildings—A Case Study in Nordic Climate. *Energies* **2019**, *12*, 485. [CrossRef]
48. EN 1264. *Water Based Surface Embedded Heating and Cooling Systems*; SIST: Bratislava, Slovakia, 2011.
49. ISO 11855. *Building Environment Design—Design, Dimensioning, Installation and Control of Embedded Radiant Heating and Cooling Systems*; ISO: Geneva, Switzerland, 2012.
50. EN ISO 6946. *Building Components and Building Elements—Thermal Resistance and Thermal Transmittance–Calculation Methods*; ISO: Geneva, Switzerland, 2007.
51. ASHRAE. *Handbook HVAC Fundamentals*; ASHRAE: New York, NY, USA, 2009.
52. François, A.; Ibos, L.; Feuillet, V.; Meulemans, J. Novel in situ measurement methods of the total heat transfer coefficient on building walls. *Energy Build.* **2020**, *219*, 110004. [CrossRef]
53. Wallentén, P. Convective heat transfer coefficients in a full-scale room with and without furniture. *Build. Environ.* **2001**, *36*, 743–751. [CrossRef]
54. Helm, I.; Jalukse, L.; Leito, I. Measurement Uncertainty Estimation in Amperometric Sensors: A Tutorial Review. *Sensors* **2010**, *10*, 4430–4455. [CrossRef]
55. Ohlsson, K.E.A.; Östin, R.; Olofsson, T. Accurate and robust measurement of the external convective heat transfer coefficient based on error analysis. *Energy Build.* **2016**, *117*, 83–90. [CrossRef]
56. Chen, J.; Chen, C. Uncertainty Analysis in Humidity Measurements by the Psychrometer Method. *Sensors* **2017**, *17*, 368. [CrossRef]
57. Chen, L.-H.; Chen, J.; Chen, C. Effect of Environmental Measurement Uncertainty on Prediction of Evapotranspiration. *Atmosphere* **2018**, *9*, 400. [CrossRef]

58. Jing, H.; Quan, Z.; Zhao, Y.; Wang, L.; Ren, R.; Liu, Z. Thermal Performance and Energy Saving Analysis of Indoor Air–Water Heat Exchanger Based on Micro Heat Pipe Array for Data Center. *Energies* **2020**, *13*, 393. [CrossRef]
59. Ali, A.; Cocchi, L.; Picchi, A.; Facchini, B. Experimental Determination of the Heat Transfer Coefficient of Real Cooled Geometry Using Linear Regression Method. *Energies* **2021**, *14*, 180. [CrossRef]
60. Ficker, T. Virtual emissivities of infrared thermometers. *Infrared Phys. Technol.* **2021**, *114*, 103656. [CrossRef]
61. Chen, H.-Y.; Chen, C. Determining the emissivity and temperature of building materials by infrared thermometer. *Constr. Build. Mater.* **2016**, *126*, 130–137. [CrossRef]
62. Höser, D.; Wallimann, R.; von Rohr, P.R. Uncertainty Analysis for Emissivity Measurement at Elevated Temperatures with an Infrared Camera. *Int. J. Thermophys.* **2016**, *37*, 14. [CrossRef]
63. Zribi, A.; Barthès, M.; Bégot, S.; Lanzetta, F.; Rauch, J.Y.; Moutarlier, V. Design, fabrication and characterization of thin film resistances for heat flux sensing application. *Sens. Actuators A Phys.* **2016**, *245*, 26–39. [CrossRef]
64. Heat Flux Plate/Heat Flux Sensor HFP01 & HFP03. User Manual. Hukseflux Manual v1721. Available online: https://www.hukseflux.com/uploads/product-documents/HFP01_HFP03_manual_v1721.pdf (accessed on 25 August 2021).
65. Evangelisti, L.; Guattari, C.; Gori, P.; de LietoVollaro, R.; Asdrubali, F. Experimental investigation of the influence of convective and radiative heat transfers on thermal transmittance measurements. *Int. Commun. Heat Mass Transf.* **2016**, *78*, 214–223. [CrossRef]
66. Bienvenido-Huertas, D.; Moyano, J.; Rodríguez-Jiménez, C.E.; Muñoz-Rubio, A.; Bermúdez Rodríguez, F.J. Quality Control of the Thermal Properties of Superstructures in Accommodation Spaces in Naval Constructions. *Sustainability* **2020**, *12*, 4194. [CrossRef]
67. Fohanno, S.; Polidori, G. Modelling of natural convective heat transfer at an internal surface. *Energy Build.* **2006**, *38*, 548–553. [CrossRef]
68. Musy, M.; Wurtz, E.; Winkelmann, F.; Allard, F. Generation of a zonal model to simulate natural convection in a room with a radiative/convective heater. *Build. Environ.* **2001**, *36*, 589–596. [CrossRef]
69. Rogers, G.F.C.; Mayhew, Y.R. *Engineering Thermodynamics. Work and Heat Transfer*; Pearson Education Ltd.: London, UK, 1992.
70. Michalak, P. The development and validation of the linear time varying Simulink-based model for the dynamic simulation of the thermal performance of buildings. *Energy Build.* **2017**, *141*, 333–340. [CrossRef]
71. Yin, C.-Y. Prediction of higher heating values of biomass from proximate and ultimate analyses. *Fuel* **2011**, *90*, 1128–1132. [CrossRef]
72. Qian, X.; Lee, S.; Soto, A.-M.; Chen, G. Regression Model to Predict the Higher Heating Value of Poultry Waste from Proximate Analysis. *Resources* **2018**, *7*, 39. [CrossRef]
73. Lauster, M.; Teichmann, J.; Fuchs, M.; Streblow, R.; Mueller, D. Low order thermal network models for dynamic simulations of buildings on city district scale. *Build. Environ.* **2014**, *73*, 223–231. [CrossRef]
74. Cui, Y.; Xie, J.; Liu, J.; Xue, P. Experimental and Theoretical Study on the Heat Transfer Coefficients of Building External Surfaces in the Tropical Island Region. *Appl. Sci.* **2019**, *9*, 1063. [CrossRef]
75. Gaši, M.; Milovanović, B.; Gumbarević, S. Comparison of Infrared Thermography and Heat Flux Method for Dynamic Thermal Transmittance Determination. *Buildings* **2019**, *9*, 132. [CrossRef]
76. Samardzioska, T.; Apostolska, R. Measurement of Heat-Flux of New Type Façade Walls. *Sustainability* **2016**, *8*, 1031. [CrossRef]
77. De Rubeis, T.; Muttillo, M.; Nardi, I.; Pantoli, L.; Stornelli, V.; Ambrosini, D. Integrated Measuring and Control System for Thermal Analysis of Buildings Components in Hot Box Experiments. *Energies* **2019**, *12*, 2053. [CrossRef]
78. Lihakanga, R.; Ding, Y.; Medero, G.M.; Chapman, S.; Goussetis, G. A High-Resolution Open Source Platform for Building Envelope Thermal Performance Assessment Using a Wireless Sensor Network. *Sensors* **2020**, *20*, 1755. [CrossRef]
79. Tejedor, B.; Gaspar, K.; Casals, M.; Gangolells, M. Analysis of the Applicability of Non-Destructive Techniques to Determine In Situ Thermal Transmittance in Passive House Façades. *Appl. Sci.* **2020**, *10*, 8337. [CrossRef]
80. Ambrožová, K.; Hrbáček, F.; Láska, K. The Summer Surface Energy Budget of the Ice-Free Area of Northern James Ross Island and Its Impact on the Ground Thermal Regime. *Atmosphere* **2020**, *11*, 877. [CrossRef]
81. Mosayebidorcheh, S.; Ganji, D.D.; Farzinpoor, M. Approximate solution of the nonlinear heat transfer equation of a fin with the power-law temperature-dependent thermal conductivity and heat transfer coefficient. *Propuls. Power Res.* **2014**, *3*, 41–47. [CrossRef]
82. Qian, X.; Xue, J.; Yang, Y.; Lee, S.W. Thermal Properties and Combustion-Related Problems Prediction of Agricultural Crop Residues. *Energies* **2021**, *14*, 4619. [CrossRef]
83. Ficco, G.; Frattolillo, A.; Malengo, A.; Puglisi, G.; Saba, F.; Zuena, F. Field verification of thermal energy meters through ultrasonic clamp-on master meters. *Measurement* **2020**, *151*, 107152. [CrossRef]
84. Ficco, G.; Iannetta, F.; Ianniello, E.; d'AmbrosioAlfano, F.R.; Dell'Isola, M. U-value in situ measurement for energy diagnosis of existing buildings. *Energy Build.* **2015**, *104*, 108–121. [CrossRef]
85. Gil, B.; Rogala, Z.; Dorosz, P. Pool Boiling Heat Transfer Coefficient of Low-Pressure Glow Plasma Treated Water at Atmospheric and Reduced Pressure. *Energies* **2020**, *13*, 69. [CrossRef]
86. Michalak, P. Annual Energy Performance of an Air Handling Unit with a Cross-Flow Heat Exchanger. *Energies* **2021**, *14*, 1519. [CrossRef]
87. Kämpf, J.H.; Robinson, D. A simplified thermal model to support analysis of urban resource flows. *Energy Build.* **2007**, *39*, 445–453. [CrossRef]

88. Ben-Nakhi, A.E.; Aasem, E.O. Development and integration of a user friendly validation module within whole building dynamic simulation. *Energy Convers. Manag.* **2003**, *44*, 53–64. [CrossRef]
89. Mahdaoui, M.; Hamdaoui, S.; Ait Msaad, A.; Kousksou, T.; El Rhafiki, T.; Jamil, A.; Ahachad, M. Building bricks with phase change material (PCM): Thermal performances. *Constr. Build. Mater.* **2021**, *269*, 121315. [CrossRef]
90. Dobrzycki, A.; Kurz, D.; Mikulski, S.; Wodnicki, G. Analysis of the Impact of Building Integrated Photovoltaics (BIPV) on Reducing the Demand for Electricity and Heat in Buildings Located in Poland. *Energies* **2020**, *13*, 2549. [CrossRef]
91. Babiarz, B.; Szymański, W. Introduction to the Dynamics of Heat Transfer in Buildings. *Energies* **2020**, *13*, 6469. [CrossRef]
92. Jezierski, W.; Sadowska, B.; Pawłowski, K. Impact of Changes in the Required Thermal Insulation of Building Envelope on Energy Demand, Heating Costs, Emissions, and Temperature in Buildings. *Energies* **2021**, *14*, 56. [CrossRef]

Article

Design and Energy Performance Analysis of a Hotel Building in a Hot and Dry Climate: A Case Study

Sultan Kobeyev [1], Serik Tokbolat [2,*] and Serdar Durdyev [3,*]

[1] Department of Civil and Environmental Engineering, Nazarbayev University, 53 Kabanbay Batyr Ave., Nur-Sultan 010000, Kazakhstan; sultan.kobeyev@nu.edu.kz
[2] School of Architecture, Design and the Built Environment, Nottingham Trent University, 50 Shakespeare Street, Nottingham Trent University, Nottingham NG1 4FQ, UK
[3] Department of Engineering and Architectural Studies, Ara Institute of Canterbury, 130 Madras Street, Christchurch 8011, New Zealand
* Correspondence: serik.tokbolat@ntu.ac.uk (S.T.); Serdar.Durdyev@ara.ac.nz (S.D.)

Citation: Kobeyev, S.; Tokbolat, S.; Durdyev, S. Design and Energy Performance Analysis of a Hotel Building in a Hot and Dry Climate: A Case Study. *Energies* **2021**, *14*, 5502. https://doi.org/10.3390/en14175502

Academic Editor: Paulo Santos

Received: 12 August 2021
Accepted: 1 September 2021
Published: 3 September 2021

Publisher's Note: MDPI stays neutral with regard to jurisdictional claims in published maps and institutional affiliations.

Copyright: © 2021 by the authors. Licensee MDPI, Basel, Switzerland. This article is an open access article distributed under the terms and conditions of the Creative Commons Attribution (CC BY) license (https://creativecommons.org/licenses/by/4.0/).

Abstract: In times of unprecedented climate change and energy scarcity, the design and delivery of energy-efficient and sustainable buildings are of utmost importance. This study aimed to design a hotel building for hot and dry climate conditions and perform its energy performance analysis using energy simulation tools. The model of the hotel building was constructed by a graphical tool OpenStudio and EnergyPlus following the ASHRAE Standard 90.1. To reduce the energy demand of the hotel, parametric analysis was conducted and building envelope parameters such as the thickness of insulation layer in the exterior wall and the roof, thermal conductivity of insulation layer, rate of infiltration, U-factor of windows, and thermal resistance of air gap in the interior walls (R-value), window-to-wall ratio, and orientation of the building were tested and the impact on the energy use of the building was analyzed. It was found that most of the design assumptions based on the ASHRAE standard were already optimal for the considered locality, however, were still optimized further to reach the highest efficiency level. Apart from this, three sustainable technologies—thermochromic windows, phase change materials, and solar panels—were incorporated into the building and their energy consumption reduction potential was estimated by energy simulations. Cumulatively, these sustainable technologies were able to reduce the total energy use from 2417 GJ to 1593 GJ (i.e., by 824 GJ or 34%). Calculation of payback period and return on investments showed that thermochromic windows and solar panels have relatively short payback periods and high return on investments, whereas PCM was found to be economically nonviable. The findings of this study are deemed to be useful for designing a sustainable and energy-efficient hotel building in a sub-tropical climate. However, the overall design and energy performance analysis algorithm could be used for various buildings with varying climate conditions.

Keywords: energy performance; design parameters; energy simulation; building envelope

1. Introduction

Globally, buildings account for about one-third of the final energy use, and the major part of that energy is consumed, among others, by heating, ventilation, and air conditioning systems [1,2]. Thus, buildings offer a significant potential to reduce the energy consumed by the building sector and thus mitigate the effects of climate change and the ongoing energy crisis [3]. Therefore, it is imperative for all the new buildings to satisfy the highest standards of energy efficiency and to operate at the lowest possible energy consumption (EC) levels. The up-to-date research findings refer to a combination of measures that could increase the energy efficiency of the buildings such as, for example, design optimization and integration of sustainable energy technologies and sustainable materials [4]. Generally, solutions for achieving net-zero energy buildings (NZEB) can be broadly categorized into three groups—passive efficiency strategies, integration of renewables, and active energy technologies [5,6].

Design optimization includes addressing such measures as the window-to-wall ratio (WWR), building orientation, shadings, cool roofs, modifications to buildings' envelope and thermal mass, etc. There is a wide consensus among the scientific community that early actions taken with optimizing the above-mentioned parameters yield substantial improvements in the efficiency of buildings, which will eventually impact the sizing of HVAC equipment and help to achieve a higher-performing building [7–9]. In this study, the following design parameters were considered: building envelope optimization (thickness and conductivity of insulation layers, infiltration rate, U-factor of windows, and thermal resistance of interior walls), WWR, and building orientation.

Several studies have been conducted by researchers to determine the most energy-efficient options for optimization of building design at a preliminary stage. Tuhus-Dubrow and Krarti [10] used a genetic algorithm to determine optimum building shapes, orientation, wall and roof construction, window type, window area, foundation, infiltration rate, insulation thickness, and shading of residential buildings. Ascione et al. [11] optimize the annual energy consumption of buildings in the Mediterranean climate zone by using optical properties of external plastering, wall insulations properties, and composition of outer walls as design parameters. Wang and Wei [12], on the other hand, focused on the composition of walls, number, type, dimensions, and material of windows to optimize energy consumption and cost of office buildings using a quantum genetic algorithm. Similarly, Carlucci et al. [13] used multi-objective optimization by the non-dominated sorting genetic algorithm to enhance the thermal comfort of a building with U-values of walls, roofs, and floors as well as U-value and visible transmittance of windows as design variables.

Apart from design optimization, the overall energy efficiency and general sustainability levels of buildings were found to be improved by integrating various sustainable energy technologies and materials. Since the hotel buildings tend to have a great number of windows due to the nature of the room allocation as well as the potential impact on solar gain and overheating, the study paid significant attention to the selection of window types. The study also considered integrating phase change materials (PCMs) and solar energy technologies.

There is a wide consensus among the scientific community and industry practitioners that windows and skylights are building elements with the lowest energy efficiency [14–16]. Owing to their relatively smaller thickness compared to exterior walls and roofs, windows have a much higher rate of thermal heat transfer [16]. As there are already a number of heat generators inside the building, such as equipment and people, extra heat gain through windows is often unwanted, especially in moderate climates.

Thermochromic windows are one of the promising technologies intended to address this issue that has drawn particular attention among researchers in recent years [17]. These are a type of smart windows with a special glazing that can passively modulate infrared transmittance from sunlight and alter its optical and thermal properties depending on the intensity of solar heat gain [16]. By reflecting infrared and near-infrared lights while still transmitting visible light, thermochromic windows can reduce cooling energy demand by up to 81.7% and maintain healthy indoor conditions with visual and thermal comfort [16].

Thermochromic windows can be manufactured from various materials, including mercury (II) iodide, chromium (III) oxide, and lead (II) oxide, but so far, the most researched material with thermochromic properties is vanadium dioxide (VO_2) [14,16,18]. Traditionally, pure VO_2-based coatings were used for manufacturing thermochromic windows as this material provides adequate luminous transmittance and solar modulation properties. However, it has a number of shortcomings, such as relatively low stability of properties, cost inefficiency, and color alterations [16]. To resolve these issues, VO_2 is usually doped with additional chemical elements or various organic and inorganic nanomaterials.

Phase change materials (PCMs) are a group of materials that could be considered as sustainable materials that help with energy savings in buildings. PCMs have a high latent heat that can passively store and release thermal energy by undergoing a solid-liquid phase transition. By doing so, PCMs can decrease heating and cooling energy consumption of buildings, reduce indoor temperature fluctuation, shift peak times of energy demand and thus reduce grid load and assist intermittent renewable energy sources, and maintain comfortable indoor temperature conditions [19]. The model of PCM in EnergyPlus was analytically validated by Tabares-Velasco et al. [20], who states that for the accuracy of the model, (1) Conduction Finite Difference algorithm should be used, (2) the number of timesteps per hour should be higher than 20, and (3) space discretization value should be set to at least 3.

Usually, due to partial charging and discharging, PCMs are not exploited to their full potential. Moreover, PCMs cause an additional problem by discharging heat into indoor spaces during cooler times of the day [21]. Therefore, various means of assisting discharge of PCMs, such as natural ventilation, night ventilation, mechanical ventilation, were actively investigated in recent years. One of them is temperature-controlled natural ventilation demonstrated by Prabhakar et al. [21] and Pisello et al. [22], which is reported to significantly improve the energy efficiency of PCMs.

Solar panels or photovoltaic PV panels are devices made of semiconductors that convert energy from photons into electricity by displacing electrons from the semiconductor material's atoms. It has such advantages as carbon neutrality, low maintenance cost, and relatively low environmental footprint among others. However, it is intermittent and significantly dependent on ambient weather conditions, set aside such challenges as large space requirements and the necessity of energy storage systems. Nevertheless, solar energy technologies are generously subsidized in many parts of the world, which, together with technological advancements, help lower their capital costs and foster their widespread installation [23]. For instance, in the United States, prices of solar panels have dropped by half since 2008, while the capacity of installed panels increased by 35 times [24].

As it can be seen from the literature review, various studies address the impacts of design parameters on the energy use of buildings in general [10–13]. Some studies separately study the impacts of the integration of renewable energy technologies [5]. However, it was also noted that most articles are concerned about residential buildings and tend to consider only impacts of either design parameters or integration of sustainable energy technologies as standalone solutions. This study, however, is aimed to adopt a systems-based approach and to provide a combined energy performance analysis of both aspects at the same time.

For the purposes of this research, a hotel building was selected. Hotel buildings are quite common around the world and a significant number of them are located near holiday destinations which are known by their tropical and sub-tropical climate conditions [25]. They usually have high comfort levels and, thus, heavily rely on HVAC systems. This motivated the authors of the research to investigate the hotel buildings as their energy optimization and energy efficiency could potentially make a significant contribution to the overall sustainability of the built environment. Thus, this study conducted an energy performance analysis of a hotel building. The primary objectives of this study are:

(1) To optimize thermodynamic properties of a hotel's building envelope and compare the resultant performance with the baseline model configured as per ASHRAE 90.1.
(2) To investigate the extent to which PV panels, PCMs, and thermochromic windows can improve the energy efficiency of a hotel and subsequently check their economic feasibility.

The results of the study would contribute to the existing body of knowledge at least in two important ways. First of all, it will provide practical insights into the effect of envelope optimization and the performance of the above-mentioned sustainable technologies on the energy efficiency of a hotel building in the subtropical climate zone, which, to the best of the authors' knowledge, has not been well studied yet. Moreover, the following paper would suggest a methodology for optimization of building envelopes and green technologies at the preliminary stages of building design. Although a case study was shown for a high-rise hotel building only, the application of the methodological flow can readily be extended to other building types as well. A simulation model of a theoretical hotel building was created using graphical and energy simulation tools, and then six potential energy performance improvement measures were considered. Subsequently, a basic cost-benefit analysis was performed to evaluate the economic viability and financial attractiveness of the building optimization options. The findings of the study were collated to a set of recommendations for designing and building an energy-efficient hotel building in a sub-tropical hot and dry climate. The steps undertaken in this study are presented in Figure 1.

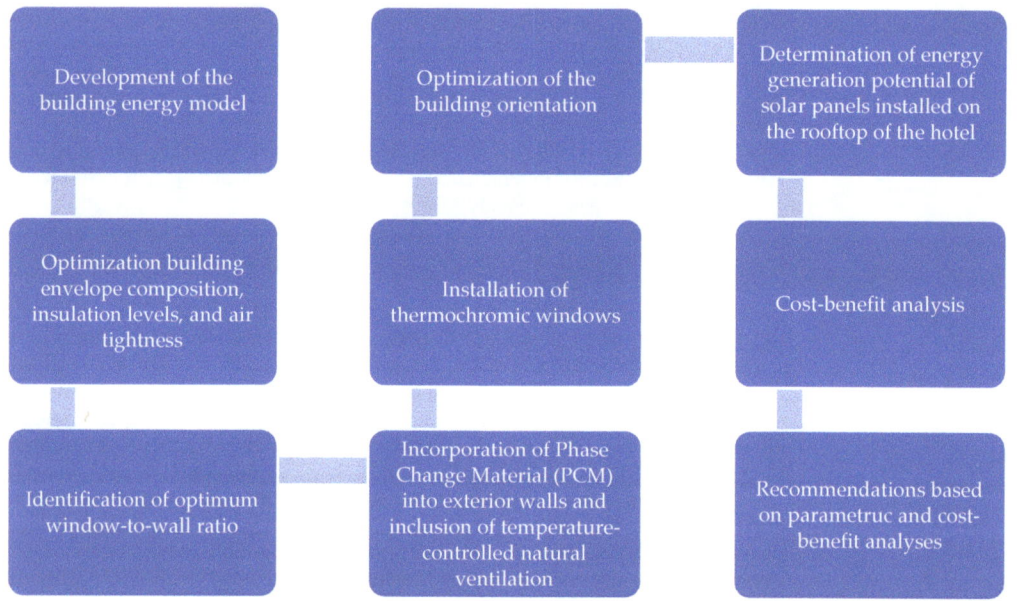

Figure 1. The flow of the research work.

2. Methodology

The adopted methodology in this study is based on energy performance simulation using a combination of graphical and energy simulation tools which were built based on the relevant standards and codes used in the United States. Building energy efficiency levels have to comply with International Energy Conservation Code. All the analysis presented in this section will be in accordance with ASHRAE [26]. To optimize building envelope characteristics, window-to-wall ratio, building orientation, and characteristics of three sustainable technologies considered in the following study, a simple parametric analysis will be performed and the optimization will be carried out sequentially considering each variable separately. The advantage of adopting this method is that it is simple and quick to perform, which may come at the cost of possibly overlooked interactions between some parameters, such as properties of the building envelope, building orientation, and PCM melting temperature. To account for this kind of interdependencies between various parameters, one may want to use more mathematically rigorous optimization methods, such as the genetic algorithm or stochastic gradient descent.

Development of the Building Energy Model

The energy model of the building was created using the OpenStudio plugin in SketchUp and improved in OpenStudio. The energy simulation of the model was then performed separately in EnergyPlus v9.3. For the validity of the PCM model, the number of timesteps per hour was set to 20 and the Conduction Finite Difference (CondFD) algorithm was used following the recommendations of Tabares-Velasco et al. [20].

Each floor was divided into separate spaces (for guest rooms, bathrooms, lobby, corridors, kitchens, etc.) and each room was set to be a distinct thermal zone. The material composition of walls, slabs, roof, and windows in the initial baseline model was assigned by using templates provided by ASHRAE Standard 90.1 [26] taking into account that the hotel is to be located in climate zone 3B according to the IECC Climate Zone Map [27]. The material composition of exterior walls and roof are shown below in Table 1.

Table 1. Material composition of exterior walls and roof.

	Thickness [m]	Conductivity [W/m·K]	Density [kg/m^3]	Specific Heat [J/kg·K]
		Exterior wall		
Stucco	0.0253	0.692	1858	837.0
Concrete	0.2032	1.311	2240	836.8
Wall Insulation	0.0623	0.049	265	836.8
Gypsum	0.0127	0.160	785	830.0
		Roof		
Roof Membrane	0.0095	0.160	1121	1460.0
Roof Insulation	0.1701	0.049	265	836.8

Moreover, space types and schedules for occupancy loads, lighting, and electric equipment were also assigned referring to ASHRAE Standard 90.1. For heating, cooling, and air conditioning of spaces in the hotel Packaged Terminal Air Conditioning (PTAC) system with direct expansion cooling and electric heating was selected due to the following reasons:

- Individual control and a separate air conditioning unit are needed for each hotel room
- The hotel comprises many small rooms and is located in a relatively moderate climate
- The surface area of the building is relatively large and thus different sides of the building may experience different air conditioning needs.

Thermostat set points of different space types have some variations, but for guest rooms, setpoints were selected to be 21 °C and 24 °C of mean air temperature for heating and cooling, respectfully, all days of the week and round-the-clock.

The model was initially built having all 24 floors (see Figure 2), but later, to reduce the simulation time, the number of residential floors was reduced and a zone multiplier of 19 was set for the eleventh floor. Floor and roof surfaces exposed because of this were set to be adiabatic. For simulation, the Conduction Transfer Function algorithm with 6 timesteps per hour was used for all retrofits except for PCMs.

In general, there is no exact scientific reason for choosing this particular type of building other than an attempt to make it representative of common hotel buildings. The hotel is rectangular with an aspect ratio of 2 with rooms arranged along its perimeter and an atrium at the center, which is a common practice for accommodating hotel floors with as many rooms as possible, while also ensuring that indoor areas are well lit.

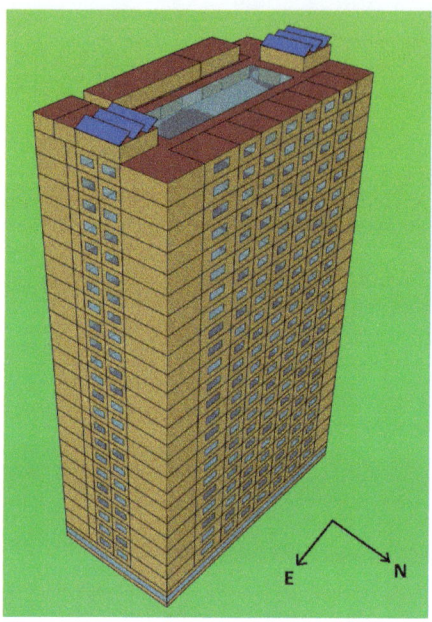

Figure 2. Energy simulation model of a hotel building.

3. Results and Discussions

3.1. Building Envelope

As was mentioned earlier, the baseline composition of exterior walls and roof were assigned based on the type of building and the climate zone. With this initial configuration, the building consumes an annual heating and cooling energy of 2557.82 GJ. To reduce this value, a parametric analysis was conducted to optimize some key parameters of the building envelope, namely thickness of insulation layer in the exterior wall and the roof ($D_{i,\,wall}$ and $D_{i,\,roof}$, respectively), the thermal conductivity of the insulation layer (k_i), rate of infiltration (I), U-factor of windows, and thermal resistance of air gap in the interior walls (R-value). All six variables control the rate of heat transfer through the envelope of a building and significantly influence its energy demand. Table 2 summarizes the range and increments of these parameters used in the parametric analyses:

Table 2. Parametric analyses for optimization of the building envelope.

Parameters	Values Used in the Parametric Analyses
$D_{i,\,wall}$, m	[0.005:0.005:0.075]
$D_{i,\,roof}$, m	[0.05:0.05:0.40]
k_i, W/m-K	[0.010:0.005:0.060]
I, m^3/s-m^2	[0.00005]^[0.0001: 0.0001: 0.0005]
U-factor, W/m^2-K	[0.3:0.5:5.8]
R-value, m^2-K/W	[0.1:0.05:0.4]^[0.4:0.1:0.8]^[1:1:4]

To reduce the number of simulations required while still accounting for interdependencies between these parameters, two cycles of "multi-objective" optimizations were conducted first involving only parameters of insulation layers, followed by the inclusion of infiltration, U-values of windows, and thermal resistance of air gap.

From Figure 3, it can be noticed that the original building envelope has already been quite energy-efficient and that in most cases, further alterations of building envelope properties tend to cause negative energy savings. The parametric analysis also reveals that for all six properties, higher values are preferable and the bottom line for all the analyses performed in this section is—the building should be well insulated yet adequately ventilated. The magnitude of energy consumption is the most sensitive to the U-factor of windows and thermal resistance (R-value). Ultimately, the building envelope was optimized as follows, Table 3.

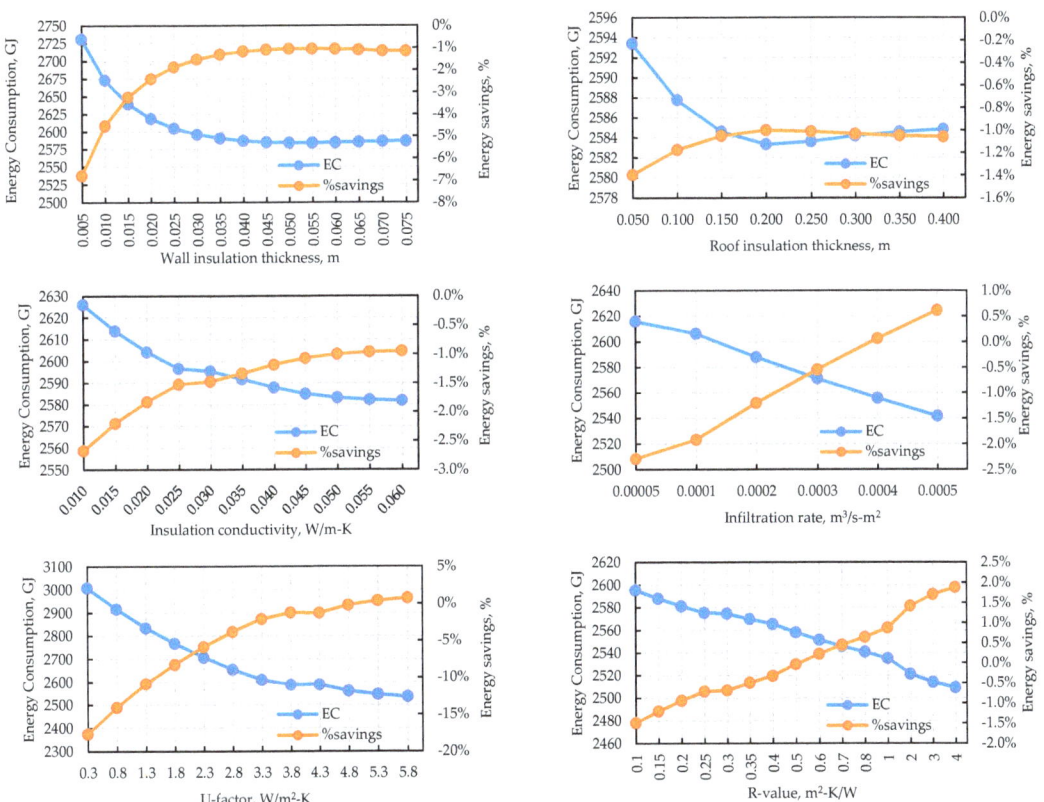

Figure 3. Parametric analysis of building envelope properties.

Table 3. Resultant building envelope parameters.

$D_{i,wall}$, m	$D_{i,roof}$, m	k_i, W/m-K	I, m³/s-m²	U-Factor, W/m²-K	R-Value, m²-K/W	EC_{total}, GJ	ES, GJ	%Savings
0.045	0.400	0.030	0.0005	5.8	4.0	2438.3	119.5	4.67%

Here, energy savings (ES) and the percentage-based savings (%savings) refer to energy savings relative to the baseline model expressed respectively in terms of absolute magnitude and percentage.

3.2. Window-to-Wall Ratio

Windows and skylights are usually components of building envelopes that have the least thermal resistance, therefore the window-to-wall ratio (WWR) plays a crucial role in improving buildings' energy efficiency. In the original design, the WWR of different sides of the building are non-uniform, but for the sake of simplicity, optimization will be carried out assuming that WWR will be the same for all sides of the hotel. Table 4 shows the original window-to-wall ratios:

Table 4. Original window-to-wall ratios of the building.

	Total	North (315°–45°)	East (45°–135°)	South (135°–225°)	West (225°–315°)
Gross Wall Area, m²	17,197.80	5712.45	2886.45	5712.45	2886.45
Window Area, m²	2400.80	882.40	265.20	988.00	265.20
Gross WWR, %	13.96	15.45	9.19	17.30	9.19

WWR was varied from 5% to 70% with an increment of 5%. Figure 4 below shows how the energy consumption of the building is affected by these changes (Note: percent savings indicated on the graph are relative to the baseline model):

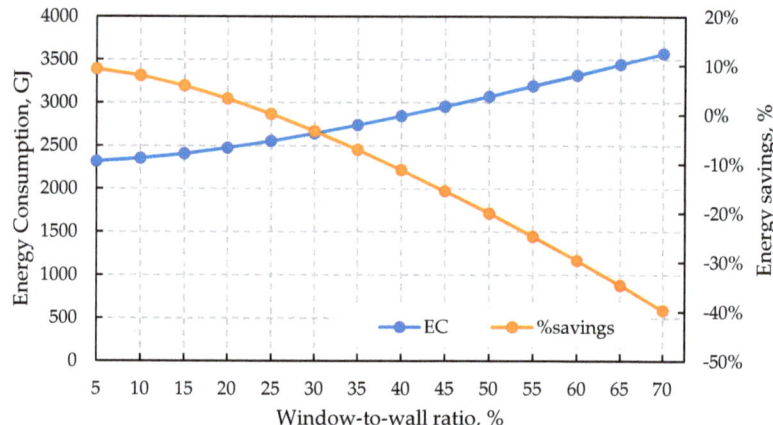

Figure 4. Effect of window-to-wall ratio on building energy consumption.

As it is shown in Figure 4, the higher WWR, the higher the energy demand of the building. However, having very small windows would be impractical for hotel buildings both from the point of view of aesthetics and the impact of insufficient daylighting on the physical and mental well-being of tenants. Therefore, considering these factors, it was decided to keep the WWR close to the original value, i.e., 15%.

3.3. Building Orientation

Since the building has an aspect ratio of 2, the orientation of the building has a significant influence on the energy consumption of the building mostly owing to the amount of solar irradiation (refer to Figure 2, to see how the building is originally orientated). In fact, from the annual sun path diagram in Figure 5, it is evidently clear that throughout the year, the northern face of the building is exposed to the direct solar irradiation less compared to other sides of the building:

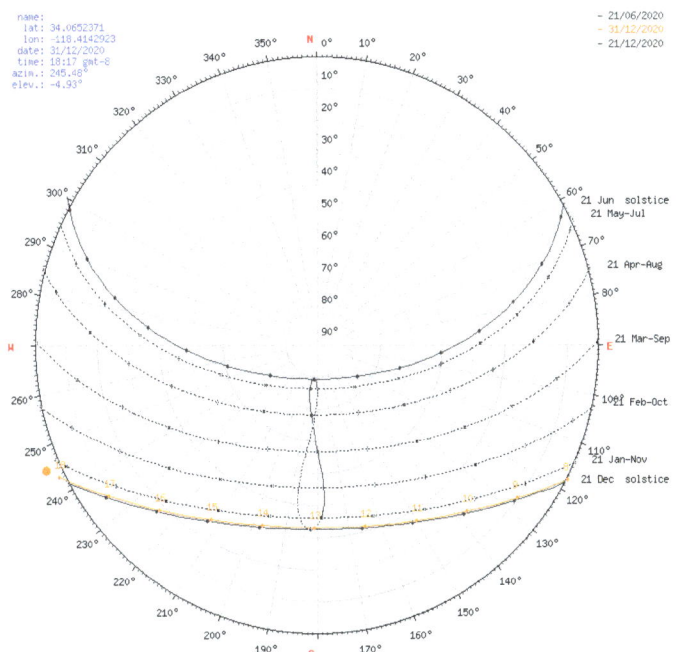

Figure 5. Sun path diagram for the city of Los Angeles, USA [28].

To find how the energy demand required for heating and cooling of the hotel is affected by the orientation, the orientation of the hotel relative to the north direction was varied from 0° to 355° with an increment of 5°. The result of these simulations is provided in Figure 6:

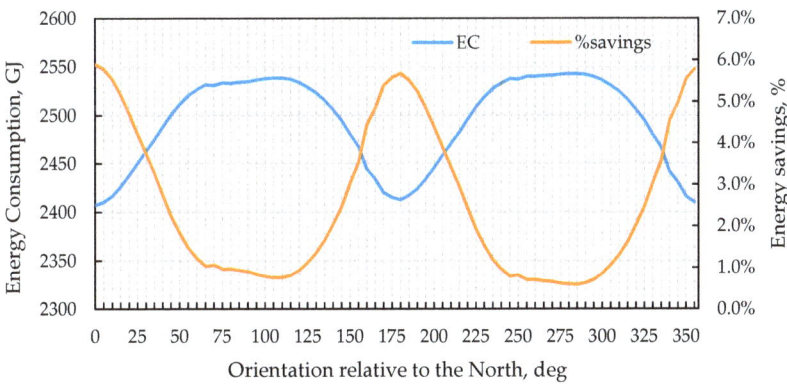

Figure 6. Effect of building orientation on heating and cooling energy demand.

As can be seen in Figure 6, the energy demand of the hotel is the lowest when its longer side is oriented perpendicular to the North direction. In terms of energy required for air conditioning, there is not much difference between orienting it at 0° or 180° relative to the north. Therefore, architectural considerations would prevail in this case: since an open-air restaurant is to be located on the top floor, it is preferable to have it oriented towards the southern direction—where several landmark buildings are located—to offer guests a better view. Thus, the building will be oriented at 180° relative to the North direction.

3.4. Thermochromic Windows

For the purposes of the hotel project, properties of double-layer VO_2 glazing covered by aluminum-doped zinc oxide (ZnO:Al) described by Kang et al. [17] were used for modeling of thermochromic windows. Kang et al. [17] and Kamalisarvestani et al. [14] reported properties of this material at 20 °C and 90 °C, therefore, referring to Giovannini et al. [18], it was assumed that optical and thermal properties vary linearly between these two temperatures (Figure 7).

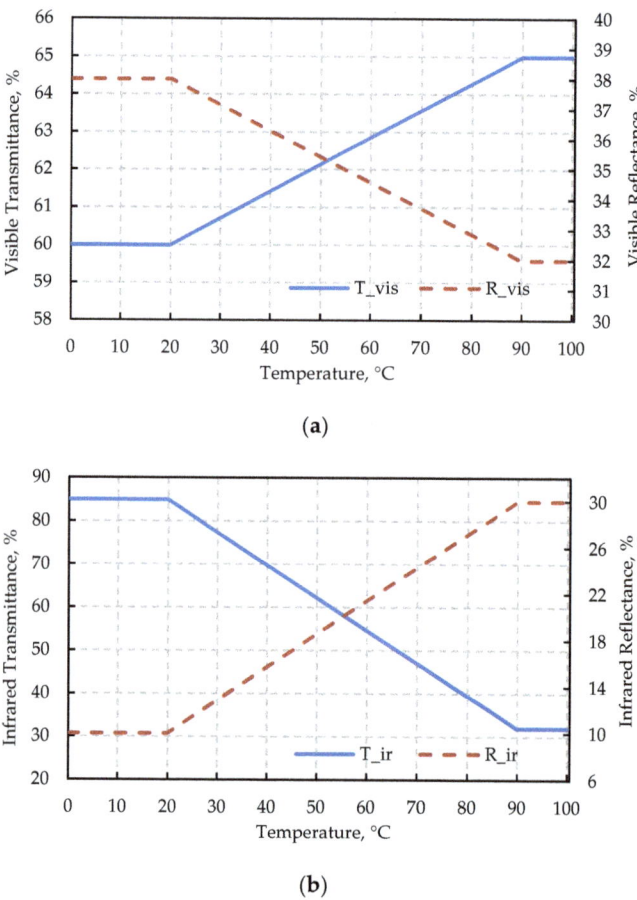

Figure 7. Cont.

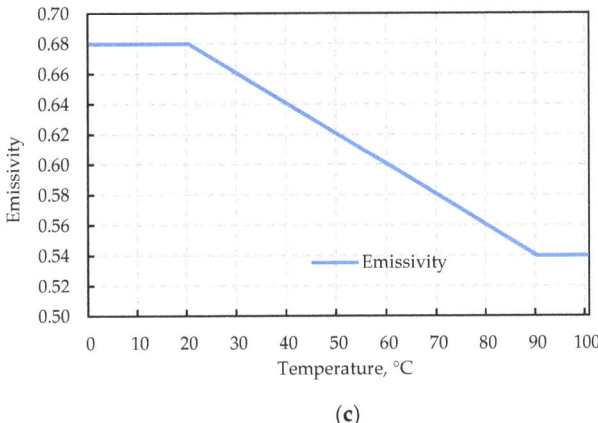

(c)

Figure 7. Optical properties of the thermochromic windows [17].

Figure 8 shows changes in energy consumed by cooling, heating, and fan of the original building and energy consumption before and after installation of thermochromic windows. As can be seen from this figure, heating energy demand decreased by 160 GJ (11.9%), and cooling energy demand increased by 33 GJ (22.4%) due to the installation of thermochromic windows. The fan energy changes proportionally to the total cooling and heating energy consumption, therefore its share in total energy consumed by the HVAC system remains practically unchanged.

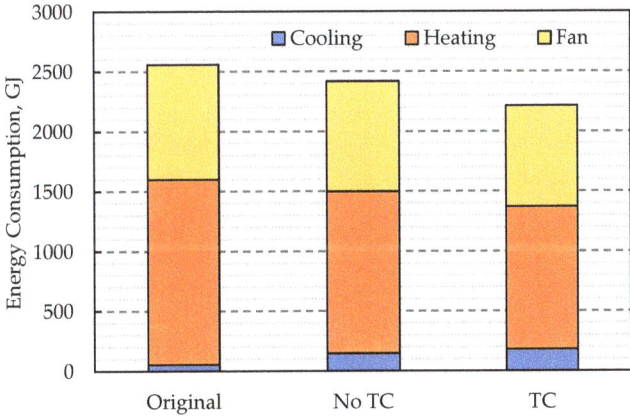

Figure 8. Energy consumption of the hotel in its original design and before and after installation of thermochromic windows.

3.5. PCM with Temperature-Controlled Natural Ventilation

The energy model of the hotel building and heat balance algorithms were configured following recommendations in the literature described in the introduction section [19,20]. A homogenous PCM layer with a 0.020 m thickness was installed into the exterior wall between the insulation layer and gypsum board (refer to Table 5). Density (ρ), conductivity (k), latent heat (H_f), and specific heat (C_p) of organic PCM A28 manufactured by PlusICE® [29] will be taken as a basis for thermophysical properties:

Table 5. Properties of PCM A28 (PlusICE®) [29].

Properties	Values
ρ [kg/m³]	789
k [W/(m·K)]	0.21
H_f [kJ/kg]	265
C_p [J/(kg·K)]	2220
T_m [°C]	20–30

To find the optimum PCM, melting temperature (T_m) will be varied from 20 °C to 30 °C and enthalpy curves will be generated using Feustel's model [30]:

$$h(T) = C_{p,const}T + \frac{H_f}{2}\left\{1 + \tanh\left[\frac{2\beta}{\tau}(T - T_m)\right]\right\}$$

where T is the temperature (°C), β is the inclination of enthalpy curve (–), τ is the width of the phase transition zone (°C). Figure 9 shows enthalpy curves of PCMs for the entire search domain used in optimization, which were generated using Feustel's model:

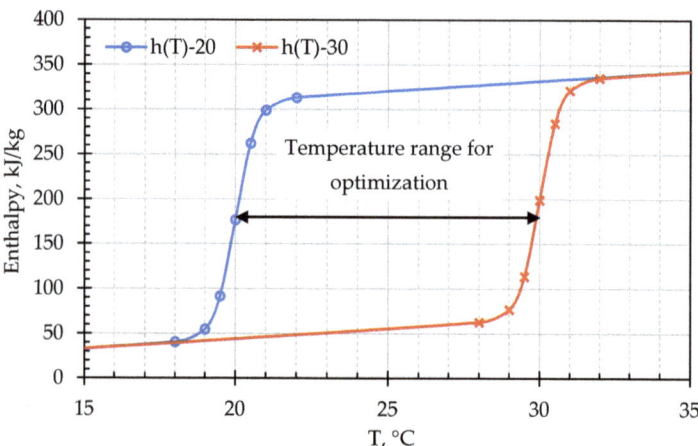

Figure 9. Enthalpy-temperature curves of PCMs for search domain used in optimization.

For the building energy analysis of the hotel, the temperature-controlled natural ventilation described by Prabhakar et al. [21] and Jacobson and Jadhav [31] was adopted with some minor modifications. To model this type of natural ventilation, the airflow network model (AFN) in EnergyPlus was used. In this model, algorithms can take into account wind, buoyancy, and pressure differences created by natural and forced movement of air for calculations of airflow [32]. In case of temperature-controlled natural ventilation, windows were opened by 50% at any time during the day and nighttime, if (1) outdoor temperature will be above HVAC heating setpoints and (2) temperature difference between indoor and outdoor temperatures were more than 1.5 °C (originally, in the above two papers, 3 °C temperature difference was used). Airflow network control was set to multizone without distribution and wind pressure coefficients were calculated by surface averages. Figure 10 below demonstrates the results of simulations:

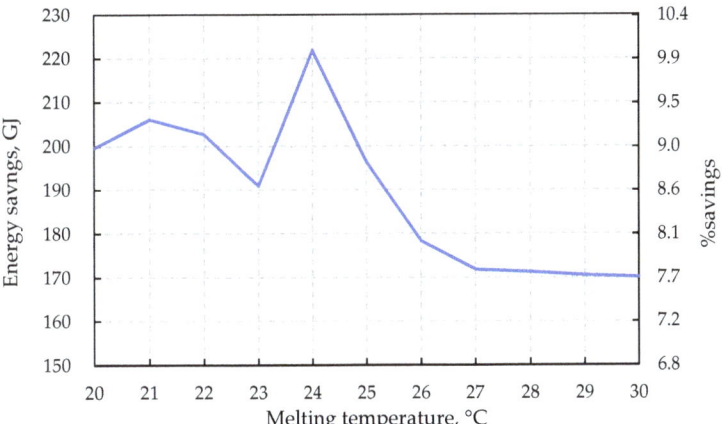

Figure 10. Optimization of PCM melting temperature.

As can be seen in this graph, the highest savings are achieved by PCM with a melting temperature of 24 °C, which yields 222 GJ (10.03%) of energy consumption reductions.

3.6. Photovoltaic Panels

Several photovoltaic (PV) panels shall be installed on the rooftop of two kitchens and bathrooms on the 24th floor. According to Jacobson and Jadhav [31], in Bakersfield (Kern County, CA, USA), a city located 112 km from Los Angeles, the optimal tilt angle of solar panels is 29°. Using this tilt angle and taking one solar panel with the unit area, the effect of panels' orientation relative to the North was checked. The results of these simulations are presented in Figure 11:

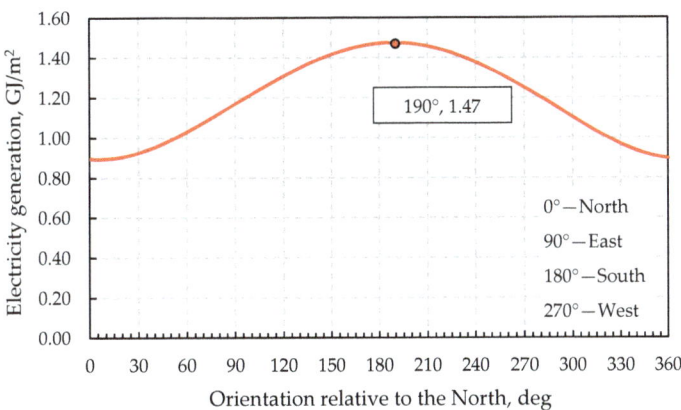

Figure 11. Effect of orientation of PV on electricity generation potential.

As can be seen from this figure, the energy generation potential of photovoltaic panels is maximum if they are oriented at 190° relative to the north direction, i.e., when the panel is facing towards the southern direction. Therefore, a total of ten PV panels were fit onto the rooftop with dimensions and interspacing as shown in Figures 12 and 13 (the total area of all solar panels is 273.0 m^2). Their annual electrical energy generation capacity was estimated via energy simulation to be 397.86 GJ.

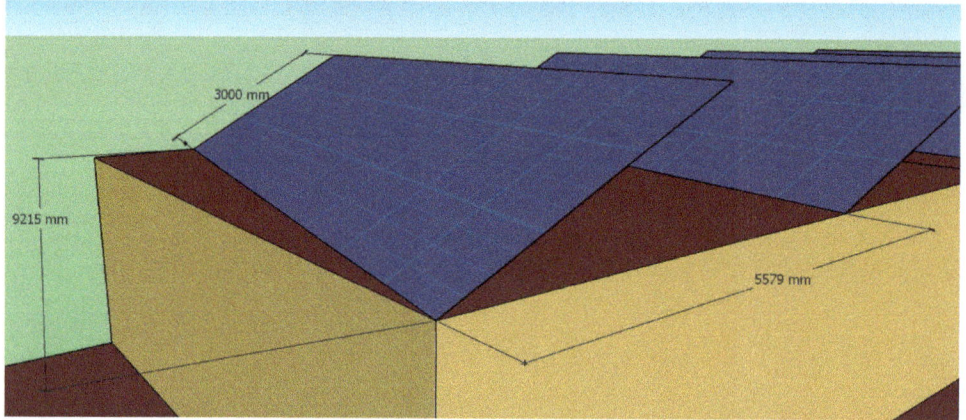

Figure 12. Dimensions and tilt angle of photovoltaic panels.

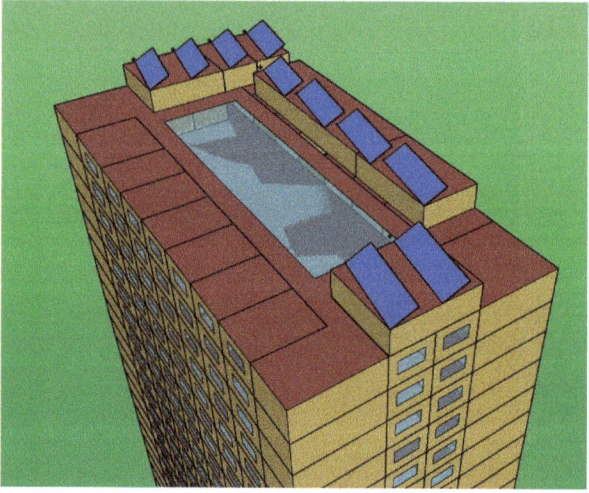

Figure 13. General overview of solar panels.

4. Cost-Benefit Analysis

To summarize the results of previous sections, due to the installation of thermochromic windows and PCMs, energy savings of 204 GJ and 222 GJ were obtained. Additionally, 398 GJ of electricity is to be generated by solar panels installed on the rooftop. Collectively, these three technologies are estimated to reduce the total energy consumption from 2417 GJ to 1593 GJ (i.e., by 824 GJ or 34%). In this section, a dynamic payback period, net present value (NPV), and a rough estimate of return on investments (ROI) due to the installation of these three sustainable energy technologies are provided. Optimization of building envelope properties, window-to-wall ratio, and building orientation were not considered during economic analysis since all these energy analyses are being considered at the preliminary stages of building's design, and as such investments required for their implementation are virtually nonexistent. However, it is still important to consider these parameters to improve the energy efficiency of the design and to reduce the operational costs required for heating and cooling the building during its lifetime.

Unit prices of thermochromic windows, PCMs, and solar panels are obtained from manufacturers. They are shown in Table 6 together with material costs, their design life, and energy-saving potential over their lifetime. For simplicity, transportation and installation costs were ignored at this stage.

Table 6. Material quantities, costs, and energy savings by thermochromic windows, PCMs, and solar panels.

Construction	Quantity	Unit Cost	Total Material Cost, USD	Design Life, Years	Energy Savings per Year	Energy Savings over Lifetime
TC window [a]	2497.5 m²	50 USD/m²	124,875	>12	204 GJ	2448 GJ
PCM [b]	344.0 m³	4572 USD/m³	1,572,768	>27	222 GJ	5994 GJ
Solar panels [c]	37.5 kW	2710 USD/kW	101,625	~25	398 GJ	9950 GJ

[a] [33], [b] [29], [c] [34].

According to the U.S. Bureau of Labor Statistics [35], the average electricity price in Los Angeles from January to September in 2020 was 54.44 USD/GJ. Assuming that the electricity prices in Los Angeles will follow the national average over the last 20 years, which, according to statista.com [36], is 2.30% per year, and taking MARR to be 6%, dynamic payback period, NPV, ROI, and revenue per available room (RevPAR) can be calculated for each of these three technologies, Table 7:

Table 7. Payback period and return on investments for TC window, PCM, and solar panels.

Construction	Dynamic Payback Period, Years	Net Present Value, USD	Return on Investments, %	RevPAR, USD/Year
TC window	14.8	−16,264	−13.0	4.9
PCM	—	−1,346,990	−85.6	254.9
Solar panels	5.6	177,866	175.0	N/A

Here, dynamic payback period, NPV, ROI, and RevPAR were calculated using the following formulas:

$$\text{Dynamic payback period} = \frac{\log\left\{\left[1 - \frac{\text{Initial Investments} \times \text{MARR}}{\text{Annual Cost Savings} - \text{Maintenance Cost}}\right]^{-1}\right\}}{\log\{1 + \text{MARR}\}}$$

$$NPV = (\text{Annual Cost Savings} - \text{Maintenance Cost})\left[\frac{(1+\text{MARR})^{\text{design life}} - 1}{\text{MARR}(1+\text{MARR})^{\text{design life}}}\right] - \text{Initial Investments}$$

$$ROI = \frac{NPV}{\text{Initial Investments}} \times 100\%$$

$$RevPAR = \frac{\text{Total Room Revenue}}{\text{Number of Available Rooms}}$$

As the above results show, to finance TC windows and PCMs, 4.9 USD and 254.9 USD have to be allocated annually from each room.

As can be seen from Table 7, TC windows and solar panels have relatively short payback periods and high returns on investments, whereas PCM seems to be not economically viable at all. Overall, among the considered technologies, only PV panels have positive NPV, meaning that only this technology yields profits by the end of its lifetime. Nevertheless, it should be taken into account that these calculations did not take into account various other overhead costs and the fact that these technologies may significantly deteriorate over their lifetime and thus have an annual energy saving potential that is cardinally different from what was estimated in this section. In addition, 25 years is a long enough period for significant climate change events to occur, therefore energy performance analysis of PCMs and solar panels under future weather conditions is advised for higher accuracy.

5. Conclusions

The study aimed to identify ways of improvement of the building energy efficiency of a newly designed hotel building developed in accordance with ASHRAE Standard 90.1. To minimize heating and cooling energy demand of the model building, building envelope properties (thickness and thermal conductivity of insulation layers, U-factor of windows, thermal resistance of air gaps, and infiltration rate), window-to-wall ratio, and building orientation were optimized. Moreover, the energy-saving potential of thermochromic windows, phase change materials, and photovoltaic panels were estimated along with the economic feasibility of their installation. Based on the performed parametric analysis, the study found that while some of the assumptions made in the design model were already optimal, other design considerations could be improved. Specifically, the parametric analysis indicated that all 6 properties were optimal in terms of energy efficiency which indicates that the ASHRAE Standard 90.1-2019 is quite strict and allows achieving good results in terms of the building envelope. The parametric analysis also indicated that the energy consumption is the most sensitive to the U-factor of windows and thermal resistance (R-value) of building envelope components. Therefore, the optimal set of values for designing hotel buildings for the climate conditions of the considered locality was provided. In addition, the energy performance analysis also showed that the window-to-wall ratio and orientation of the hotel building had an impact on the energy use levels. It was found that the higher the WWR is, the higher the energy demand of the building was. Keeping WWR close to the original value, i.e., as 15% was also the most optimal solution. In terms of orientation, the energy demand of the hotel was found to be the lowest when its longer side was oriented perpendicular to the North direction.

The integration of thermochromic windows, phase change materials, and photovoltaic panels allowed to increase the energy efficiency of the building. After installing the thermographic windows heating energy demand decreased by 160 GJ (11.9%) and cooling energy demand increased by 33 GJ (22.4%) due to the installation of thermochromic windows. The energy-saving is significant. It was also the case with the PCMs as the highest savings were achieved with a melting temperature of 24 °C, which yielded 222 GJ (10.03%) of energy consumption reductions. Finally, the solar panels were able to generate 398 GJ of electricity. Cumulatively, these three technologies were able to reduce the total energy use from 2417 GJ to 1593 GJ (i.e., by 824 GJ or 34%). These findings indicate that the initial design considerations were optimal after minor adjustments but adding extra sustainable energy solutions brought notable energy use reductions. Since design parameters stayed close to the original values, the cost-benefit analysis of these three technologies only was carried out. It was found that windows and solar panels have relatively short payback periods and high returns on investments, whereas PCM was found to be economically nonviable in the given climate zone. Therefore, the designers and relevant stakeholders should weigh the benefits and costs of the sustainable measures and adopt only feasible ones.

These analyses may be extended further to include HVAC with model predictive control and analyses of the future energy demand of the building.

Author Contributions: Conceptualization, S.K.; methodology, S.K.; software, S.K.; formal analysis, S.K. and S.T.; resources, S.K., S.T. and S.D.; data curation, S.K.; writing—original draft preparation, S.K.; writing—review and editing, S.K., S.T. and S.D.; supervision, S.K., S.T. and S.D. All authors have read and agreed to the published version of the manuscript.

Funding: Research received no external funding.

Institutional Review Board Statement: Not applicable.

Informed Consent Statement: Not applicable.

Data Availability Statement: The data presented in this study are available on request from the corresponding author.

Conflicts of Interest: The authors declare no conflict of interest.

Abbreviation

AFN	Airflow network model
ASHRAE	American Society of Heating, Refrigerating and Air-Conditioning Engineers
CondFD	Conduction Finite Difference
EC	Energy consumption
ES	Energy savings
HVAC	Heating, Ventilation, and Air Conditioning
IECC	International Energy Conservation Code
MARR	Minimum Acceptable Rate of Return
NPV	Net present value
NZEB	Net-zero energy buildings
PCM	Phase change materials
PTAC	Packaged Terminal Air Conditioning
RevPAR	Revenue Per Available Room
ROI	Return on investments
TC	Thermochromic (windows)
WWR	Window-to-wall ratio

References

1. European Environment Agency. *IPCC Fifth Assessment Report: Climate Change 2014 (AR5)*; European Environment Agency: København, Denmark, 2014.
2. Tokbolat, S.; Karaca, F.; Durdyev, S.; Calay, R.K. Construction professionals' perspectives on drivers and barriers of sustainable construction. *Environ. Dev. Sustain.* **2019**, *22*, 4361–4378. [CrossRef]
3. Durdyev, S.; Ismail, S.; Ihtiyar, A.; Abu Bakar, N.F.S.; Darko, A. A partial least squares structural equation modeling (PLS-SEM) of barriers to sustainable construction in Malaysia. *J. Clean. Prod.* **2018**, *204*, 564–572. [CrossRef]
4. Durdyev, S.; Zavadskas, E.K.; Thurnell, D.; Banaitis, A.; Ihtiyar, A. Sustainable Construction Industry in Cambodia: Awareness, Drivers and Barriers. *Sustainability* **2018**, *10*, 392. [CrossRef]
5. Harkouss, F.; Fardoun, F.; Biwole, P.H. Optimal design of renewable energy solution sets for net zero energy buildings. *Energy* **2019**, *179*, 1155–1175. [CrossRef]
6. Rodriguez-Ubinas, E.; Montero, C.; Porteros, M.; Vega, S.; Navarro, I.; Castillo-Cagigal, M.; Matallanas, E.; Gutiérrez, A. Passive design strategies and performance of Net Energy Plus Houses. *Energy Build.* **2014**, *83*, 10–22. [CrossRef]
7. Saikia, P.; Pancholi, M.; Sood, D.; Rakshit, D. Dynamic optimization of multi-retrofit building envelope for enhanced energy performance with a case study in hot Indian climate. *Energy* **2020**, *197*, 117263. [CrossRef]
8. Kheiri, F. A review on optimization methods applied in energy-efficient building geometry and envelope design. *Renew. Sustain. Energy Rev.* **2018**, *92*, 897–920. [CrossRef]
9. Echenagucia, T.M.; Capozzoli, A.; Cascone, Y.; Sassone, M. The early design stage of a building envelope: Multi-objective search through heating, cooling and lighting energy performance analysis. *Appl. Energy* **2015**, *154*, 577–591. [CrossRef]
10. Tuhus-Dubrow, D.; Krarti, M. Genetic-algorithm based approach to optimize building envelope design for residential buildings. *Build. Environ.* **2010**, *45*, 1574–1581. [CrossRef]
11. Ascione, F.; Bianco, N.; De Masi, R.F.; Mauro, G.M.; Vanoli, G.P. Design of the Building Envelope: A Novel Multi-Objective Approach for the Optimization of Energy Performance and Thermal Comfort. *Sustainability* **2015**, *7*, 10809–10836. [CrossRef]
12. Wang, Y.; Wei, C. Design optimization of office building envelope based on quantum genetic algorithm for energy conservation. *J. Build. Eng.* **2021**, *35*, 102048. [CrossRef]
13. Carlucci, S.; Cattarin, G.; Causone, F.; Pagliano, L. Multi-objective optimization of a nearly zero-energy building based on thermal and visual discomfort minimization using a non-dominated sorting genetic algorithm (NSGA-II). *Energy Build.* **2015**, *104*, 378–394. [CrossRef]
14. Kamalisarvestani, M.; Saidur, R.; Mekhilef, S.; Javadi, F. Performance, materials and coating technologies of thermochromic thin films on smart windows. *Renew. Sustain. Energy Rev.* **2013**, *26*, 353–364. [CrossRef]
15. Baetens, R.; Jelle, B.P.; Gustavsen, A. Properties, requirements and possibilities of smart windows for dynamic daylight and solar energy control in buildings: A state-of-the-art review. *Sol. Energy Mater. Sol. Cells* **2010**, *94*, 87–105. [CrossRef]
16. Aburas, M.; Soebarto, V.; Williamson, T.; Liang, R.; Ebendorff-Heidepriem, H.; Wu, Y. Thermochromic smart window technologies for building application: A review. *Appl. Energy* **2019**, *255*, 113522. [CrossRef]
17. Kang, L.; Gao, Y.; Luo, H.; Wang, J.; Zhu, B.; Zhang, Z.; Du, J.; Kanehira, M.; Zhang, Y. Thermochromic properties and low emissivity of ZnO: Al/VO2 double-layered films with a lowered phase transition temperature. *Sol. Energy Mater. Sol. Cells* **2011**, *95*, 3189–3194. [CrossRef]
18. Giovannini, L.; Favoino, F.; Pellegrino, A.; Verso, V.R.M.L.; Serra, V.; Zinzi, M. Thermochromic glazing performance: From component experimental characterisation to whole building performance evaluation. *Appl. Energy* **2019**, *251*, 113335. [CrossRef]

19. Kalnæs, S.E.; Jelle, B.P. Phase change materials and products for building applications: A state-of-the-art review and future research opportunities. *Energy Build.* **2015**, *94*, 150–176. [CrossRef]
20. Tabares-Velasco, P.C.; Christensen, C.; Bianchi, M. Verification and validation of EnergyPlus phase change material model for opaque wall assemblies. *Build. Environ.* **2012**, *54*, 186–196. [CrossRef]
21. Prabhakar, M.; Saffari, M.; de Gracia, A.; Cabeza, L.F. Improving the energy efficiency of passive PCM system using controlled natural ventilation. *Energy Build.* **2020**, *228*, 110483. [CrossRef]
22. Pisello, A.L.; Castaldo, V.L.; Rosso, F.; Piselli, C.; Ferrero, M.; Cotana, F. Traditional and Innovative Materials for Energy Efficiency in Buildings. *Key Eng. Mater.* **2016**, *678*, 14–34. [CrossRef]
23. Burnett, J.W.; Hefner, F. Solar energy adoption: A case study of South Carolina. *Electr. J.* **2021**, *34*, 106958. [CrossRef]
24. U.S. Energy Information Administration. *Levelized Costs of New Generation Resources in the Annual Energy Outlook 2021*; U.S. Energy Information Administration: Washington, DC, USA, 2021.
25. Dibene-Arriola, L.M.; Carrillo-González, F.M.; Quijas, S.; Rodríguez-Uribe, M.C. Energy Efficiency Indicators for Hotel Buildings. *Sustainability* **2021**, *13*, 1754. [CrossRef]
26. ASHRAE. *ANSI/ASHRAE/IES Standard 90.1-2019-Energy Standard for Buildings except Low-Rise Residential Buildings*; ASHRAE: Atlanta, GA USA, 2019.
27. Baechler, M.C.; Williamson, J.L.; Cole, P.C.; Hefty, M.G.; Love, P.M.; Gilbride, T.L. Guide to Determining Climate Regions by County. In *Building America Best Practices Series*; Pacific Northwest National Lab.(PNNL): Richland, WA, USA, 2010; Volume 7.1.
28. SunEarthTools. Sun Path Diagram for the City of Los Angeles. 2021. Available online: https://www.sunearthtools.com/dp/tools/pos_sun.php (accessed on 18 July 2021).
29. PlusICE. *PlusICE Organic (A) Range*; PlusICE: Lynge, Denmark, 2021.
30. Feustel, H. *Simplified Numerical Description of Latent Storage Characteristics for Phase Change Wallboard*; Indoor Environment Program, Energy and Environment Division: Berkeley, CA, USA, 1995.
31. Jacobson, M.Z.; Jadhav, V. World estimates of PV optimal tilt angles and ratios of sunlight incident upon tilted and tracked PV panels relative to horizontal panels. *Sol. Energy* **2018**, *169*, 55–66. [CrossRef]
32. U.S. Department of Energy. *EnergyPlus™, Version 9.3.0*; Documentation: Engineering Reference; U.S. Department of Energy: Washington, DC, USA, 2020.
33. Arutjunjan, R.E.; Markova, T.S.; Halopenen, I.Y.; Maksimov, I.K.; Tutunnikov, A.I.; Yanush, O.V. Thermochromic Glazing for "Zero Net Energy" House. In Proceedings of the Glass Processing Days Conference Proceedings, Tampere, Finland, 15–18 June 2003.
34. Energysage. Solar Panels in Los Angeles County, CA. 2021. Available online: https://www.energysage.com/local-data/solar-panel-cost/ca/los-angeles-county/ (accessed on 15 June 2021).
35. U.S. Bureau of Labor Statistics. Average Energy Prices, Los Angeles-Long Beach-Anaheim–September 2020. 2020. Available online: https://www.bls.gov/regions/west/news-release/2020/averageenergyprices_losangeles_20201015.htm (accessed on 14 June 2021).
36. Annual Year-On-Year Growth in Residential Electricity Prices in the United States from 2000 to 2020, with a Forecast Until 2022. 2020. Available online: Statista.com (accessed on 14 June 2021).

Article

Experimental Evaluation of Energy-Efficiency in a Holistically Designed Building

Raluca Buzatu [1], Viorel Ungureanu [2,3,*], Adrian Ciutina [4], Mihăiță Gireadă [1], Daniel Vitan [5] and Ioan Petran [6]

1. Research Institute for Renewable Energy—ICER, Politehnica University of Timisoara, 300774 Timisoara, Romania; ralu.buzatu@gmail.com (R.B.); mihaita.gireada@student.upt.ro (M.G.)
2. Laboratory of Steel Structures, Romanian Academy, Timisoara Branch, 300223 Timisoara, Romania
3. CMMC Department, Politehnica University of Timisoara, 300224 Timisoara, Romania
4. CCTFC Department, Politehnica University of Timisoara, 300224 Timisoara, Romania; adrian.ciutina@upt.ro
5. EI Department, Politehnica University of Timisoara, 300223 Timisoara, Romania; vitan.danut@gmail.com
6. Department of Structures, Technical University of Cluj-Napoca, 400027 Cluj-Napoca, Romania; ioan.petran@dst.utcluj.ro
* Correspondence: viorel.ungureanu@upt.ro

Citation: Buzatu, R.; Ungureanu, V.; Ciutina, A.; Gireadă, M.; Vitan, D.; Petran, I. Experimental Evaluation of Energy-Efficiency in a Holistically Designed Building. *Energies* **2021**, *14*, 5061. https://doi.org/10.3390/en14165061

Academic Editor: Paulo Santos

Received: 5 July 2021
Accepted: 9 August 2021
Published: 17 August 2021

Publisher's Note: MDPI stays neutral with regard to jurisdictional claims in published maps and institutional affiliations.

Copyright: © 2021 by the authors. Licensee MDPI, Basel, Switzerland. This article is an open access article distributed under the terms and conditions of the Creative Commons Attribution (CC BY) license (https://creativecommons.org/licenses/by/4.0/).

Abstract: The building sector continues to register a significant rise in energy demand and environmental impact, notably in developing countries. A considerable proportion of this energy is required during the operational phase of buildings for interior heating and cooling, leading to a necessity of building performance improvement. A holistic approach in building design and construction represents a step to moderate construction costs in conjunction with reduced long-term operating costs and a low impact on the environment. The present paper presents an experimental evaluation of the energy efficiency of a building under real climate conditions; the building, which represents a holistically designed modular laboratory, is located in a moderate continental temperate climate, characteristic of the south-eastern part of the Pannonian Depression, with some sub-Mediterranean influences. Considerations for the holistic design of the building, including multi-object optimization and integrated design with a high regard for technology and operational life are described. The paper provides a genuine overview of the energy efficiency response of the building during six months of operational use through a monitored energy management system. The energetic analysis presented in the paper represents an intermediary stage as not all the energetic users were installed nor all the energetic suppliers. However, the results showed a reliable thermal response in the behaviour of recycled-PET thermal wadding used as insulation material in the building and for the intermediary stage in which the building has only secondary energy users, the energetic balance proves its efficiency, keeping the buffer stock of energy high values over 90%.

Keywords: holistic; energy management system; sustainable; building performance; thermal performance; indoor comfort

1. Introduction

1.1. Context

The built environment with its different forms (residential buildings, workplaces, educational buildings, hospitals, libraries, community centres, and other public buildings) is the largest energy consumer and one of the largest emitters of carbon dioxide (CO_2) in the European Union (EU). Buildings caused 41.3% of the EU27 final energy consumption in the last decade (Figure 1), and are responsible for approximately 36% of the EU's greenhouse gas emissions [1]. Aiming to help address these issues, the EU agreed with new rules for the energy performance of buildings directive; in 2010 it established a legislative framework that includes the Energy Performance of Buildings Directive 2010/31/EU (EPBD) [2] and later, in 2012, the Energy Efficiency Directive 2012/27/EU [3], promoting policies that help to achieve a highly energy-efficient and zero-emission building stock in the EU by 2050, to combat energy poverty, and to encourage more automation and control systems in order to make buildings operate more efficiently. Later, in 2018 and 2019, both directives were amended, as part of the new energy rulebook called the Clean Energy for all Europeans package (2018/844/EU) [4], through which the EU improved its energy policy framework to encourage the migration from fossil fuels to cleaner energy, while also delivering on the EU's Paris Agreement [5] commitments in reducing greenhouse gas emissions and tackling global warming. At the same time, building and renovating is part of the European Green Deal [6]; an action plan striving for Europe to be the first carbon-neutral continent.

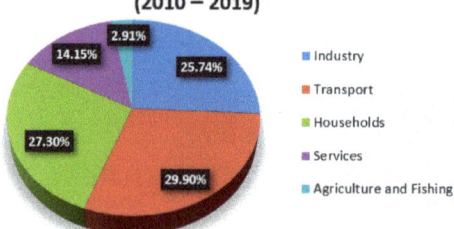

Figure 1. Final Energy consumption by sector in 27 Member States of the European Union (average from final energy consumption registered in 2010–2019). Adapted from [1].

Two issues need to be addressed to make Europe's building sector compatible with the Paris Agreement: reducing the energy demand by employing energy efficiency measures, alongside increasing the use of renewable energy sources.

Besides the building's envelope, human behaviour is also a key factor in defining energy demand in a building. Both intelligent use of building automation technologies and improved awareness-raising contribute to diminished energy consumption [7].

Implementing building automation technologies, adopting renewable energy sources, and providing energy-efficient envelopes are deficient in meeting important sustainability objectives, if the design stages of the buildings are contrived successively and independently, leading to an unalterable variable selection starting with the first steps of the design process, which highly shortens the ability to find optimal solutions of a sustainable approach in the end [8]. By consequence, embodying a holistic approach in building design, considering cross-disciplinary analysis and multi-object optimization is essential in the building sector [9]. Addressing concerns such as embodied GHG emissions (GHG emissions from the energy that is used to extract raw materials, produce and transport materials and components during production and construction phases, as well as the energy used for the maintenance, renovation and building's deconstruction/demolish) and operational GHG emissions (GHG emissions from the energy consumed in buildings during operation phase) are equally important [10]. Many studies [11,12] emphasize the circumstance that reaching net-zero emissions of carbon dioxide is an impossible objective to attain simply by reducing the energy consumption of end-users, which necessitates a more holistic approach in the construction field by means of using environmentally friendly materials on a large scale and adopting a mix of solutions in replacing the fossil demand.

Improving the connection between the end-of-life (EOL) of the products and the beginning-of-life (BOL) of the products despite the barriers encountered by the circular economy (CE), such as developing circular product design guidelines or identifying the needed assortment, categorizing and recovery infrastructure is another important process in reducing the carbon dioxide emissions [13].

1.2. Aim of the Research

The achievement of energy-efficient buildings requires an integrated design concerning various factors such as climate, occupant behaviour, technology, operation and maintenance, etc [14].

The literature review [15,16] shows that the current body of knowledge shows attention towards the economic values of sustainable construction and towards case studies (from the methodological point of view), which demands additional research in the environmental and social context of constructions, as well as in the experimental and quantitative research. The present work aims to investigate and confirm multiple sustainable factors gathered in a holistically designed building through an experimental evaluation of the energy efficiency of a modular laboratory.

2. Building and Equipment

2.1. Site and Climate

The case study is located in Timișoara, the capital city of Timiș County, western Romania (Figure 2).

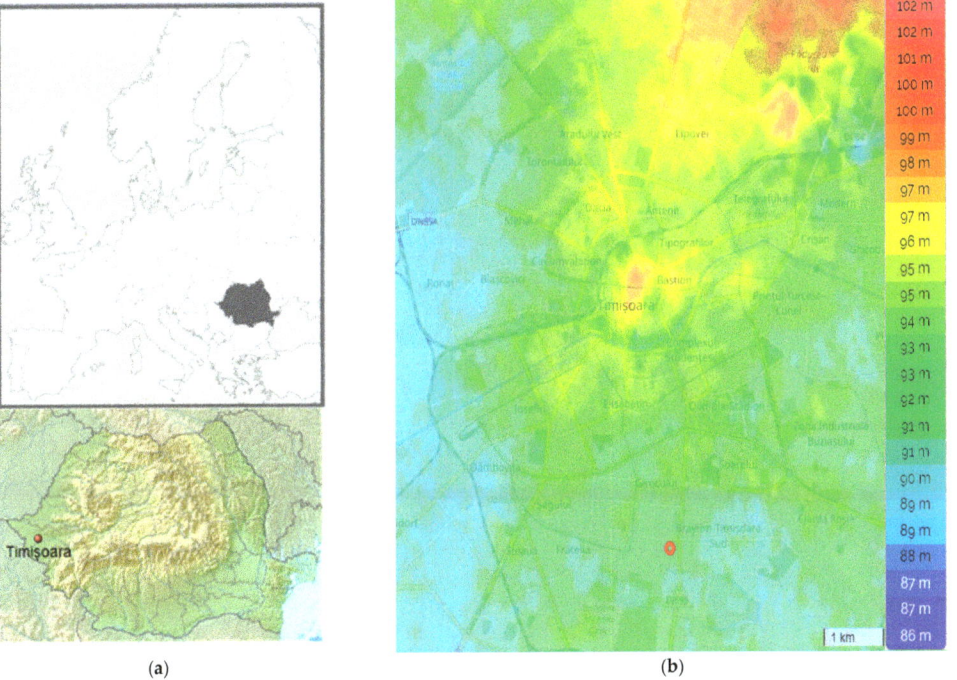

Figure 2. Location of the case study: (a) country context, (b) Timișoara's urban layout on topographic map and location of the Experimental Module. Reprinted from [17].

Located on the Bega River, the city of Timișoara is considered the informal capital of the historical Banat region, being the country's third most populous city, with almost 320,000 inhabitants and close to half a million inhabitants in its metropolitan area [18]. At a geographical level, Timișoara is located at the intersection of the 21st meridian east with the 45th parallel north, being almost an equal distance from the north pole and the equator and in the eastern hemisphere. Timișoara lies at an altitude of 86–102 m (Figure 2b) on the southeast edge of the Banat Plain which is part of the Pannonian Plain.

According to the Köppen-Geiger climate classification [19], the Banat region exhibits a Cfb climate; a marine climate with mild summers and cool but not cold winters. The average annual temperature in Timișoara is 11.1 °C, having the warmest month, on average, in July, with an average temperature of 21.7 °C (average high 27.8 °C) and the coolest month on average, January, with an average temperature of −1.7 °C (average low −4.8 °C) [20,21]. Figure 3 shows calculated values for the dry bulb temperature ranges for each month and the full year, enclosing the recorded high and low temperature (round dots), the design high and low temperatures (top and bottom of green bars), average high and low temperatures (top and bottom of yellow bars), and average temperature (open slot). The majority of the recorded hours are below the comfort zone, both during the warm and cold periods of the year.

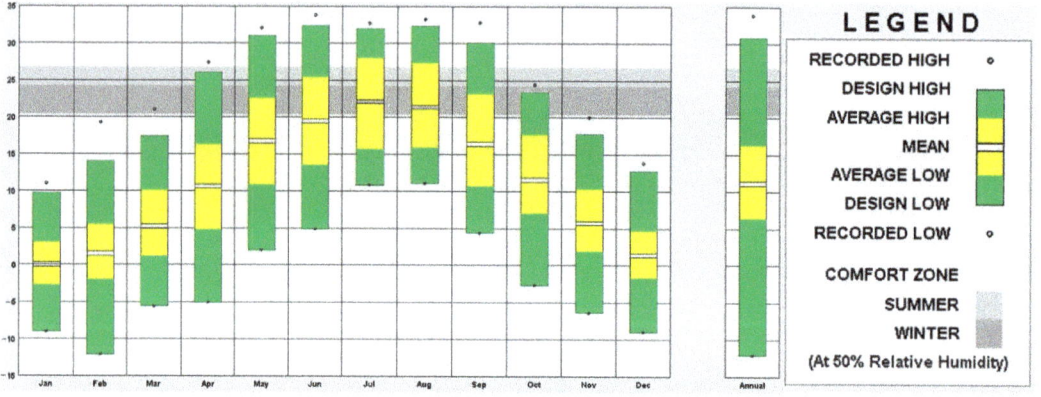

Figure 3. Temperature range for Timișoara (IWEC Data, 152,470 WMO Station). Reprinted from [21,22].

The annual average relative humidity is 80% in Timișoara, where June is the month with the highest rainfall (76 mm average rainfall) and February is the driest month (36 mm average rainfall) [20].

Recent studies [23–25] over climate and bioclimatic conditions in Romania show changes in the bioclimatic indices over the period 1961–2016 in terms of frequency of occurrence considering the number of days for each class of bioclimatic indices and in terms of duration of their occurrence period. For the stated period, bioclimatic indices such as the universal thermal climate index (UTCI), the effective temperature (ET), the equivalent temperature (TeK), the temperature-humidity index (THI), and the cooling power (H) reveal a shift from cold stress conditions to warm and hot conditions, as the climate in the big cities of Romania (Timișoara being among them) became hotter during the warm periods of the year and milder during the cold season. In terms of thermal sensation, a general negative trend was noticed in the number of comfortable days [21]. Figure 4 shows a psychrometric chart for the Timișoara location, based on IWEC weather data [18] and ASHRAE 55 standard [23] and shows that in residential buildings only 14% of the hours (1226 h) during a year are indoor comfortable hours for a human being when no design strategies (such as cooling, heating, humidification, dehumidification, sun shading of windows, natural ventilation cooling, fan-forced ventilation cooling, etc.) are considered. Every hour of registered climate data is shown as a dot on this chart.

The colour of each dot represents whether the hour is comfortable (green dots) or uncomfortable (red dots). To reach more than 90% of indoor comfortable hours during a year, one must consider design strategies such as heating and humidification for 7047 h (from a total of 8760 h annually) and cooling along with dehumidification (when needed) for 387 h annually which leads to significant energy use during the year and for the building's life span. In this specific location, the same achievement of more than 90% of indoor comfortable hours during a year can be reached when integrating holistic and passive design strategies in building design, such as internal heat gain, sun shading of windows, direct gain passive solar, night flushing of high thermal mass, etc. This reduces heating and humidification needs to 4424 h annually (almost 38% less heating hours annually) and cooling and dehumidification needs to 31 h annually (92% less cooling hours annually), as shown in Figure 5.

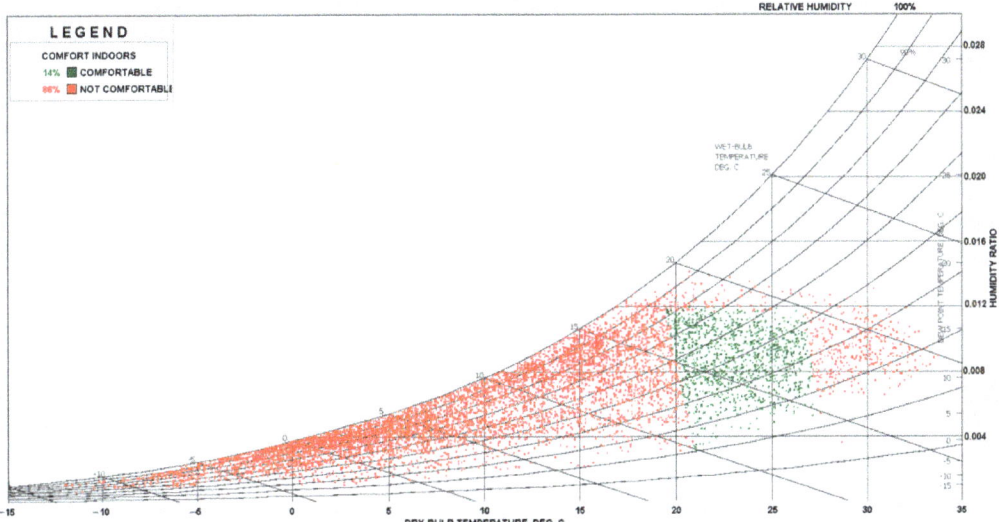

Figure 4. Psychrometric chart for Timișoara location (IWEC Data, 152,470 WMO Station): comfort indoors without design strategies. Reprinted from [22].

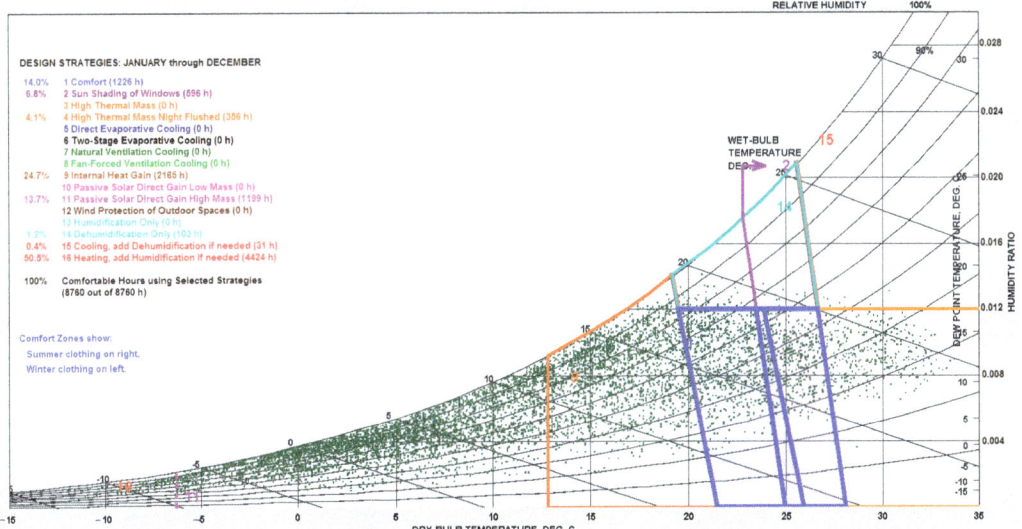

Figure 5. Psychrometric chart for Timișoara location (IWEC Data, 152,470 WMO Station): comfortable indoors hours using both active and passive design strategies. Reprinted from [22].

Integrating passive design strategies in building design and concurrent engineering (CE) overall is the necessary to meet the climate change milestones related to keeping the global temperature rise for this century well-below two degree Celsius, to achieve a climate neutral world by mid-century within zero-carbon solution targets [5] in current bioclimatic conditions with the context of a future weather shift and to provide a more resilient future for our built environment.

Based on IWEC data [21], the representative concentration pathway 4.5 [26] emissions scenario (RCP of an additional 4.5 W/m² of heating in 2100 compared with preindustrial conditions representing moderately aggressive mitigation that requires that carbon dioxide (CO_2) emissions start declining by approximately 2045) and a warming percentile of 50%, the local weather previsions over the course of the 21st century due to the impact of climate change, shows a continuous shift in decreasing number of colder days in a typical year and an increasing number of hotter days (Figure 6). For example, the number of days with an average temperature of 26.9 °C will increase from 3, registered at the present, to 10 days by 2035, to 21 days by 2065 and will reach a number of 30 days annually by 2090, while the number of days with an average temperature of −0.2 °C will decrease from 70, which are registered at present, to 57 days by 2035, to 52 days by 2065 and 47 days annually by the year 2090.

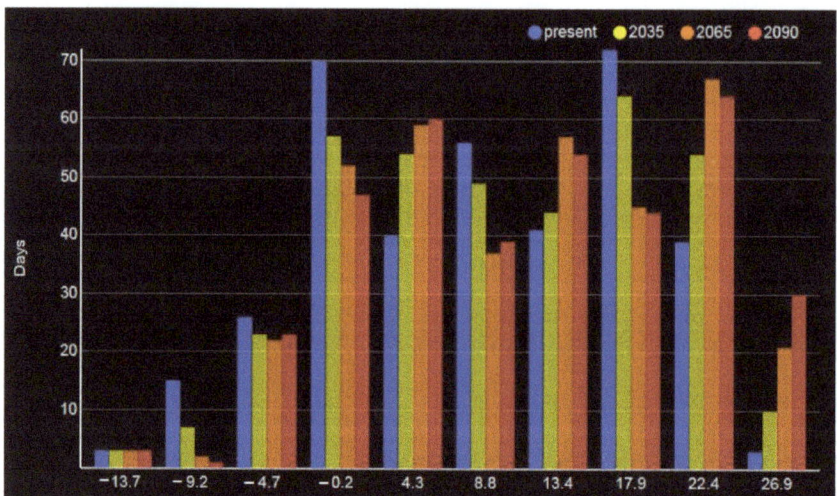

Figure 6. Projected weather data for Timișoara location based on RCP 4.5 and 50% warming percentile representing the shift of the number of days of average daily temperature. Reprinted from [27].

As RCP 1.9 is the pathway that limits global warming to below 2 °C, as the Paris Agreement specifies, which is significantly below the greenhouse gas concentration trajectory of RCP 4.5, which is considered to be a possible scenario for 2100 (in which global temperature rises between 2 and 3 °C, over the 21st century and many plants and animal species will be unable to adapt to its effects). Integrating a holistic concept in sustainable building design proves to be important.

2.2. Construction of the Experimental Module

The modular laboratory, illustrated in Figure 7, on which the experimental measurements were performed was constructed based on a selection of structural systems and materials under constituent factors of sustainable building principles, such as material efficiency, resource efficiency, health and well-being or cost-efficiency.

Figure 7. LSF experimental module.

The structure is a lightweight steel-framed (LSF) construction with cold-formed elements. The structural system was chosen on the account of the sustainable characteristics of steel, essentially, small weight with high mechanical strength, tremendous potential for recycling, deconstruction and future reuse, onsite reduced severance, speed of construction, flexible structural system for modular design, economic in transportation and handling, reduced foundation costs, [28–30]. The LSF structure is a two-story, modular construction, with a 5 m long span, 5 m long bay, 3.80 m eave height (on the southern side), 6.10 m eave height (on the northern side), and 6.95 m ridge height.

The eastern façade has two 0.76 m × 0.96 m window openings, the southern façade integrates a 3.56 m × 2.73 m glass curtain opening, while the western façade has a 0.76 m × 0.96 m window opening and a 0.97 m × 2.73 m door opening. There are no openings on the northern side of the building. The access to the second floor is ensured by a 1 m × 1 m attic scuttle door.

Using an LSF structure allowed the adoption of a precast wedge foundation system, designed as a quick foundation system, easy to handle and install, fully recoverable at the end-of-Life of the building and suitable for reuse [31]. The foundations' design was part of the holistic approach design of the experimental module, adopted regarding environmentally conscious design, modular and standardized design, reusable/recyclable element design, life cycle design, waste generation assessment, environment-friendly demolition method, working conditions, safety design and consideration of costs for materials, waste disposal and life cycle [9].

The southern side of the roof was designed with a roof pitch of 42°, in the pursuit of gaining an optimal performance of a roof-mounted solar energy system.

The materials used in the experimental module's construction were selected in the same approach of holistic design and ease for deconstruction and future reuse of the components. Table 1 presents the thermal conductivities of the materials used in the LSF experimental module.

The structure is proper for various envelope configurations. The current envelope configuration (Figure 8) was carefully selected with consideration for the local sourcing of building materials to keep transport emissions and associated costs to a minimum.

Table 1. Thermal conductivity (λ) of the materials used in the LSF experimental module.

Material	λ [W/(m·K)/]	Specific Heat [J/kg·°C]	Density [kg/m³]
Steel profiles (C150/2, C200/1.5)	50.00	420	7800
OSB [1]	0.130	1700	620
Recycled-PET [2] thermal wadding	0.048	1350	20
Wood fibreboard	0.050	2100	270
Vapor barrier	0.22	1700	130
Aluminium sheet	160	880	2800
XPS [3]	0.035	1450	35
PIR [4] sandwich panel	0.023	1400	30
Glass (door and windows)	0.024	840	838

[1] OSB: oriented strand board; [2] PET: polyethylene terephthalate; [3] XPS: extruded polystyrene; [4] PIR: polyisocyanurate.

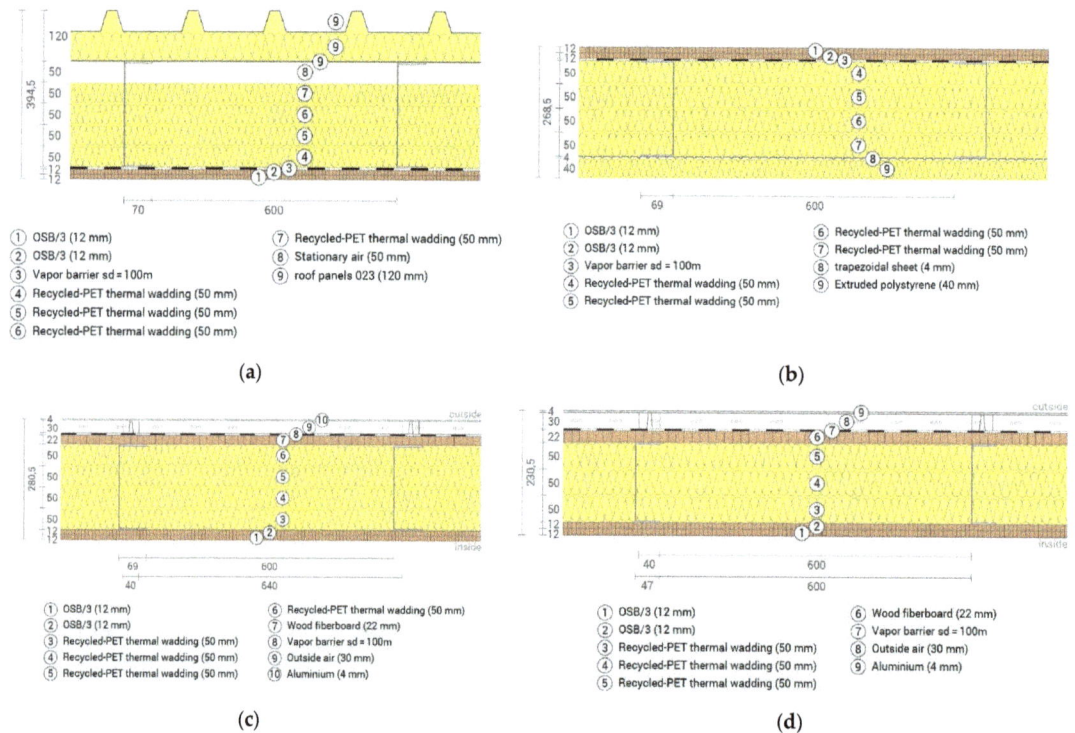

Figure 8. LSF construction elements stratification: (a) roof; (b) floor; (c) northern wall; (d) eastern and western wall.

As an inner sheathing layer of walls, ceiling, and floor, the LSF experimental module was designed to have oriented strand board (OSB) panels (24 mm thick). In between the steel frame, recycled-PET thermal wadding (150 mm or 200 mm thick, by the case) was used as batt insulation. For walls, the thermal insulation system was completed in the exterior with an overlaid layer of wood fibreboards (22 mm thick) and finished by a layer of rectangular aluminium panels (4 mm thick). In order to avoid moisture from the ground, the floor was 400 mm elevated. In between the steel frame of the floor, recycled PET thermal wadding was used (200 mm thick) as batt insulation. Below the thermal insulation wadding, was installed a layer consisting of trapezoidal steel sheets (4 mm thick), and beneath, an exterior continuous layer (40 mm) of extruded polystyrene (XPS). Both the floor and roof were waterproofed by poly-vinyl chloride (PVC) membranes. On the roof, the thermal insulation system was completed in the exterior with PIR sandwich panels (120 mm thick).

The LSF envelope elements (materials, thicknesses, number of layers) are displayed in Table 2.

Table 2. Materials, thicknesses (d) and thermal transmittances (U) of the experimental module elements.

Element	Material (Layers from Inside to Outside)	d [mm]	U-Value [W/(m²·K)]
Floor	OSB	24	0.272
	Vapor barrier	0.5	
	Recycled-PET thermal wadding	200	
	Steel sheet	4	
	XPS	40	
	Total thickness	268.5	
Walls (north)	OSB	24	0.314
	Recycled-PET thermal wadding	200	
	Wood fibreboard	22	
	Vapor barrier	0.5	
	Rear ventilated level (outside air)	30	
	Aluminium cladding	4	
	Total thickness	280.5	
Walls (east and west)	OSB	24	0.355
	Recycled-PET thermal wadding	150	
	Wood fibreboard	22	
	Vapor barrier	0.5	
	Rear ventilated level (outside air)	30	
	Aluminium cladding	4	
	Total thickness	230.5	
Roof	OSB	24	0.192
	Vapor barrier	0.5	
	Recycled-PET thermal wadding	200	
	Stationary air	50	
	PIR sandwich panel	120	
	Total thickness	394.5	
Door and windows	Glass with argon filling	24	0.880
	PVC casement	92	
Glass Curtain	Glass with argon filling	44	0.740
	PVC casement	92	

Thermal Insulation Fabricated from Recycled Post-Consumer PET Bottles

The thermal insulation layers of the envelope elements (Figure 9), consisting of a thermal insulation wadding, are fabricated of polyester fibre, recycled from post-consumer polyethylene terephthalate (PET) bottles. The insulation material is produced entirely from recycled PET bottles, which withholds CO_2 emissions and ensures environmental benefits. Besides the significantly low environmental impacts shown by the product [32], the recycled-PET thermal wadding provides high mechanical and physical properties [33], which remain unaffected by time passing and ensures acoustic insulation properties as well.

Since there are no chemical or textile agents used in the production process, the product contains no harmful substances for human health [34]. Another property of the recycled-PET thermal wadding is the material circularity: at the end-of-life of the building where it was installed, the product can be recycled in a proportion of 100% and used as a raw material for new thermal insulation wadding. The eco-efficiency of this specific thermal insulation is also from the proximity of the production place to the construction site of the laboratory: a transportation distance of only 15 km contributed to the created value of the product system, along with other factors, such as reusing post-consumer PET bottles as a raw material in the production stage, the absence of chemicals in the production process, the lack of wastes resulted from production or installation of the product.

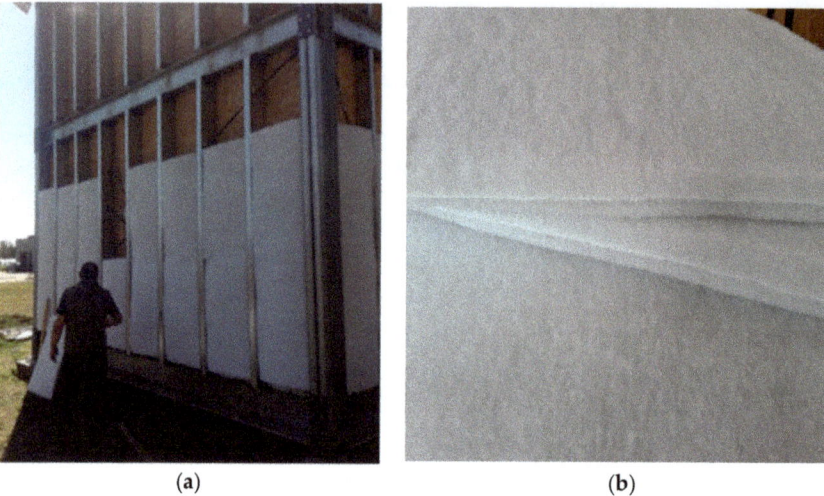

Figure 9. Recycled-PET thermal wadding: (**a**) installation phase at the construction site; (**b**) layers of insulation before installation.

2.3. Experimental Installation and Data Acquisition

The primary function of a building is to provide a suitable, comfortable, inner environment, according to the building's functions. A holistic design of an energy-efficient building regards the installation of renewable energy sources and energy conservation and an integrated design with regard towards technology, operation, and maintenance. In a building's lifetime, the greatest amount of energy is required during the operational phase, therefore the building's envelope has a pivotal impact on the building's behaviour.

2.3.1. Passive Design Strategies

The holistic design of the building regarded a series of passive strategies for the design of the LSF experimental module. Natural illumination is granted by a 3.56 m × 2.73 m glass curtain, installed on the south façade of the building, which also provides passive solar heating during daylight. When additional artificial light is necessary, LED light sources are available. The sun shading of the glass curtain, provided by external photo-voltaic shading lamellae, ensures passive cooling of the first floor (not yet installed during the six months of monitoring).

The renewable sources of energy are based on harvesting solar and wind energy: twelve 250 W polycrystalline cell panels intake solar energy, with an estimated amount of solar energy produced on-site of 1269 kWh/year (the potential production of the installed polycrystalline cell panels under ideal conditions is 3427.29 kWh/year [32]), and a 1 kW vertical wind turbine.

2.3.2. Monitored Energy Management System

The design of the LSF experimental module, in pursuance of having an authentic, factual overview of the building's performance during the operational phase, included a monitored energy management system. The LSF experimental module is a non-grid connected building, matching its own energy needs by on-site generation, fully based on renewables. The monitored energy management system consists of an electric power distribution representing a direct current (DC) grid, similar to a smart nano-grid (SN).

The electric power distribution integrates wind and solar sources of energy, elements for conversion and storage of the electrical energy, and a distributed control and energy management system through a SCADA system. Common electrical appliances (fridge, TV, PC) are used and adapted for DC supply, in order to reproduce residential application.

The architecture of the SN, presented in Figure 10, consists of a high voltage DC bus (HVDC), with a value of 350 V, and a low voltage DC bus (LVDC), with a level of 24 V. For alternating current (AC) loads and as a backup solution, the SN owns an AC bus with a voltage of 230 VRMS.

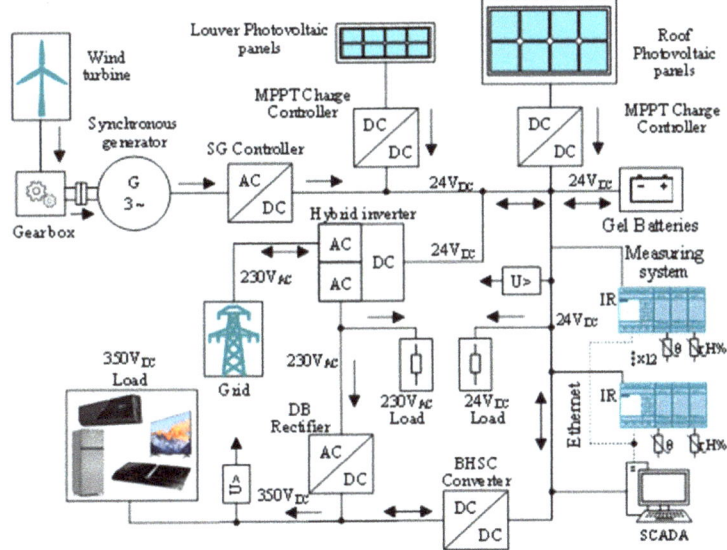

Figure 10. The architecture of the implemented smart nano-grid (adapted from [9]).

A synchronous generator (SG) coupled through a gearbox ensures the harvest of the wind energy from the vertical wind turbine. The electrical power provided by the SG is injected into the LVDC bus using the SG controller. A maximum power point tracking (MPPT) charge controller through which the LVDC is connected to the photovoltaic panels, helps to convert solar energy into electrical energy. A smaller MPPT charge controller is used for the louver photovoltaic panels. The energy is stored in four 12 V/220 Ah valve regulated lead-acid gel batteries, which can store 10 kWh of electrical energy, enough for 2–3 days of usual household operation without recharging. The connection between the HVDC bus and LVDC bus is performed through a bidirectional hybrid switched capacitors converter (BHSC) [35].

High efficiency and low cost of high ratio voltage conversion are viable due to the BHSC converter's capabilities. The entire flow of electrical energy is controlled by a SCADA system which ensures the data acquisition of all parameters.

2.3.3. Data Acquisition Infrastructure

The LSF experimental module's data acquisition infrastructure consists of 3 CO_2 sensors, 14 humidity sensors and 53 temperature sensors distributed as presented in Figure 11. A measuring station, composed from 12 so-called intelligent relays (IR) is used for acquiring the data from the sensors [36], providing digital inputs and outputs, which can be used in small automation such as residential automation [9]. The sensors (Figure 12a) distributed on the walls are located on the outer face of the interior walls, between the insulation layers and on the inner face of the exterior walls, as illustrated in Figure 12b.

The SCADA interface was designed with the LabView 2021 software development platform provided by National Instrument (11500 N Mopac Expwy Austin, TX 78759-3504, USA), and it is supported by a dedicated station server. For redundancy, a second SCADA system was designed with the Logo Web Editor V1.0 software (Siemens AG, P.O. Box 48 48 90026 Nuremberg, Germany) development platform [36] which is supported by the IR. Unlike other SCADA systems which run over a dedicated station (server or desktop), this second SCADA system is accessible using a web page. The acquired data are stored on the server station and backup is also stored on the IR which is equipped with a micro-SD card.

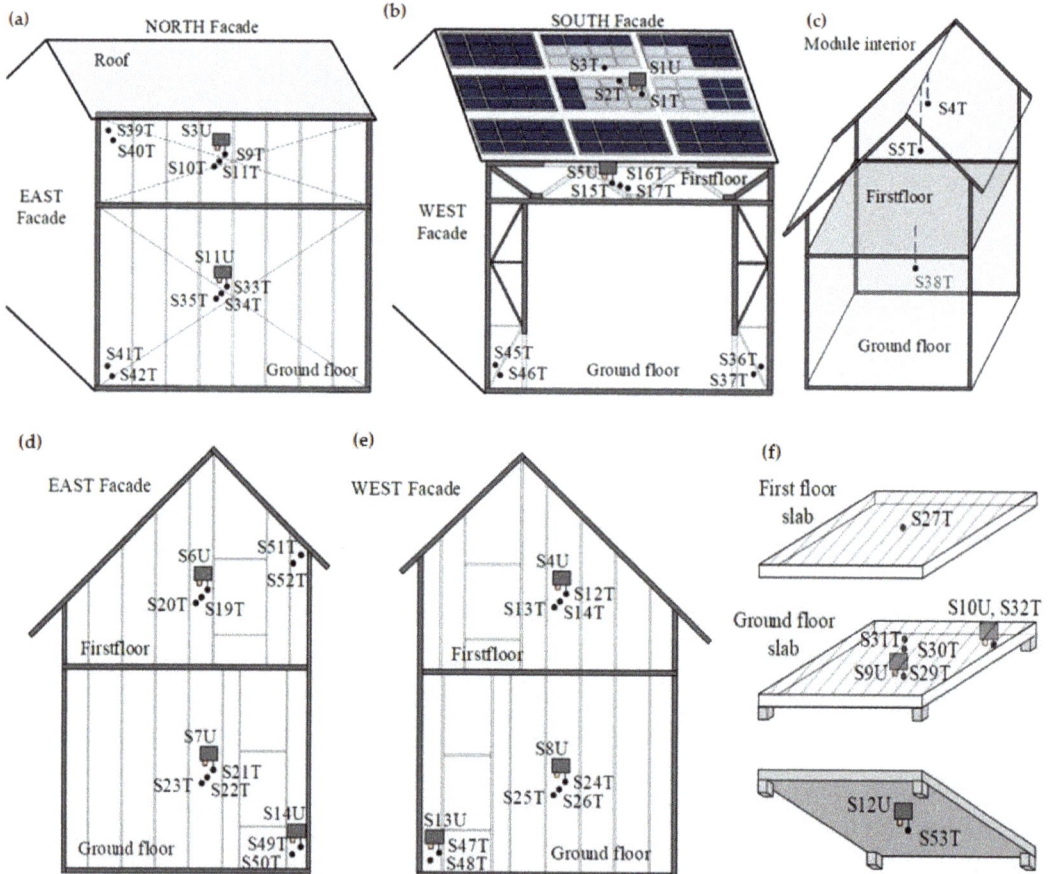

Figure 11. Sensors' distribution on LSF experimental module: (**a**) north façade (**b**) southern façade and roof (**c**) interior (**d**) east façade (**e**) west façade (**f**) in slabs (adapted from [9]).

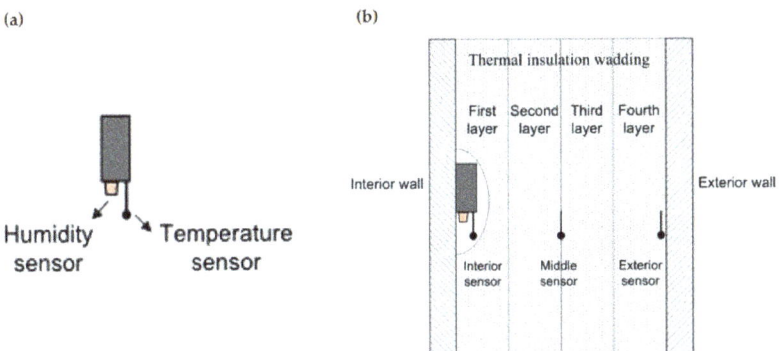

Figure 12. (a) Humidity and temperature sensor configuration (b) sensors' distribution between thermal insulation layers.

3. Results and Discussion

3.1. Thermal Monitoring

Figures 13–18, illustrated below, show the information provided by the monitoring management system registered during a supervision interval of six months (02 December 2020–17 May 2021). The recordings transferred from the sensors reveal the behaviour of the experimental module's envelope and indoor comfort conditions. In the temperature graphics, data provided from the sensors located on the outer face of the interior walls are shown in yellow, data provided from the sensors located between the insulation layers are shown in blue while data provided from the sensors located on the inner face of the exterior walls are shown in purple (Figure 12b). It should be noted that at the time of monitoring the external photo-voltaic shading lamellae were not installed yet, nor any other HVAC system; therefore, no mechanically cooling, heating or dehumidification system contributed to the indoor comfort. The interior temperature was influenced only by solar gain, electrical appliances, and human interaction during maintenance and observation interference.

No doubt, the building occupancy has a direct influence on the thermal performance of the building. The building was only used sporadically during the recording period, human interaction during maintenance and observation was the only interaction.

During the winter period (from 1 December 2020 to 21 March 2021), the outdoor minimum air temperature was ranging from −1 to 2 °C, while the maximum one from 5 to 9 °C. During the spring period (from 22 March 2021 to 15 May 2021), the outdoor minimum air temperature was ranging from 3 to 8 °C, while the maximum one from 11 to 21 °C. In both periods, the daily maximum and minimum outdoor air temperatures had significant variations during the monitoring period. Looking to the middle and interior sensors, it is very easy to see they are following those from the exterior with a difference of 2–3 °C (Figures 13–16). The differences are a bit higher in the case of north and west facades, especially from the sensors located on the ground floor. The temperature profiles for both floors were quite similar, but due to the presence of the larger glazed area on the southern facade, higher temperatures were recorded for the interior sensor. These highlight the effect of the glazed area as a strategy to capture solar gains. In the spring, a difference of 2–3 °C was observed between the indoor air temperatures of the space located on the ground floor and the one the upper floor. Figures 13–16 show thew indoor temperature does not remain stable in the rooms. The situation was expected to be like this due to the missing systems for indoor comfort and human interaction.

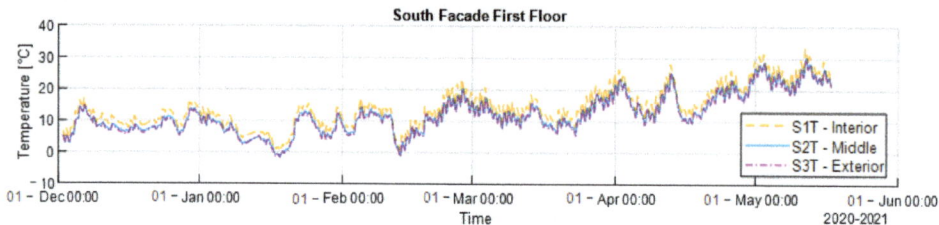

Figure 13. (a) Temperature data provided by the sensors for southern façade first floor.

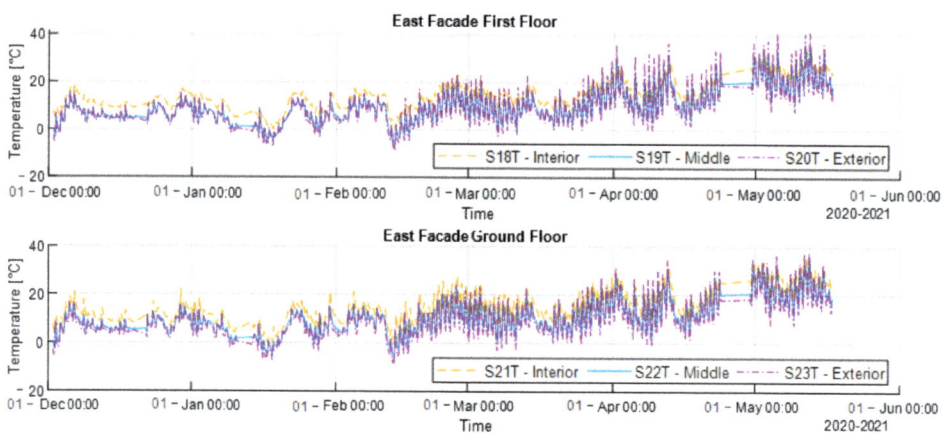

Figure 14. Temperature data provided by the sensors for the eastern façade first floor (above) and ground floor (below).

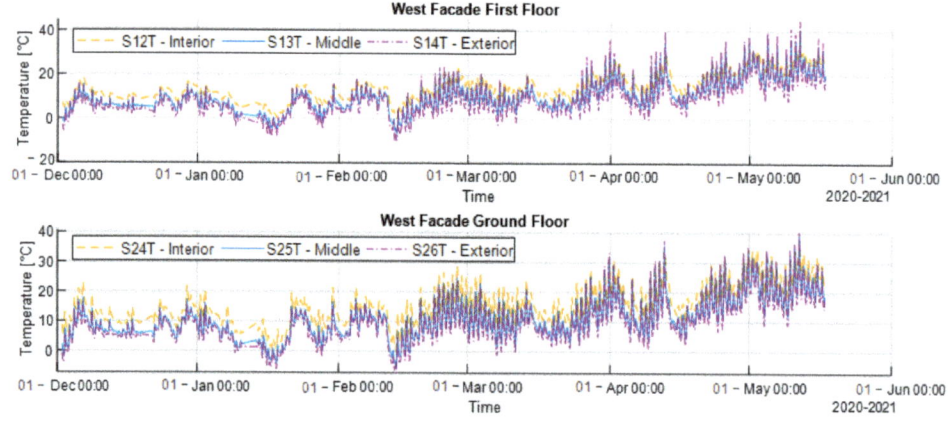

Figure 15. Temperature data provided by the sensors for the western façade first floor (above) and ground floor (below).

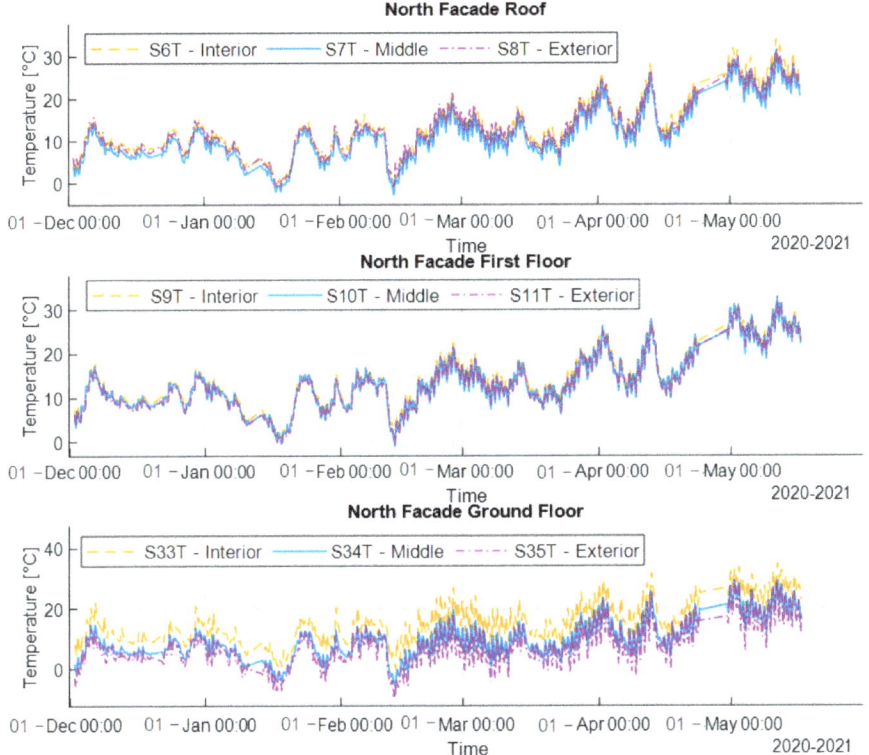

Figure 16. Temperature data provided by the sensors for the northern side of the building: roof (above), first floor façade (middle), and ground floor façade (below).

Regarding the outdoor relative humidity, two different periods can be identified, i.e., from 1 December 2020 to 21 March 2021, where the outdoor relative humidity varies between 90 and 95% and from 22 March 2021 to 15 May 2021, where the outdoor relative humidity varies between 74 and 78% (Figure 17).

The humidity sensors placed on the internal sides of the walls shows there was a 10–20% daily variation, reaching maximum values of around 70% during the night and lower figures of 40% during the day in the winter, while in spring values from 45 to 25% are observed. Slightly higher values for the relative humidity were obtained for the sensors S13U placed on the west ground floor corner and S14U placed on the east ground floor corner, close to the north facade. The reduced ventilation rate, due to the lack of occupancy, might be the main reason for the high indoor relative humidity levels.

The LSF module is also equipped with a CO_2 sensor, whose provided data are reflected in Figure 18. The carbon dioxide (CO_2) concentration was evaluated and classified according to the categories defined by EN 15,251 [37]. The CO_2 concentrations were measured during the same period of time. Higher values of CO_2, between 300 and 350 parts per million (ppm) were recorded during human interference in the building for maintenance or observation. However, even the top values of CO_2 concentration remain in the normal CO_2 concentration of air quality, which corresponds to category I according to EN 15251.

As the LSF experimental module is completely off-grid and during the monitorisation period the wind turbine was not yet installed, there were two intervals (10 January 2021–08:02 AM to 14 January 2021–01:42 PM and 23 April 2021–07:18 AM to 30 April 2021–

02:11 PM) in which the energy production of the roof PV was insufficient (due to heavy cloud cover), and the sensors could not provide data (as the graphics show).

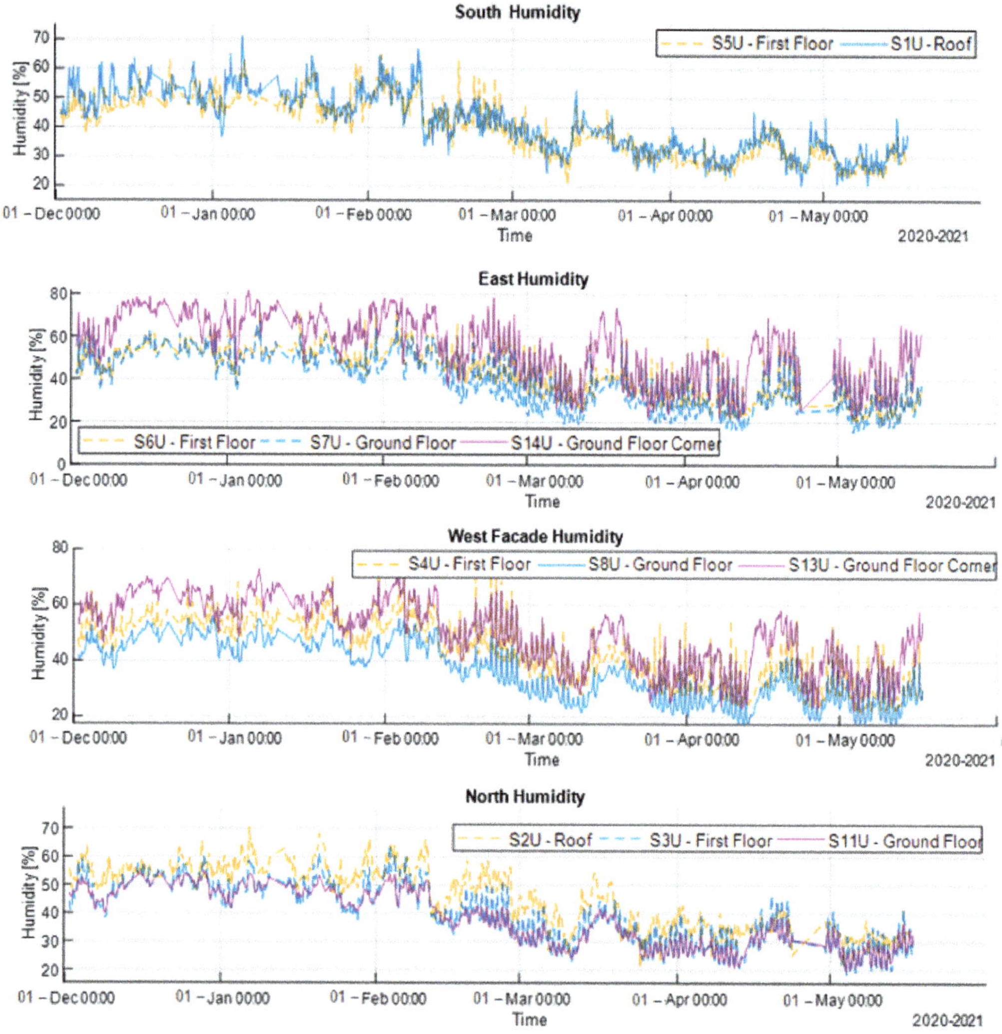

Figure 17. Humidity data provided by the sensors from various locations of each façade.

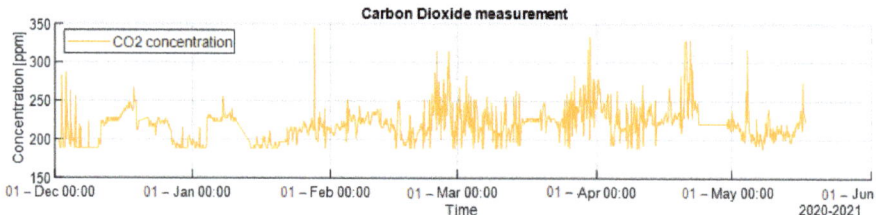

Figure 18. Carbon dioxide (CO_2) concentration within the experimental module.

3.2. Analysis of the Energy Production

The next section presents an energy analysis report of the LSF module. The energy shown in the following diagrams is provided only by the roof PV. The wind turbine and PV louver were not integrated into the physical system during the monitoring period. For comparison, a winter month, December (Figure 19), and a final spring month, May (Figure 20), were chosen. The blue line represents the state of charge of the storage system, the orange bars represent the energy production of the roof PV, while the red bars represent the energy consumptions by the LSF module. Against expectations, the higher energy production is in December, due to the necessary energy to charge the batteries.

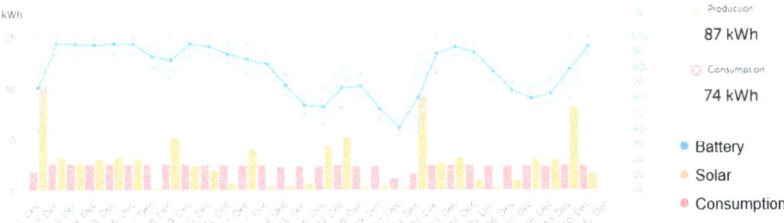

Figure 19. Energy analysis report of the LSF module during December 2020.

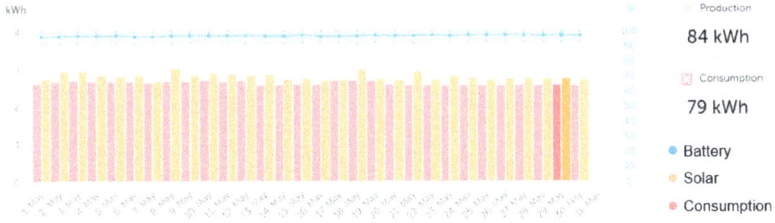

Figure 20. Energy analysis report of the LSF module during May 2021.

It can be observed that there are periods of up to 10 kWh energy production/day, which compensate for cloudy and snowy days when the energy is assured from the batteries. In normal operation, the LSF module energy consumption is constant and is approximately 2.6 kWh/day (Figure 21); however, to not discharge the batteries more than 40% to extend the batteries life, the consumption was reduced and only the essential equipment was powered. Figure 21 presents the hourly energy analysis for two summer days. Over the nights, the batteries are discharged up to 92–93%, which covers eight–nine hours without solar radiation. The essential equipment consists of the SCADA system and the measuring system. In the end, if we want to assume the total energy that can be generated by the three renewable energy sources (roof photovoltaic panels, louver photovoltaic panels, and wind turbine), we can state that the energy provided is around 5 kWh during peak production.

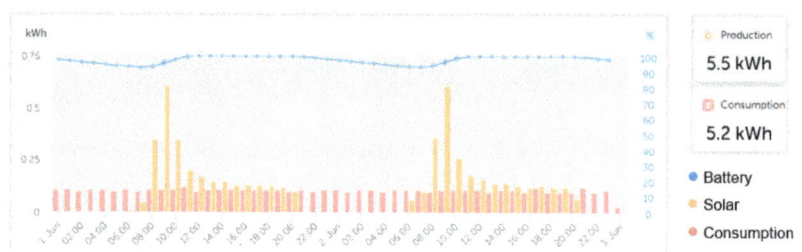

Figure 21. Energy analysis report of the LSF module, two days overview.

3.3. Conditions and Limitations of the Study

The outcomes of this study are based on the analysis of only six months of thermal behaviour, it was not possible to statistically analyse and compare the behaviour of this building during large periods of time. The results presented are particular to the Banat zone due to the particular type of climate. However, the benefits of holistically designed buildings and of the recycled-PET thermal wadding insulation can be extrapolated to other areas.

Another limitation of this study is the fact that the building is an experimental laboratory that was not constantly inhabited during the monitorisation period. Since this building is mainly used for short periods of time (maintenance or observation), potential actions of building occupants who could alter in any way the indoor environmental quality were not addressed.

Furthermore, at the time of monitorisation, the external photovoltaic shading lamellae were not installed, a fact which led to the lack of sun shading of the glass curtain and a lower rate of indoor comfortable hours in the days with a clear sky and outside temperatures above 20 °C. Another equipment that was not yet installed at the time of the monitorisation period was the wind turbine, which could have been helpful with the energy production during the two periods of heavy cloud cover of the sky when the energy production of the roof PV was insufficient.

4. Conclusions

Given the EU's commitment in the Paris Agreement to limit the increase in global average temperature to less than 1.5 °C above pre-industrial levels and the significant contribution of GHG emissions of the building sector, it is imperative to minimize both the embodied GHG emissions and the operating GHG emissions from the construction and renovation of buildings. The weight of embodied GHG emissions varies with the design, the origin of energy, the mix of materials used, and with the construction of the buildings, while the operating GHG emissions are determined by the building performance and the amount of renewable energy in building energy consumption in correlation with fossil-based energy sources.

To achieve buildings with a reduced impact on the environment (either from the construction or operational phase) and moderate construction costs, one needs the embody a holistic approach, integrating cross-disciplinary analysis and multi-object optimization.

The holistic design approach of the LSF experimental module presented in the paper involved the adoption of various criteria regarding a sustainable building, such as resource efficiency, material efficiency, ecology preservation, environmentally conscious design, life cycle design, reusable/recyclable materials, modular and standardized design, environment-friendly demolition method, waste recycling and reuse, safety design, consideration of life cycle cost, materials cost and health and well-being. Besides assigning renewable energy sources, conservation sources of energy, and inclusion of passive design strategies, to meet energy efficiency targets, the holistic design of the modular laboratory required an integrated design with consideration for technology and operation.

The monitored energy system included in the design of the LSF experimental module brings an important contribution in having a genuine overview of the building's performance during the operational phase.

Thermal performance monitoring included the assessment of air temperature and relative humidity. These parameters were evaluated over six months, from 1 December 2020 to 15 May 2021. The results showed that it is difficult to obtain adequate thermal comfort conditions without an active heating system.

Despite the fact that the building does not have any mechanically cooling, heating or dehumidification systems to augment the indoor comfort conditions, the recordings showed for the monitored period that during mid-season, the rooms had adequate comfort conditions.

However, the presented results highlight the importance of the occupants' actions that will influence the thermal performance of the building through heating/cooling and ventilation, useful for removing air pollutants and heat loads.

Not controlling the solar radiation (as the shading PV lamellae were not installed yet) increased the risk of overheating hours, as the results showed for the last two weeks of monitorisation. The future use of an external solar shading device will be more efficient in reaching thermal comfort conditions within comfort limits, reducing the risk of excessive solar gains and overheating. The use of an external solar shading device, such as the PV louver, will be more effective to reduce the risk of excessive solar gains and overheating during summer.

Furthermore, additional studies are needed to complement and understand the benefits of improving thermal comfort conditions and reducing energy requirements for heating, to validate the effectiveness of the research presented and to disseminate the assets on a holistic design approach. In future works, a main issue of the model should be addressed, i.e., human behaviour interaction with the building in defining energy demand. Another important aspect is the energy production from renewable sources. The wind turbine and PV louver are now integrated into the physical system. From September 2021, a one-year monitoring period will start including all these missing components.

Author Contributions: Conceptualization, R.B. and A.C.; methodology, R.B.; software, R.B., M.G. and D.V.; validation, V.U. and A.C.; resources, R.B., V.U., M.G and D.V.; writing—original draft preparation, R.B.; writing—review and editing, M.G., A.C. and V.U.; data acquisition M.G., D.V. and I.P.; visualization, V.U.; supervision, A.C. and V.U.; project administration, V.U.; funding acquisition, V.U. All authors have read and agreed to the published version of the manuscript.

Funding: This research was funded by the Romanian Ministry of Research and Innovation, CCCDI—UEFISCDI, project number PN-III-P1-1.2-PCCDI-2017-0391.

Institutional Review Board Statement: Not applicable.

Informed Consent Statement: Not applicable.

Data Availability Statement: The data presented in this study are available on request from the corresponding author.

Acknowledgments: This work was supported by a grant of the Romanian Ministry of Research and Innovation, CCCDI—UEFISCDI, project number PN-III-P1-1.2-PCCDI-2017-0391/CIA_CLIM –Smart buildings adaptable to the climate change effects, within PNCDI III.

Conflicts of Interest: The authors declare no conflict of interest. The funders had no role in the design of the study; in the collection, analyses, or interpretation of data; in the writing of the manuscript, or in the decision to publish the results.

References

1. EU Commission; DG Energy. *EU Energy Statistical Pocketbook and Country Datasheets (UPDATED JUNE 2021)*; EU Commission: Brussels, Belgium, 2019.
2. EU Commission. *Directive 2010/31/EU of the European Parliament and of the Council of 19 May 2010 on the Energy Performance of Buildings*; EU Commission: Brussels, Belgium, 2010.

3. EU Commission. *Directive 2012/27/EU of the European Parliament and of the Council of 25 October 2012 on Energy Efficiency, Amending Directives 2009/125/EC and 2010/30/EU and Repealing Directives 2004/8/EC and 2006/32/ECText with EEA Relevance*; EU Commission: Brussels, Belgium, 2012.
4. EU Commission. *Clean Energy for All Europeans Package*; EU Commission: Brussels, Belgium, 2017.
5. *Paris Agreement*; United Nations Framework Convention on Climate Change: Paris, France, 2015.
6. European Commission. *The European Green Deal*; EU Commission: Brussels, Belgium, 2019.
7. *Study on Energy Savings Scenarios 2050*; Fraunhofer Institute for Systems and Innovation Research ISI: Karlsruhe, Germany, 2019.
8. Gagnon, R.; Gosselin, L.; Armand Decker, S. Performance of a sequential versus holistic building design approach using multi-objective optimization. *J. Build. Eng.* **2019**, *26*, 100883. [CrossRef]
9. Buzatu, R.; Muntean, D.; Ungureanu, V.; Ciutina, A.; Gireadă, M.; Vitan, D. Holistic energy efficient design approach to sustainable building using monitored energy management system. *IOP Conf. Ser. Earth Environ. Sci.* **2021**, *664*, 012037. [CrossRef]
10. European Academies Science Advisory Council, Secretariat. *Decarbonisation of Buildings: For Climate, Health and Jobs*; German National Academy of Sciences Leopoldina: Halle, Germany, 2021; EASAC Policy Report 43; ISBN 978-3-8047-4263-5.
11. Gaujena, B.; Agapovs, V.; Borodinecs, A.; Strelets, K. Analysis of Thermal Parameters of Hemp Fiber Insulation. *Energies* **2020**, *13*, 6385. [CrossRef]
12. Limpens, G.; Jeanmart, H.; Maréchal, F. Belgian Energy Transition: What Are the Options? *Energies* **2020**, *13*, 261. [CrossRef]
13. Mangers, J.; Minoufekr, M.; Plapper, P.; Kolla, S. An Innovative Strategy Allowing a Holistic System Change towards Circular Economy within Supply-Chains. *Energies* **2021**, *14*, 4375. [CrossRef]
14. Lumpkin, D.R.; Horton, W.T.; Sinfield, J.V. Holistic synergy analysis for building subsystem performance and innovation opportunities. *Build. Environ.* **2020**, *178*, 106908. [CrossRef]
15. Solaimani, S.; Sedighi, M. Toward a holistic view on lean sustainable construction: A literature review. *J. Clean. Prod.* **2020**, *248*, 119213. [CrossRef]
16. Kiani Mavi, R.; Gengatharen, D.; Kiani Mavi, N.; Hughes, R.; Campbell, A.; Yates, R. Sustainability in Construction Projects: A Systematic Literature Review. *Sustainability* **2021**, *13*, 1932. [CrossRef]
17. Timișoara Topographic Map, Elevation, Relief. *Topographic-MapCom*. Available online: https://en-au.topographic-map.com/maps/9h9q/Timi%C8%99oara/ (accessed on 20 May 2021).
18. Volumul I: Populația Stabilă (Rezidentă)—Structura Demografică. Recensamant. 2011. Available online: http://www.recensamantromania.ro/noutati/volumul/ (accessed on 20 May 2021).
19. Climate of the World: Romania. Available online: https://www.weatheronline.co.uk/reports/climate/Romania.htm (accessed on 20 May 2021).
20. Timișoara, Romania—Detailed Climate Information and Monthly Weather Forecast. Weather Atlas. Available online: https://www.weather-atlas.com/en/romania/timisoara-climate (accessed on 21 May 2021).
21. U.S. Department of Energy. Weather Data by Location. EnergyPlus. Available online: https://energyplus.net/weather-location/europe_wmo_region_6/ROU//ROU_Timisoara.152470_IWEC (accessed on 21 May 2021).
22. US Department of Energy. Climate Consultant 6.0 Software, USA (Build. 16). Available online: https://climate-consultant.informer.com/versions/ (accessed on 20 May 2015).
23. Bojariu, R.; Bîrsan, M.-V.; Cică, R.; Velea, L.; Burcea, S.; Dumitrescu, A.; Cărbunaru, F.; Lenuța, M. *Schimbările Climatice: De la Bazele Fizice la Riscuri și Adaptare*; Administrația Națională de Meteorologie: Bucharest, Romania, 2015.
24. Banc, Ș.; Croitoru, A.-E.; David, N.A.; Scripcă, A.S. Changes Detected in Five Bioclimatic Indices in Large Romanian Cities over the Period 1961–2016. *Atmosphere* **2020**, *11*, 819. [CrossRef]
25. Ionac, N.; Ciulache, S. *Atlasul bioclimatic al României*; Ars Docendi: Bucharest, Romania, 2008.
26. Intergovernmental Panel on Climate Change. *AR5 Climate Change 2014: Impacts, Adaptation, and Vulnerability—IPCC*; IPCC: Geneva, Switzerland, 2014.
27. Arup North America Ltd., Argos Analytics LLC. WeatherShiftTM 2.0 Software, San Francisco, California. Available online: https://www.weathershift.com/ (accessed on 20 May 2021).
28. Veljkovic, M.; Johansson, B. Light steel framing for residential buildings. *Thin Walled Struct.* **2006**, *44*, 1272–1279. [CrossRef]
29. Gorgolewski, M. Developing a simplified method of calculating U-values in light steel framing. *Build. Environ.* **2007**, *42*, 230–236. [CrossRef]
30. Santos, P.; Martins, C.; da Silva, L.S. Thermal performance of lightweight steel-framed construction systems. *Metall. Res. Technol.* **2014**, *111*, 329–338. [CrossRef]
31. Ciutina, A.; Mirea, M.; Ciopec, A.; Ungureanu, V.; Buzatu, R.; Morovan, R. Behaviour of wedge foundations under axial compression. *IOP Conf. Ser. Earth Environ. Sci.* **2021**, *664*, 012036. [CrossRef]
32. Buzatu, R.; Muntean, D.; Ciutina, A.; Ungureanu, V. Thermal Performance and Energy Efficiency of Lightweight Steel Buildings: A Case-Study. *IOP Conf. Ser. Mater. Sci. Eng.* **2020**, *960*, 032099. [CrossRef]
33. Intini, F.; Kühtz, S. Recycling in buildings: An LCA case study of a thermal insulation panel made of polyester fiber, recycled from post-consumer PET bottles. *Int. J. Life Cycle Assess.* **2011**, *16*, 306–315. [CrossRef]
34. URBAN-INCERC. *Saltele de Vatelina Pentru Izolare Termica si Fonica—Tip Softex*; Agrement Tehnic 001ST-03/039-2018; URBAN-INCERC: Timisoara, Romania, 2018.

35. Hulea, D.; Muntean, N.; Gireada, M.; Cornea, O. A Bidirectional Hybrid Switched-Capacitor DC-DC Converter with a High Voltage Gain. In Proceedings of the 2019 International Aegean Conference on Electrical Machines and Power Electronics (ACEMP) & 2019 International Conference on Optimization of Electrical and Electronic Equipment (OPTIM), Istanbul, Turkey, 2–4 September 2019. [CrossRef]
36. LOGO! Logic Module. *SiemensCom Global*. Available online: https://new.siemens.com/global/en/products/automation/systems/industrial/plc/logo.html (accessed on 2 June 2021).
37. EN15251. *Indoor Environmental Input Parameters for Design and Assessment of Energy Performance of Buildings-Addressing Indoor Air Quality, Thermal Environment, Lighting and Acoustics*; BSI: London, UK, 2007.

Article

The Potential of the Reed as a Regenerative Building Material—Characterisation of Its Durability, Physical, and Thermal Performances

Raphaele Malheiro [1], Adriana Ansolin [1], Christiane Guarnier [2], Jorge Fernandes [1], Maria Teresa Amorim [3], Sandra Monteiro Silva [1] and Ricardo Mateus [1,*]

[1] Institute for Sustainability and Innovation in Structural Engineering (ISISE), University of Minho, 4800-058 Guimarães, Portugal; raphamalheiro@gmail.com (R.M.); pg37064@alunos.uminho.pt (A.A.); jepfernandes@me.com (J.F.); sms@civil.uminho.pt (S.M.S.)
[2] Federal Center for Technological Education "Celso Suckow da Fonseca" (CEFET/RJ), Rio de Janeiro, RJ 20271-110, Brazil; christiane.guarnier@cefet-rj.br
[3] Centre for Textile Science and Technology (2C2T), University of Minho, 4800-058 Guimarães, Portugal; mtamorim@det.uminho.pt
* Correspondence: ricardomateus@civil.uminho.pt

Citation: Malheiro, R.; Ansolin, A.; Guarnier, C.; Fernandes, J.; Amorim, M.T.; Silva, S.M.; Mateus, R. The Potential of the Reed as a Regenerative Building Material—Characterisation of Its Durability, Physical, and Thermal Performances. *Energies* **2021**, *14*, 4276. https://doi.org/10.3390/en14144276

Academic Editor: Paulo Santos

Received: 9 June 2021
Accepted: 12 July 2021
Published: 15 July 2021

Publisher's Note: MDPI stays neutral with regard to jurisdictional claims in published maps and institutional affiliations.

Copyright: © 2021 by the authors. Licensee MDPI, Basel, Switzerland. This article is an open access article distributed under the terms and conditions of the Creative Commons Attribution (CC BY) license (https://creativecommons.org/licenses/by/4.0/).

Abstract: Knowing the properties of vernacular materials is crucial to heritage conservation and to develop innovative solutions. Reed, considered to be a carbon-neutral and a carbon dioxide sink material, has been used for centuries for diverse uses. Its high availability and properties made it a popular building material, including in Portuguese vernacular architecture. An experimental investigation was conducted to evaluate the physical performance, thermal performance, and durability of the reed found in Portugal since the characterisation of this material was not found in previous studies. The influence of geometric characteristics and the presence of nodes on these properties were also analysed, and the results showed that they are irrelevant. The studied reeds were found to have an adequate thermal performance to be used as thermal insulation. Their thermal resistance (1.8 m^2·°C/W) and thermal conductivity (0.06 W/m·°C) are under the requirements defined by Portuguese regulations on thermal insulation materials. Overall, the physical characteristics (moisture content, density, and retraction) are compatible to its use in the construction. Concerning durability, there was only a trend for mould growth in particular environments. The results provide valuable data to be considered in the development of new construction products based on this natural and renewable material. Additionally, considering the studied samples, the reed found in Portugal has characteristics suitable for use as a building material, especially as a thermal insulation material.

Keywords: reed (*Arundo donax*); material characterisation; sustainability; natural materials; vernacular architecture

1. Introduction

Reed has been used for thousands of years in diverse uses by many cultures [1–3]. The reed has been a conventional construction material since ancient times. It was used to make baskets, fences, windbreakers, building walls, roofs, floors, shading barriers, and temporary shelters for men and animals; music instruments; paper; and bio-fuel. Its characteristics, such as its high availability, lightweight stem, and fair mechanical strength and high flexibility (due to the tubular shape of the stem), has allowed for different uses of reed and made it a popular component as a construction material [2–5].

Its low mechanical strength and easy combustion have made it difficult to use in building structures, being more commonly used in ceilings or supports for covering roofs, wall panelling to improve their thermal performance, or to complement the earthquake resistance of internal and external walls [3]. However, the solution adopted in the construction of the external walls of earthquake-resistant buildings with mats of *Arundo donax* L. in Calabria, Italy, has shown good durability since many examples are still perfectly intact [3].

Around the world, it is possible to see vernacular buildings entirely built out of reeds, just with some construction elements (walls and ceilings), and using reeds together with other materials [6]. In Portuguese vernacular architecture, the reed was also used as a construction material, particularly in the southern region, probably due to its high availability. However, since it is spread throughout the Portuguese territory [7], its use in vernacular architecture can also be found in other regions of mainland and island Portugal [8,9].

Reed was used mainly as an element for thermal insulation in walls [10] and roofs. The encaniçado is an example of the use of reeds in the construction of roofs (Figure 1). This technique allows for a roof to achieve better thermal performance [11], and it is widely used in the vernacular buildings of the Alentejo and Algarve regions in noble and common buildings. The encaniçado consists of rows of reeds tied together [12] and to the structural timber beams. In some cases, a layer of mortar is applied over the reeds to flatten the surface and connect them to the roof tiles that are placed over the mortar.

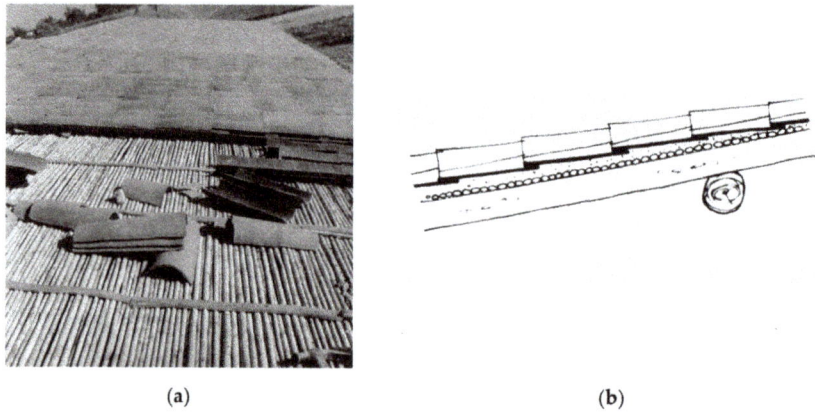

(a)   (b)

Figure 1. Use of reed for roof construction: (**a**) common gable building [8] and (**b**) section of a roof [8].

Concerning the use of reed as an element for thermal insulation in walls, the walls of palheiros, and the tabique technique are important examples of Portuguese vernacular architecture using reeds. Palheiros are palafitic timber buildings, and although their timber walls usually just have an air cavity, there are records of palheiros in Leirosa (near Figueira da Foz) in which the cavity of the external wall was filled with reeds to improve thermal insulation [8,13] (Figure 2). Reed is also found in earth-filled timber frame walls such as the tabique. The tabique technique can be found almost everywhere in Portugal [14]. It is part of the Portuguese heritage, and similar techniques are applied worldwide [15]. In brief, a tabique is formed by a regular timber frame covered with an earth mortar [14,15]. In Algarve, more specifically in the cities of Lagos and São Brás de Alportel, there are examples where the wooden structure of the tabique walls was replaced by reed panels [16].

Figure 2. Use of reed for wall construction: palheiro in Leirosa [13].

Similar solutions are used worldwide. For example, in Italy, there is a building system uses a timber-framed supporting structure and a sheathing composed of two mats made of *Arundo donax* L. fixed to the walls and then covered with a layer of lime and cement plaster [3].

Knowing the properties of the natural materials used in vernacular architecture is crucial to ensure successful heritage conservation, optimise the use of these materials, and develop innovative solutions. In general, the characterisation of natural materials is a challenge for researchers. The high variability in their properties and the absence of specific standards are the main barriers to these materials characterisation [17]. In the particular case of reeds, from the literature review, it was verified that only a few studies have presented their characteristics.

Regarding physical characteristics, the research results are even scarcer. *Phragmites australis* (PA) (common reed) and *Arundo donax* (AD) (giant reed) are frequently mentioned in these studies on natural building materials. However, it is important to note that these species have a marked difference in their geometry, particularly in average diameter: around 1.0–2.5 cm for PA [18,19] and 2.5–5.0 cm for AD [19,20]. In this sense, the comparison of results from different studies should be made with caution.

Concerning thermo-acoustic properties, different approaches to characterisation have been identified. Some studies have evaluated the thermo-acoustic potential of panels where the reed is the main material, and others have evaluated panels made only with reed [6,21–23]. Both approaches highlight the thermal insulation potential of the reed. Regarding the panels made only with the reed, the review research carried out by Asdrubali et al. [24] showed thermal conductivity values for the reed of between 0.045 and 0.056 W/m·°C. Asdrubali et al. [22] presented a thermo-acoustic characterisation of reed panels (PA) while considering different geometries, densities, humidity rates, and stems shapes in experimental research. The maximum diameter of the used reed was 1.5 cm. The thermal characterisation was carried out in a guarded hot plate and hotbox apparatus. According to the authors [22], the layout and characteristics of the reeds did not strongly influence the equivalent thermal conductivity, reaching values between 0.055 and 0.065 W/m·°C. However, the acoustic behaviour was strongly affected by the stem configuration. Since there was a significant difference in the average diameter of the different reed species, it is essential to understand whether these conclusions apply, for example, to panels made with AD.

Regarding a panel where the reed is the main material, the AD harvested in Portugal was used in a prototype of a building solution based on earth and reeds (stem and fibres) [25]. The prototype built in Lisbon had its indoor and outdoor temperatures monitored during the different seasons. The researchers [25] concluded that the solution contributed to controlling the interior air temperature, given the thermal amplitudes that were registered outside.

Molari et al. [26] used the same bamboo standards to evaluate the mechanical properties of reed (AD). According to the results, the reed's compressive strength (127.4 MPa) was 20% higher than the tensile strength (103.7 MPa) and almost twice the axial compressive strength (57.0 MPa). On the other hand, the axial Young's modulus under tension (15.3 MPa) was similar to that under compression (13.4 MPa) and about three times higher than the shear modulus (2.96 MPa). According to the authors [26], the results showed good mechanical properties of the reed that were similar to those of various species of bamboo used in constructions. However, it is important to note that, given the small reed dimensions compared to bamboo, variations concerning the bamboo standards were carried out. Therefore, caution is needed when comparing these results with the results of other research.

Furthermore, the water content was not considered in the discussion of the results. Conte et al. [27] studied the effect of water content on the structural and mechanical properties of reed (AD). The researchers used a combination of different analytical techniques (such as calorimetry and fast field cycling NMR relaxometry). They concluded that reed bending properties are strongly affected by the presence of bound water. Thus, it is important to consider the physical properties of the reed when studying its mechanical properties.

There is an important gap in the knowledge of the physical properties of reed. In recent literature, there has been no research focused on these properties. Some research with mainly thermo-mechanical objectives included the determination of some physical characteristics of the reed. Soliman [28] studied reed (AD) use in thermal insulation and, in parallel, also determined its water absorption and moisture content. After three months of drying the reed, they found 12.11% of moisture and 52.60% of water absorption (after immersion in water for 24 h at a room temperature of 23 °C). The moisture content was higher than values achieved in other research [26,29]. This difference could have been related to the test methodology.

In their study focused on mechanical characterisation, Molari et al. [26] determined the moisture content of reed, reaching values between 7.09% and 8.96%. Ortuño [29] carried out a reed (AD) characterisation for use as a construction material and determined its moisture content and density. After one year of drying, they determined the reed density as received (583 kg/m^3), the density of anhydrous reed (537 kg/m^3), and the density of saturated reed (1040 kg/m^3). Since the reed used as a construction material does not receive any treatment, they considered the density equal to 583 kg/m^3 and determined its moisture content to be 8.63%. This value was similar to moisture content value found by Molari et al. [26].

Since the properties of natural materials can be site-dependent [29,30] and no characterisation of reed from Portugal was found in previous studies, this research fills a gap in this area of knowledge, thus assuming an innovative position. In the present paper, the characterisation results of Portuguese reed are presented and discussed. This experimental investigation assessed the physical and thermal behaviour of reed and its durability, aiming to produce knowledge that can be used to design new solutions for more sustainable construction and the conservation of vernacular buildings.

2. Materials and Methods

2.1. Specimens

The giant reed (*Arundo donax*) used in this research was harvested in Serpa, inland Southern Portugal. The location was chosen due to the abundance of the reed [7] and proximity to areas where the reed has an important presence in vernacular architecture, as mentioned in the previous section.

The harvest took place in winter. In winter, the plant has less sap (it is drier), and, therefore, it can regenerate quickly. The cut stems have less moisture, dry faster, and are less susceptible to biological agents. According to Alentejo's locals, "reed harvested in January lasts all year" [18].

The harvested reeds were stored in a laboratory environment (20 ± 2 °C and 50 ± 2% RH) in a vertical position, and they remained there for six months in the drying process. After the dying process, the reeds with cracks or other anomalies were discarded, and the others were identified and classified according to their external diameter. To characterise the reeds, they were divided into three groups: G0—diameter less than 11 mm; G1—diameter between 11 and 15 mm (average thickness of 1.6 mm); and G2—diameter between 15 and 22 mm (average thickness of 2.5 mm). G0 was discarded because it showed insufficient mechanical resistance, often breaking during the works. G1 and G2 were characterised in three ways: physical properties, thermal properties, and durability.

2.2. Physical Properties

Since there are no normative procedures for the characterisation of reeds, the methods used were adapted from other materials. Due to the differences between the complex reed node structure and the reed stem, it was decided to assess the node influence on the results. Thus, in the physical tests, reed samples with and without nodes were studied. For each test, five samples were used (5-cm-length stems).

2.2.1. Moisture Content

The moisture content (MC) test was carried out based on the Portuguese Standard NP—614 [31] for wood. The test consisted of drying the sample in an oven (103 ± 2 °C) until it reached a constant mass. The moisture content was the difference between the wet mass (m_0 (g)) and the dry mass (m_1 (g)) divided by the dry mass, according to Equation (1).

$$MC = \frac{(m_0 - m_1)}{m_1} \times 100, \quad (1)$$

2.2.2. Apparent Density

The apparent density (ρ) test was carried out based on the Portuguese Standard NP—616 [32] for wood. The test consisted of determining the specimen mass and volume at a specific moisture content. The graphic program AutoCAD was used to determine the specimen's volume. The apparent density was the mass (m (g)) divided by the volume (v (m^3)), according to Equation (2).

$$\rho = \frac{m}{v}, \quad (2)$$

2.2.3. Dimensional Stability—Retraction

The linear retraction (ε) test was carried out based on NP—615 [33] for wood. The test consisted of determining the variation in studied dimensions after saturation and drying. The linear retraction was the difference between the saturated specimen dimensions (l_1 (mm)) and the air-dried specimen dimensions (l_2 (mm)) divided by the oven-dried specimen dimensions (l_3 (mm)), according to Equation (3).

$$\varepsilon = \frac{(l_1 - l_2)}{l_3} \times 100, \quad (3)$$

Reed walls' dimensional variations in length, diameter, and thickness were monitored during the test. Concerning diameter, two directions were monitored: 1–3, 2–4 (Figure 3a). Concerning thickness, four points were monitored: 1, 2, 3, and 4 (Figure 3a). Concerning the length, four directions were monitored: 1–1, 2–2, 3–3, and 4–4 (Figure 3b).

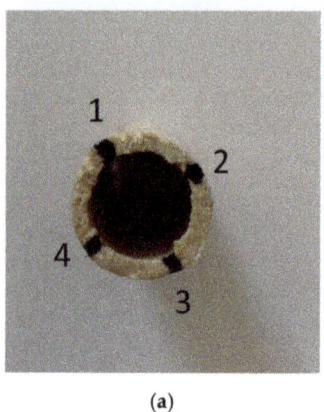
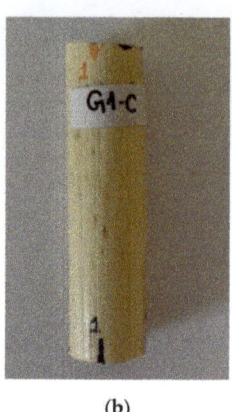

(a)   (b)

Figure 3. Markings used in the reeds to carry out the retraction test: (**a**) diameter and (**b**) length.

2.2.4. Capillary Water Absorption

The principle of the capillary water absorption (A) test consisted of determining the amount of water absorbed by porous materials over a certain time. A dry specimen was placed in an oven at a controlled temperature, and the amount of water absorbed through only one surface of a non-saturated specimen immersed in a water film of 5 ± 1 mm was measured. The absorption of water through capillarity was the difference between wet mass (m_i (g)) that had one surface in contact with water during a time (t_i) and the dry mass (m (g)) divided by the superficial area that was in contact with water (a (mm^2)), according to Equation (4).

$$A_i = \frac{m_i - m}{a}, \qquad (4)$$

The samples were dried over 24 h in an oven at 100 ± 5 °C. Afterwards, the reed base (external and internal wall) was isolated with silicone to ensure a unidimensional penetration of the water (Figure 4a). A steel wire support was built to prevent the reeds from floating during the test (Figure 4b). The samples were immersed in a water film of 5 ± 1 mm, and the water level was monitored (Figure 4c). The amount of absorbed water was measured after 15, 30, 60, 90, 120, 150, 180, 210, 240, 300, 360, 1440, and 2880 min (Figure 4d).

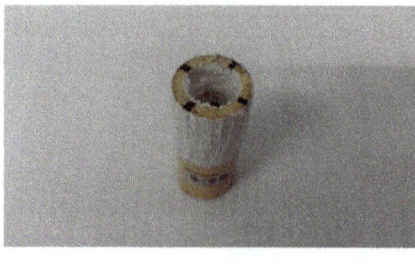
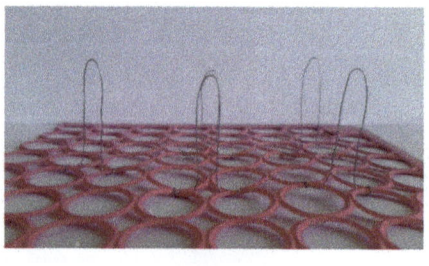

(a)   (b)

Figure 4. *Cont.*

(c)   (d)

Figure 4. Capillary water absorption test steps: (**a**) sample preparation, (**b**) steel wire support, (**c**) samples immersed in water, and (**d**) weighing the sample on a precision balance.

2.3. Thermal Properties

The characterisation of the thermal properties of the reed was based on the analysis of its thermal transmittance, thermal resistance, and thermal conductivity. Reed panels were built to analyse these properties using the hotbox test.

2.3.1. Reed Panels

Reed panels of 15 cm × 15 cm (Figure 5a) were built to carry out the calibrated hotbox tests. The reed stems were overlapped to create a 10-cm-thick panel (Figure 5b). In order to contain the reed stems, the panels were tied with steel wire. A steel wire was applied in the borders of the panel to decrease the influence on the heat flux during the hotbox test. Two types of panels were built: type 1 (only G1 reeds stalks) and type 2 (only G2 reeds stem). Their main characteristics are presented in Table 1.

(a)   (b)

Figure 5. Reed panels used in hotbox test: (**a**) horizontal dimensions and (**b**) thickness.

Table 1. Characteristics of the panels used in the hotbox test.

ID	Quantity of Reed Stems	Reed Stems with Nodes (%)	Quantity of Nodes	Average Diameter (mm)	SD [1]	Average Thickness (mm)	SD	Steel Wire Mass (g)	Panel Mass: Reed and Wire (g)	Density (kg/m^3)
Type 1	84	92.31%	103	13.25	0.93	1.70	0.29	19.50	489.70	208.98
Type 2	56	100%	65	16.83	1.57	2.29	0.52	19.50	479.34	204.37

[1] Standard deviation (SD).

2.3.2. Hotbox Test

The thermal properties of the reed were evaluated in a calibrated hotbox [34] designed and built at the Department of Civil Engineering of the University of Minho, following the recommendations of ASTM C1363-11:2014 [35]. The hotbox comprised two five-sided chambers (external dimensions: 2.0 m × 1.4 m × 1.6 m)—a cold one and a hot one—and one mounting ring placed between the two chambers. The envelope was well insulated and made of extruded polystyrene (thickness: 0.20 m; U = 0.21 W/(m²·°C)).

The reed panel was placed in the mounting ring set between the two chambers. It was enclosed between two medium density fibreboards (MDFs) in order to provide a flat surface for installing the flux meter and thermocouples (Figure 6). The reed panels were tested in a horizontal position because the reed accommodation in this position achieved a lower thermal transmittance than the vertical position [21].

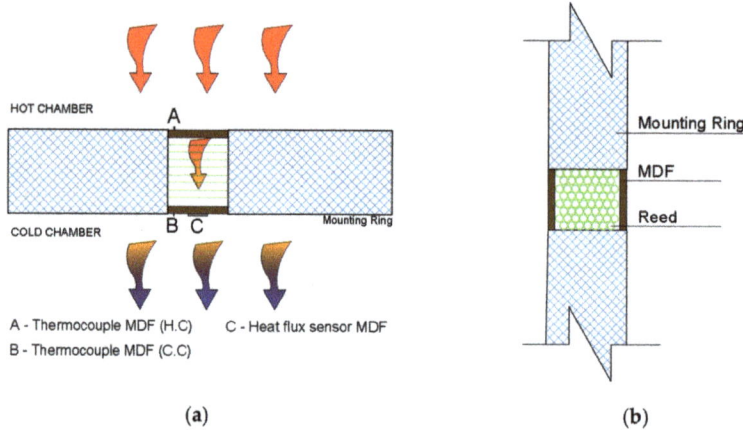

(a) (b)

Figure 6. Set up of reed panel in hotbox test: (a) horizontal view and (b) vertical view.

The temperature difference between the hot and cold chambers was maintained at around 10 °C. The tests were carried out by the thermal flux meter methodology, according to ISO 9869-1:2014 [36], where the heat flux was measured through a heat flux sensor installed in the central part of the reed panel and the temperature was measured by four thermocouples (two in each chamber: one in the middle of the chamber and the other on the surface of the reed panel). With the values of the heat flux (q) and the surface temperatures (T) while using Equation (5), it was possible to determine the thermal resistance (Re_{set}) of the set (reed panel and MDF), where ΔT is the difference between the surface MDF temperature in the hot and cold chambers.

$$Re_{set}\left[m^2{}^\circ C/W\right] = \frac{\Delta T}{q}, \qquad (5)$$

The thermal resistance of the reed panel (Re_{reed}) was determined using Equation (6), where Re_{MDF} is the thermal resistance of the MDF used.

$$Re_{reed}\left[m^2{}^\circ C/W\right] = Re_{set} - (2*Re_{MDF}), \qquad (6)$$

The thermal transmittance (U_{reed}) and thermal conductivity (λ_{reed}) of the reed panel were determined using Equations (7) and (8), where e is the thickness of the panel.

$$U_{reed}\left[W/m^2{}^\circ C\right] = \frac{1}{Re_{reed}}, \qquad (7)$$

$$\lambda_{reed}[W/m°C] = \frac{e}{Re_{reed}}, \qquad (8)$$

2.4. Durability—The Mould Development Test

Reed is an organic material. As a natural building material, reed becomes mouldy like wood or wood-based products in favourable conditions [37]. Mould is often an early indication of increased moisture levels in buildings. Problems caused by mould are mainly discolouration, odours, and health problems.

In order to evaluate the resistance of the reed for mould growth, an exploratory overstress test was carried out for the samples. The assessment of the emergence and development of mould in the reed samples was made visually with images collected using binocular materials microscope LEICA DM 750 M from Leica Microsystems, a 5 MP HD Microscope Camera Leica MC170 HD (sourced by Leica Microsistemas Lda., Microscopia e Histologia, Carnaxide Portugal) and Leica Application Suite (version 4.12.0).

Four samples in each reed group (G1 and G2) were placed in a Petri dish, visually evaluated, and laid in ARALAB FitoClima 1000EC45 climatic chamber ($22 \pm 2\ °C$ and $90 \pm 5\%$ HR) for 42 days. The chamber conditions were appropriate for mould development in reed [38]. One reed dried in an oven ($103 \pm 2\ °C$) until reaching a constant weight (G1—Dry and G2—Dry), was studied too. The samples were evaluated weekly. To reduce the time the samples were outside the chamber and to avoid changes in the final results, the visual analysis was performed in the shortest possible time of less than 20 min.

During the visual analysis, the quantification of mould growth was based on the mould index used in the experiments for visual inspection (Table 2). To minimise the margin of error, the visual inspection was made independently by two persons.

Table 2. Mould indexing classifications (adapted from [39]).

Mould Index	Coverage	Description of Classification
0	0	No growth
1	0	Some growth detected only with microscopy
2	>10%	Moderated growth detected with microscopy
3	0–10%	Some growth detected visually
4	10–50%	Visually detected
5	50–80%	Visually detected
6	80–100%	Visually detected

3. Results and Discussion
3.1. Physical Properties
3.1.1. Moisture Content

Figure 7 shows the mass loss as a time function. Similar behaviour was observed in all studied samples, regardless of node presence and group. The samples reached a constant mass after around 240 min. There was a marked mass loss during the first 30 min of the test, followed by a residual loss up to 24 h. The verified mass loss was related to the water evaporating during the drying process.

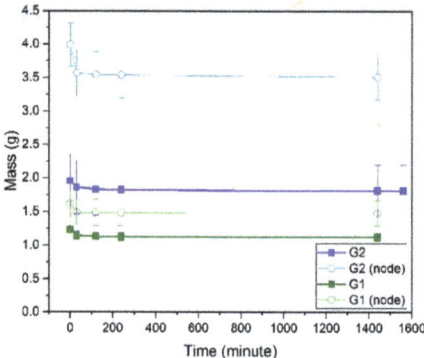

Figure 7. Mass loss of reed as a function of test time.

The results of the moisture content test are presented in Table 3. After six months in the natural drying process (Section 2.1), the reeds showed similar moisture content, regardless of group (G1 or G2). According to the results, the reed diameter (and thickness) had no significant influence on the moisture content. Despite the difference between used methodologies, Molari et al. [26] and Ortuño [29] studied reed (AD) and found similar moisture content values of 8.96% and 8.63%, respectively. Nevertheless, they did not evaluate the differences between regions with and without nodes. Concerning node presence, G2 with node showed a higher moisture content than G2 without node. That was not the case with the G1 reed. The presence of the node likely only had an influence on the moisture content for larger diameters. Conte et al. [27] studied node influence on the moisture content of reed (AD), and they also found a higher moisture content in reeds with nodes. The research did not show the diameter of the studied reed. According to the authors, the largest moisture content of the reed with a node could be explained by the presence of a larger amorphous cellulose fraction inside the node compared with the reed without the node.

Table 3. Moisture content of reeds.

Identification	Moisture Content (%)	SD [1]
G1	9.463	0.195
G1—node	9.432	0.056
G2	9.495	0.242
G2—node	10.28	0.150

[1] Standard deviation.

The values present in Table 3 are close to the moisture values for wood showed in Portuguese standard (12% [31]) and those expected for bamboo after the natural drying process, i.e., between 10% and 15% [40]. In this sense, it is possible to say that, after six months of natural drying, regardless of the diameter and node presence, the reed could reach moisture levels close to the reference values for wood and bamboo.

3.1.2. Apparent Density

Density is related to most physical properties of materials. Therefore, it becomes an indicator of the quality of natural materials and their possible applications. The higher the density, the higher the resistance and the better the quality [29]. Reed density can vary between its base and its tip, and it depends on its age: the more mature the reed, the denser it is [41]. The studied stems were from the middle part of the reed and were between one and two years of age. The graph presented in Figure 8 shows the density values found in this study and their comparison with bamboo and other wood materials used in construction.

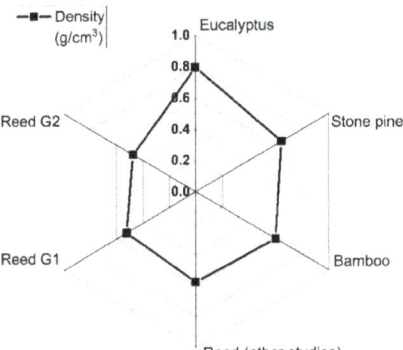

Figure 8. Density of the studied reed (G1 and G2), bamboo, and other wood materials used in the construction.

From the comparison of the different groups, it is possible to observe a slight difference between the density values. G1 at 0.524 g/cm³ (SD 0.063) showed a greater density than G2 at 0.476 g/cm³ (SD 0.011). These values were in accordance with other studies, namely that of Molari et al. [26], 0.577 g/cm³, and Ortuño [29], 0.583 g/cm³. Regarding bamboo, for example, the density varies according to the species (0.6 and 1.0 g/cm³ [40]). No studies were found on the relationship between reed' physical properties and its different species. Regarding the wood used in construction, reed density (G1 and G2) is close to the density of stone pine: 0.645 g/cm³ [42].

3.1.3. Dimensional Stability—Retraction

Because it is a hygroscopic material, reed changes its moisture content according to the relative air humidity, expanding with increasing humidity levels and retracting with decreasing levels. Thus, the dimensional stability of reed plays an important role in its use in construction, whether in the renovation interventions of heritage or new applications.

The results of the linear retraction in length, diameter, and thickness of the reed are shown in Figure 9. The observed retraction values revealed a specific behaviour for each studied part from the reed stem. For example, regardless of the group and node presence, the thickness of the reed had the highest retraction percentage and the length had the lowest, always less than 1%. Concerning length, the found values were similar to those of bamboo from Portugal retraction (0.23%) obtained in [43].

Figure 9 shows that the presence of the node had a more significant influence on G2 than G1, particularly in the reed thickness retraction. The node's influence was more significant in point "1" (G2), where the reed with node showed a retraction in thickness that was 1.80 times greater than the reed without node. Regarding the retraction in diameter, the node presence had no influence in G1 but had a slight influence in G2. Regarding the retraction in length, the values were negligible regardless of the group.

In general, the results showed that the reeds with larger diameters tended to present greater retraction, despite their nodes being less vulnerable to this phenomenon. The thickness retraction reached significant values for G1 and G2. Nevertheless, this behaviour must be considered when reeds are used in situations that require greater geometric stability. This high retraction can also influence cases where reeds are used with joints and accessories.

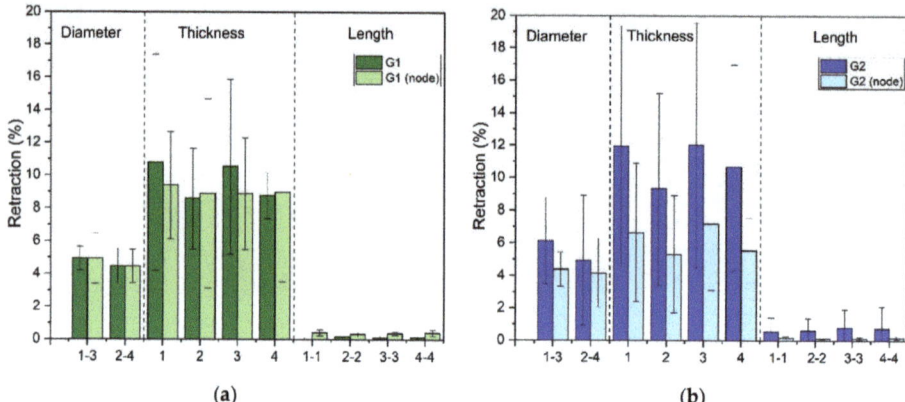

Figure 9. Linear retraction percentage: (a) G1 and (b) G2.

3.1.4. Capillary Water Absorption

The curves shown in Figure 10a illustrate the variation of capillary water absorption as a function of the square root of test time. Figure 10b shows the first stage of the test (first 360 min) in detail.

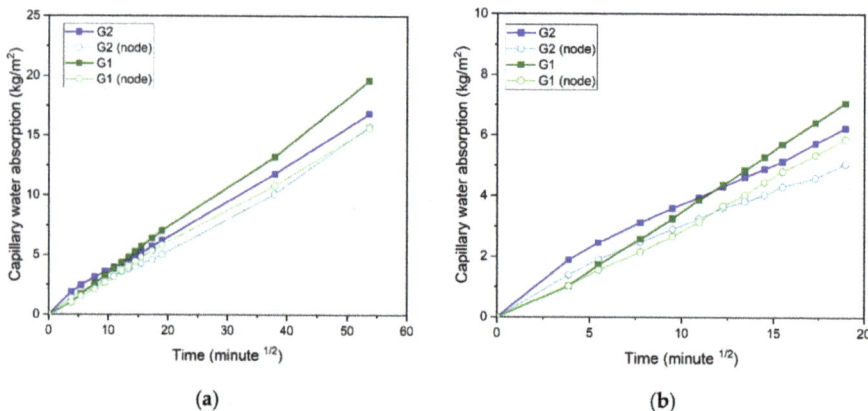

Figure 10. Capillary water absorption as a function of the square root of test time: (a) complete curves and (b) detail.

According to Figure 10a, until 48 h, none of the samples reached a saturation stage. This behaviour may have been related to the hygroscopic nature of the material and showed that the reed could absorb a high amount of water. This high water absorption was in accordance with the results of a study by Soliman [28]. Though they used another method to assess water absorption (total immersion of reed in water during 24 h), they also found a high water absorption value of 52.60%.

Concerning the node presence, it was observed that the samples with nodes absorbed less water than samples without nodes. This "difficulty" in water transport may be related to the dense and complex structure of the node. Reed nodes are likely to be similar to bamboo nodes. In bamboo, the fibrovascular bundles (which transport water and nutrients) are parallel in the stem region, but have a random distribution in the node region [44]. This complex arrangement can make the transportation of water difficult. The retraction test also presented results that corroborated this difference in behaviour between regions without and with nodes.

Concerning groups, G2 absorbed less water than G1. In addition, the lower slope of the curves (Figure 10b) shows that the capillary absorption of G2 was also the slowest. As such, it is likely that the fibrovascular bundles of G2 were smaller than those of G1. Though the capillary water transport in G1 was faster than G2, the transport occurred throughout the reed stem regardless of diameter and resulted in a very similar moisture content for both groups of around 9% (Table 3).

3.2. Thermal Properties

An infrared camera (ThermaCAMTM T400, FLIR Systems AB, Danderyd, Sweden) was used to evaluate the distribution and homogeneity of the temperatures in the hotbox during the test, as well as to detect the presence of thermal bridges or discontinuities. Figure 11 shows the uniformity of temperatures in the cold chamber and no thermal bridges. It is also possible to observe the hot air outlet in the upper ventilation grid.

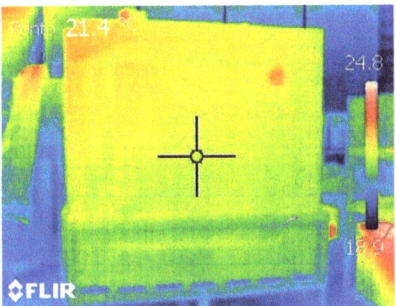

Figure 11. Infrared image from the cold side of hotbox during the test with the reed panel.

The parameters monitored during the hotbox test are shown in Figure 12. Figure 12a refers to the results of panel type 1 (G1), and Figure 12b refers to the results of panel type 2 (G2).

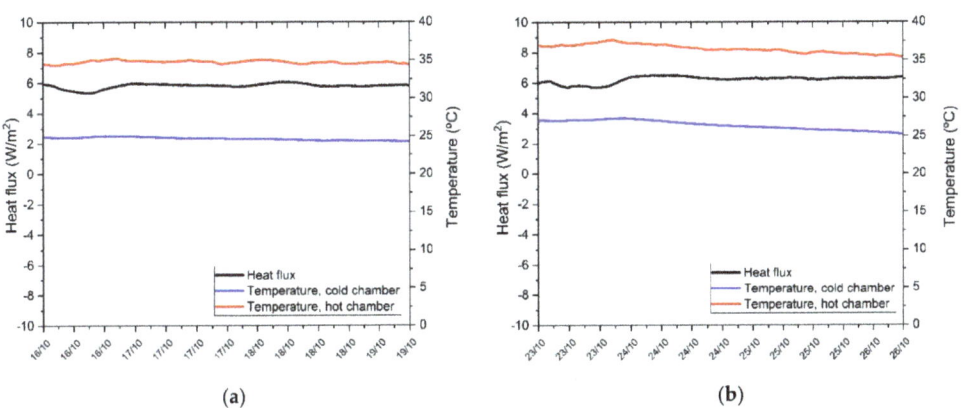

Figure 12. Measurements of the heat flux and hot and cold superficial temperatures in the set: (a) panel reed G1 and MDF; (b) panel reed G2 and MDF.

From Figure 12, it can be seen that despite the heterogeneity of this natural material, the temperature and the heat flux remained stable during the test period (72 h). In the type 1 (G1) test, the cold chamber remained very stable, maintaining a temperature of 24 °C and the hot chamber temperature remained near 34 °C. In the type 2 (G2) test, the cold chamber showed an average temperature near 26 °C and the hot chamber showed an average temperature near 36 °C.

Based on the results from Figure 12 and the application of the equations presented in Section 2.3.2, Table 4 summarises the average values obtained for the thermal properties of the studied reed panels.

Table 4. Parameters monitored during the hotbox and the thermal properties of the studied reed panels.

Panel	Superficial Temperature (°C)		Heat Flux (W/m²)	Re_{set} (m²·°C/W)	Re_{MDF} (m²·°C/W)	Re_{reed} (m²·°C/W)	λ_{reed} (W/m·°C)
	Hot Chamber	Cold Chamber					
Type 1	34.66	24.66	5.61	1.834	0.147	1.540	0.064
Type 2	36.37	26.37	5.88	1.886	0.147	1.592	0.063

According to Table 4, type 1 (G1) and 2 (G2) panels showed similar thermal behaviour. From the results of the performed measurements, two main aspects can be discussed. First, the type 1 (G1) panel had almost twice as many nodes as the type 2 (G2) panel (Table 1). Thus, the number of nodes had no significant influence on the thermal behaviour of the reed panels. Second, the panels were built with similar densities to assess the influence of reed diameter (Table 1). For the studied diameter range (11–22 mm), the reed diameter had no significant influence on the thermal properties of the reed panels. These results were in accordance with those of Asdruballi et al. [22], a study where reeds (PA) with diameters between 3.6 and 8.8 mm were analysed using a guarded hot plate and hotbox test, and conclusions showed that the diameter did not influence thermal conductivity [22]. The studied reed panels showed a good thermal behaviour regardless of the studied group.

Considering the thickness of the reed panels (100 mm), their thermal resistance values represented 60% of the thermal resistance value of some commercially used insulation materials such as rock wool, XPS, and cork (approximately 2.60 m²·°C/W). The obtained thermal conductivity values were similar to those presented in other studies (using reed from different origins) [22,44] and corroborated this type of reed potential for use as a thermal insulation material.

3.3. Durability—The Mould Growth Test

To analyse the moisture content's influence on the durability of the natural materials, reed (without nodes) was subjected to two drying processes: the natural process (G1 and G2) and the oven drying process (G1—Dry and G2—Dry). The quantification of mould growth is shown in Figure 13.

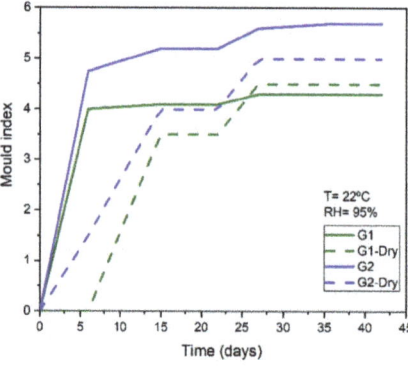

Figure 13. The mould index of the reed as a function of the test time.

From Figure 13, it can be seen that the growth of mould was detected in all studied situations. The mould growth intensity was different depending on the reed group and drying process. G2 had a more intense mould growth than G1 during the entire test period, regardless of the drying process. This behaviour could have been related to the G2 thickness. Its greater thickness than G1 could provide a larger surface for mould development. Since the moisture content of the two groups was similar (Table 3), it is believed that this variable did not influence this specific behaviour.

Concerning the drying process, it is possible to say that reference samples (G1—Dry and G2—Dry) had a less intense mould growth than those subjected to the natural drying process (G1 and G2). In this specific behaviour, moisture content was found to play an important role since the higher moisture content (G1 and G2) provided a more favourable environment for mould development. Tables 5 and 6 show mould growth observations for the larger (G2) and smaller (G1) mould growth intensities observed in this study. In Tables 5 and 6, it is possible to compare the mould index and the images of the samples at four different test moments: 0, 14, 28, and 42 days.

Table 5. Development of surface moulds in G1 dry reed.

Test Period	Mould Index	Bare Eyes	Microscope (10×)
14 days	3.5		
28 days	4.5		
42 days	4.5		

Table 6. Development of surface moulds in G2 reed.

Test Period	Mould Index	Bare Eyes	Microscope (10×)
0 day	0		
14 days	5		
28 days	5.5		
42 days	6		

The general aspects of the samples presented in Tables 5 and 6 are quite different. In the photographs corresponding to the bare-eyes analysis, it is possible to observe intense dark areas in the G2 sample, while in the G1—Dry sample, a generalised brown colour was observed. The microscope images show the evolution of the mould in the sample at the same point of the test.

In the G2 sample, the fifth stage of growth (index 5) was reached after 14 days of testing. The bare-eyes analysis showed that almost 80% of the sample was covered with grey and black mould. After 14 days, the mould growth was moderate, and the last stage of mould growth (index 6) was reached at the end of the test. In this stage, it was possible to see, in the microscope image, a high density of mould in the sample. In the G1—Dry sample, the third stage of growth (index 3) was reached after 14 days of testing, and the maximum growth stage did not exceed index 4. Grey and black mould was observed in the microscope image only after 28 days, and the sample showed a low density of mould growth until the end of the test.

These results show the trend in the development of mould in the studied reeds. The interior surfaces of the reeds were studied individually, but the results were in accordance with the study of [37]. In this research, [37] reeds tied with steel wire were subjected to the overstress test (22 °C and 90% HR), and at the end of 42 days, the external surfaces of the reeds showed a mould development index between 5 and 6. The results of this study also showed the important role played by moisture content in the durability context. Drier reeds lead to less intense and slower mould growth.

These results confirmed the trend in the development of mould in the studied reed. However, these results must be carefully analysed. There are two main aspects to consider. First, the chamber conditions were the ones that maximise mould growth in reed [38]. Similar conditions could occur in particular natural environments. In Portugal, for example, only the coastal region reaches temperature and humidity conditions similar to those studied. This situation can happen during the summer for a short time, usually at night [10]. Second, the reed was studied while considering the resistance to the mould of the inner face, i.e., the most vulnerable one. This area is the less dense side of the reed [41] and is therefore more vulnerable to aggressive agents.

4. Conclusions

To reduce the environmental impact of the construction sector, preserve and renovate vernacular buildings, and create commercial value for locally available natural materials such as the giant reed (*Arundo donax*), it is necessary to know a material's physical and thermal proprieties.

Reed is considered to a carbon-neutral raw material and a carbon dioxide sink, and it has been used for centuries for diverse uses. Its high availability and properties made it a popular building material, including in the construction of Portuguese vernacular buildings.

In the present paper, an experimental study was carried out to characterise the giant reed (*Arundo donax*) that is most common in Portugal. Considering the uses of reed in Portuguese vernacular architecture (as for thermal insulation) and the absence of data on the physical properties of the reed, its thermal and physical properties were assessed in this study. The durability of natural materials plays an important role in this context, so it was also evaluated in this study. The results were presented and discussed in the previous sections, and the main conclusions are as follows:

(1) Based on the studied thermal parameters (thermal resistance, thermal transmittance, and thermal conductivity), it is possible to conclude that reed from Portugal, under the studied conditions, has an adequate thermal performance. Furthermore, its thermal resistance (1.8 $m^2 \cdot °C/W$) and thermal conductivity (0.06 $W/m \cdot °C$) were found to be in accordance with the requirements defined by Portuguese law for thermal insulation materials. The thermal resistance of reed is almost 60% of the insulation materials used in Portugal (e.g., rock wool, XPS, and cork). The geometric characteristics (diameter) and node presence were found to have no influence on the studied thermal properties.

(2) The physical characteristics of reed from Portugal are compatible with its use as a construction material. The reed's density and water content were found to be similar to the organic materials conventionally used in the construction sector. The reed has satisfactory dimensional stability, making it compatible with rigid connection accessories. However, attention should be paid to aspects such as retraction in its thickness. The hygroscopic nature of reed could be related to the high water absorption reached in the tests.

(3) There is a trend for mould growth under favourable conditions (22 ± 2 °C and 90 ± 5% HR). Nevertheless, these specific temperature and humidity conditions are uncommon in Portuguese climatic conditions. Therefore, the durability of this material will not hinder its use as a building material.

The presented characterisation provides valuable data to be considered in the renovation of vernacular buildings. In addition, regarding the studied samples, the reed found in Portugal has characteristics suitable for use as a building material, especially as a thermal insulation material. Additionally, considering the abundance of reed throughout the Portuguese territory, this is a sustainable, eco-friendly, and low-cost option.

Author Contributions: R.M. (Raphaele Malheiro) and A.A. undertook the main part of the research that was the base of this article. They developed the research method and analysed the results with the contribution of C.G. and J.F. R.M. (Raphaele Malheiro) wrote the document with the input of S.M.S. and R.M. (Ricardo Mateus). S.M.S. and R.M. (Ricardo Mateus) helped to develop the discussion sections of the paper and provided critical judgment on the undertaken research. Additionally, they supervised all the works and revised the document. M.T.A. helped in the durability tests (development and data analyse). All authors have read and agreed to the published version of the manuscript.

Funding: The authors would like to acknowledge the support granted by the FEDER funds through the Competitively and Internationalization Operational Programme (POCI) and by national funds through FCT (the Foundation for Science and Technology) within the scope of the project with the reference POCI-01-0145-FEDER-029328, reVer+.

Institutional Review Board Statement: Not applicable.

Informed Consent Statement: Not applicable.

Acknowledgments: The authors would like to acknowledge the support granted by DANOSA "Derivados asfálticos normalizados, S.A." industry for the hotbox construction by providing all the necessary insulation material.

Conflicts of Interest: The authors declare no conflict of interest.

References

1. Allirand, J.-M.; Gosse, G. An above-ground biomass production model for a common reed (*Phragmites communis* Trin.) stand. *Biomass Bioenergy* **1995**, *9*, 441–448. [CrossRef]
2. Köbbing, J.F.; Thevs, N.; Zerbe, S. The utilisation of reed (*Phragmites australis*): A review. *Int. Mire Conserv. Group Int. Peat Soc.* **2013**, *13*, 1–14.
3. Barreca, F. Use of giant reed *Arundo Donax* L. in rural constructions. *Agric. Eng. Int. CIGR J.* **2012**, *14*, 46–52.
4. Zámolyi, F.; Herbig, U. Reed as building material-renaissance of vernacular techniques. In *International Symposium on Advanced Methods of Monitoring Reed Habitats*; Csaplovics, E., Schmidt, J., Eds.; Rhombos Verlag Berlin: Berlin, Germany, 2011; p. 83.
5. Speck, O.; Spatz, H.-C. Damped oscillations of the giant reed *Arundo donax* (Poaceae). *Am. J. Bot.* **2004**, *91*, 789–796. [CrossRef]
6. Barreca, F.; Martinez, A.; Flores, J.A.; Pastor, J.J. Innovative use of giant reed and cork residues for panels of buildings in Mediterranean area. *Resour. Conserv. Recycl.* **2019**, *140*, 259–266. [CrossRef]
7. *Arundo donax* Flora-On. Available online: https://flora-on.pt/#/1Arundo+donax (accessed on 11 March 2021).
8. AAVV. *Arquitectura Popular em Portugal*; Associação dos Arquitetos Portugueses: Lisboa, Portugal, 1988.
9. Mestre, V. *Arquitetura Popular da Madeira*; Argumentum: Lisboa, Portugal, 2002.
10. Fernandes, J.; Malheiro, R.; Castro, M.D.F.; Gervásio, H.; Silva, S.M.; Mateus, R. Thermal Performance and Comfort Condition Analysis in a Vernacular Building with a Glazed Balcony. *Energies* **2020**, *13*, 624. [CrossRef]
11. Fernandes, J. O Contributo da Arquitectura Vernacular Portuguesa para a Sustentabilidade dos Edifícios. Ph.D. Thesis, University of Minho, Guimarães, Portugal, 2012.
12. Cravinho, A. *Pedra & Cal n.24*; Rua Pedro Nunes: Lisboa, Portugal, 2004; pp. 12–13.
13. Fernandes, J.; Mateus, R.; Gervásio, H.; Silva, S.; Branco, J.; Almeida, M. Thermal Performance and Comfort Conditions Analysis of a Vernacular Palafitic Timber Building in Portuguese Coastline Context. *Sustainability* **2020**, *12*, 10484. [CrossRef]
14. Ferreira, D.M.; Araújo, A.; Fonseca, E.M.; Piloto, P.; Pinto, J. Behaviour of non-loadbearing tabique wall subjected to fire—Experimental and numerical analysis. *J. Build. Eng.* **2017**, *9*, 164–176. [CrossRef]
15. Sá, A.B.; Pereira, S.; Soares, N.; Pinto, J.; Lanzinha, J.C.; Paiva, A.; Cunha, S.P.D.S. An approach on the thermal behaviour assessment of tabique walls coated with schist tiles: Experimental analysis. *Energy Build.* **2016**, *117*, 11–19. [CrossRef]
16. Fonseca, I.F.P. *Permanência e Transformação: Contributos para a Utilização de Materiais Modernos na Conservação do Património Arquitetónico. Os Produtos Técnicos de Madeira*; University of Évora: Évora, Portugal, 2014.
17. Barroso, C.E.; Oliveira, D.V.; Ramos, L.F. Physical and mechanical characterization of vernacular dry stone heritage materials: Schist and granite from Northwest Portugal. *Constr. Build. Mater.* **2020**, *259*, 119705. [CrossRef]
18. Ribeiro, V. (Ed.) *Materiais, Sistemas e Técnicas de Construção Tradicional-Contributo para o Estudo d Arquitetura Vernácula da Região Oriental da Serra do Caldeirão*; Edições Afrontamento e CCDR Algarve: Faro, Portugal, 2008.
19. *Arundo donax* vs. *Phragmites australis*. Available online: http://desertfishes.org/cuatroc/organisms/arundo-vs-phragmites.html (accessed on 24 March 2021).
20. Speck, O.; Spatz, H.-C. Mechanical Properties of the Rhizome of *Arundo donax* L. *Plant Biol.* **2003**, *5*, 661–669. [CrossRef]
21. Miljan, M.; Miljan, M.-J.; Miljan, J.; Akermann, K.; Karja, K. Thermal transmittance of reed-insulated walls in a purpose-built test house. *Mires Peat* **2014**, *13*, 1–12.

22. Asdrubali, F.; Bianchi, F.; Cotana, F.; D'Alessandro, F.; Pertosa, M.; Pisello, A.L.; Schiavoni, S. Experimental thermo-acoustic characterization of innovative common reed bio-based panels for building envelope. *Build. Environ.* **2016**, *102*, 217–229. [CrossRef]
23. Barreca, F.; Fichera, C.R. Wall panels of *Arundo donax* L. for environmentally sustainable agriculture buildings: Thermal performance evaluation. *J. Food Agric. Environ.* **2013**, *11*, 1353–1357.
24. Asdrubali, F.; D'Alessandro, F.; Schiavoni, S. A review of unconventional sustainable building insulation materials. *Sustain. Mater. Technol.* **2015**, *4*, 1–17. [CrossRef]
25. Carneiro, P.; Jerónimo, A.; Faria, P. Reed-Cob: Tecnologia Inovadora de Baixo Carbono para Construção de Pequeno Porte. In *II Encontro Nacional Sobre Reabilitação Urbana e Sustentabilidade*; iiSBE Portugal & Universidade do Minho: Lisboa, Portugal, 2017.
26. Molari, L.; Coppolino, F.S.; García, J.J. *Arundo donax*: A widespread plant with great potential as sustainable structural material. *Constr. Build. Mater.* **2021**, *268*, 121143. [CrossRef]
27. Conte, P.; Fiore, V.; Valenza, A. Structural and Mechanical Modification Induced by Water Content in Giant Wild Reed (*A. donax* L.). *ACS Omega* **2018**, *3*, 18510–18517. [CrossRef]
28. Soliman, M. *Arundo donax L. and Its Use in the Thermal Insulation in Architecture to Decrease the Environmntal Pollution*; Ain Shams University: Cairo, Egypt, 2009.
29. García-Ortuño, T. *Caracterización de la Caña Común (Arundo donax L.) para su Uso Como Material de Construcción*; Universidad Miguel Hernández: Alicante, Spain, 2003.
30. Volf, M.; Diviš, J.; Havlík, F. Thermal, Moisture and Biological Behaviour of Natural Insulating Materials. *Energy Procedia* **2015**, *78*, 1599–1604. [CrossRef]
31. IPQ NP-614. *Wood-Determination of Water Content (in Portuguese)*; IPQ NP-614: Lisboa, Portugal, 1973; p. 2.
32. IPQ NP-616. *Wood-Determination of Density (in Portuguese)*; IPQ NP-616: Lisboa, Portugal, 1973; p. 2.
33. IPQ NP-615. *Wood-Determination of Retraction (in Portuguese)*; IPQ NP-615: Lissboa, Portugal, 1973.
34. Teixeira, E.R.; Machado, G.; De Adilson, P.; Guarnier, C.; Fernandes, J.; Silva, S.M.; Mateus, R. Mechanical and thermal performance characterisation of compressed earth blocks. *Energies* **2020**, *13*, 2978. [CrossRef]
35. ASTM International. ASTM C1363—11 Standard Test Method for Thermal Performance of Building Materials and Envelope Assemblies by Means of a Hot Box Apparatus 1. *Am. Soc. Test. Mater.* **2014**, *90*, 1–44.
36. ISO-9869. *Thermal Insulation-Building Elements-In-Situ Measurement Thermal Resistance and Thermal Transmittance*; International Organization for Standardization: Geneva, Switzerland, 1994; Volume 994, p. 24.
37. Lautkankare, R. The mold test. In *Guidebook of Reed Business*; Ülo, K., Ed.; Tallinn University of Technology: Tallinn, Estonia, 2013; pp. 42–45. ISBN 978-9949-484-91-1.
38. Bergholm, J. *Susceptibility to Microbial Growth of Common Reed and Other Construction Materials*; Turku University of Applied Sciences: Turku, Finland, 2012.
39. Hukka, A.; Viitanen, H.A. A mathematical model of mould growth on wooden material. *Wood Sci. Technol.* **1999**, *33*, 475–485. [CrossRef]
40. Liese, W. *The Anatomy of Bamboo Culms*; International Network for Bamboo and Rattan: Beijing, China, 1998; ISBN 8186247262.
41. Couvreur, L.; Alejandro Buzo, R. *Construir con Caña-Estudio del uso de la Caña en la Arquitectura Tradicional y de su Recuperación para la Construcción Contemporánea*; Catálogo de publicaciones del Ministerio: Madrid, Spain, 2019. Available online: www.culturaydeporte.gob.es (accessed on 13 June 2021).
42. Dias, J.R. *Caracterization of the Stone Pine Timber (in Portuguese)*; University of Coimbra: Coimbra, Portugal, 2019.
43. Freitas, M.P. *Análise Prática das Propriedades Físicas e Mecânicas do Bambu Phyllostachys edulis, Cultivado em Portugal*; University of Minho: Guimarães, Portugal, 2019.
44. Rusch, F.; Éverton, H.; Ceolin, G.B. Anatomia de hastes adultas de bambu: Uma revisão. *Pesqui. Florest. Bras.* **2018**, *38*, 38. [CrossRef]

Article

Energy Intensity Reduction in Large-Scale Non-Residential Buildings by Dynamic Control of HVAC with Heat Pumps

Alessandro Franco *, Lorenzo Miserocchi and Daniele Testi

Department of Energy, Systems, Territory, and Constructions Engineering (DESTEC), University of Pisa, Largo Lucio Lazzarino, 56122 Pisa, Italy; lorenzo.miserocchi@gmail.com (L.M.); daniele.testi@unipi.it (D.T.)
* Correspondence: alessandro.franco@ing.unipi.it

Abstract: One of the main elements for increasing energy efficiency in large-scale buildings is identified in the correct management and control of the Heating Ventilation and Air Conditioning (HVAC) systems, particularly those with Heat Pumps (HPs). The present study aimed to evaluate the perspective of energy savings achievable with the implementation of an optimal control of the HVAC with HPs. The proposed measures involve the use of a variable air volume system, demand-controlled ventilation, an energy-aware control of the heat recovery equipment, and an improved control of the heat pump and chiller supply water temperature. The analysis has been applied to an academic building located in Pisa and is carried out by means of dynamic simulation. The achieved energy saving can approach values of more than 80% if compared with actual plants based on fossil fuel technologies. A major part of this energy saving is linked to the use of heat pumps as thermal generators as well as to the implementation of an energy efficient ventilation, emphasizing the importance of such straightforward measures in reducing the energy intensity of large-scale buildings.

Keywords: energy efficiency improvement; HVAC operation; dynamic optimization; sustainability

Citation: Franco, A.; Miserocchi, L.; Testi, D. Energy Intensity Reduction in Large-Scale Non-Residential Buildings by Dynamic Control of HVAC with Heat Pumps. *Energies* **2021**, *14*, 3878. https://doi.org/10.3390/en14133878

Academic Editor: Fabrizio Ascione

Received: 9 June 2021
Accepted: 25 June 2021
Published: 28 June 2021

Publisher's Note: MDPI stays neutral with regard to jurisdictional claims in published maps and institutional affiliations.

Copyright: © 2021 by the authors. Licensee MDPI, Basel, Switzerland. This article is an open access article distributed under the terms and conditions of the Creative Commons Attribution (CC BY) license (https://creativecommons.org/licenses/by/4.0/).

1. Introduction

The civil and residential sector is responsible for about 40% of the total final energy consumption and for about 30% of greenhouse gas (GHG) emissions. The major possibilities of reducing the energy consumption lie in non-residential buildings for public use, which are characterized by higher energy intensities (EI) compared to residential ones [1]. The specific consumption of these buildings is strongly dependent on climate but also on other variables such as the building's use and the energy systems in place. A lot of scientific contributions have demonstrated this occurrence. In a paper recently published [2], the authors have shown the significant variability of the energy intensity associated to different public non-residential buildings located in the same city. Comparing the total energy consumption intensity of several large public buildings (such as offices, educational buildings, hotels, shopping malls), the authors evidenced the range of available values from 50 kWh/m^2 per year up to 500 kWh/m^2 per year, and a range from 50 kWh/m^2 up to 200 kWh/m^2 in the case of office and educational buildings. This relevant variability suggests the evidence of an "energy performance gap", that is due to faults in building design, construction, but, in particular, in the operation stages of the Heating, Ventilation and Air Conditioning (HVAC) systems [3]. To reduce this problem, improvements both in the building structure and in the energy systems' management and control are possible [4], pursued with the aim of energy efficiency enhancement and the integration of renewable energy sources (RES) in the pool of energy systems serving the building. Both these approaches can play a relevant role, as shown in recent papers considering strategic perspectives [5,6].

Considering the HVAC system, measures for reducing the energy intensity are correlated both with the systematic use of specific components, like the Heat Pump, and to the optimal control of HVAC systems. Heat pump (HP) systems are considered a strategic technology with which to increase the efficiency of heat generation and the penetration of electricity produced with renewable energy systems in buildings, as was discussed in some recent papers [7–9]. Moreover, HPs are considered relevant elements in smart microgrids and nearly Zero Energy Buildings (nZEB), adding flexibility to the system and allowing shifts in energy demand from heat to electricity, as well as from fossil fuels to renewables [10,11].

Concerning the problem of management and control, it has been estimated that about 90% of the Heating, Ventilation and Air Conditioning are not operated optimally [12]. In fact, while the single components are carefully controlled to satisfy the building demands, a supervisory control able to provide the optimal solution has still not been developed on a full-scale [13]. The development of supervisory control of HVAC systems could lead to the possibility of both reducing the energy consumption and improving occupants' comfort. Supervisory control could shift the HVAC operation from the simple pursuit of acceptable environmental conditions, considered as constraints as in technical standards, to an accurate consideration of all the targets involved. For example, in [14], the authors of this paper achieved an improvement of 25% in terms of the occupants' comfort, while simultaneously reducing the building ideal energy demand up to 30% by optimizing ambient air setpoints in an academic building.

The development of advanced Building Energy Management Systems (BEMS), available thanks to the widespread use of smart metering systems and pieces of equipment able to modulate their output without degrading their performance, is leading buildings into a transition phase in which they can interact with occupants, with other buildings, and with a grid [15,16].

Another possibility of obtaining supervisory control is represented by Model Predictive Control (MPC), which is based on the selection of an optimal time sequence of operation [17–20]. The advantages that can be obtained with this technology are directly dependent on the test case they are applied to, and especially on the type of control they are compared to. In [18], the authors provided possible reduction in energy usage between 30% and 80%, while in [19] the authors provided energy savings up to 75%.

Moreover, the development of supervisory control is paramount when considering the integration of RES in the HVAC system. On one side, it can help to successfully manage the different energy sources of the building and, on the other, it can provide the optimal operation of storages, thanks to evaluations made on future predictions. These elements can both be seen in the operation of electrically driven HP, which have been identified as one of the favored technical solutions, in combination with photovoltaics (PV), to obtain all-electric solutions [21] and achieve the feature of nearly-Zero Energy Buildings [22]. In fact, the combination with PVs requires an accurate selection of the energy source favoring renewable electricity when available, to reduce costs and environmental impact connected to the energy use, while the presence of thermal storages requires an accurate prediction of energy demands. Thus, thanks to an advanced control, these machines are expected to become fundamental in future HVAC systems [23].

The aim of this paper was both to highlight the gap in the current operation of HVACs and to analyze the energy savings that can be achieved in the field of large-scale, non-residential buildings, with a shift from fossil fuel-based technologies to the application of HVAC systems based on a systematic use of HPs. The study focused both on measures which do not require supervisory control and that are already available on the market, but are not sufficiently implemented on the field, such as Demand-Controlled Ventilation (DCV), as in [24], and on opportunities given from a supervisory control, as an energy-aware control of the heat recovery (HR) equipment or the optimal control of the heat pump and chiller supply water temperature. The methodology was applied to a test case represented by an educational building in Pisa, Italy.

The novelty of this study is that of emphasizing a hierarchy in the importance of the various measures that can be applied to HVAC systems to improve their overall performance in addition to highlighting that the main energy savings can be achieved with the use of low-complexity and at-hand interventions, represented especially by the use of HPs.

2. Energy Intensity of Large-Scale Public Buildings: The Perspectives of the Use of HVAC with HP

In a recent paper [2], the authors presented a method to evaluate building energy consumption based on an energy use index of different functional sectors in China. The study testified that the average value of buildings for public use is in general quite high, but the relevant spread among similar structures demonstrates how it could be possible to pursue relevant reduction of energy consumption.

Considering the large-scale public buildings, in general, the relevant part of energy consumption is connected to the lighting systems, to the various electricity loads, but in particular, to the operation of the HVAC system, used for maintaining adequate internal conditions inside buildings. HVAC systems, developed in various layouts, involve many components that can be arranged in several ways, obtaining different configurations [25]. In general, five different processes can be considered to achieve all the air-conditioning services [26], namely heating, cooling, humidifying, dehumidifying, and ventilation [27].

To realize these processes, typical HVAC systems can be thought of as divided in five different sections, shown in Figure 1, which in turn can be divided in localized and non-localized sections. The pre-treatment, the treatment, and the generation sections belong to the former type, and can therefore be represented showing their interconnections; the transport and the control section belong to the latter type, as they rather represent functional sections.

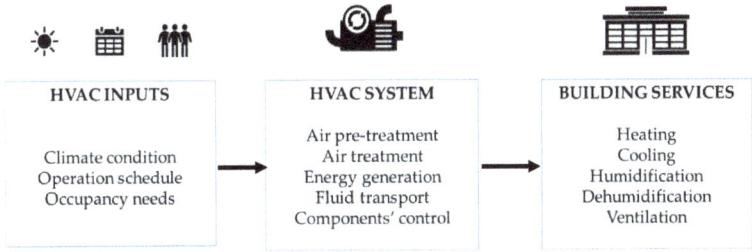

Figure 1. HVAC functional sections.

At the beginning, the incoming air is sent in the pre-treatment section, in which energy recovery measures and mixing with exhaust air are obtained. Later, in the treatment section, air is processed until it reaches the desired supply conditions; then, the generation section is present. The transport and the control section are also present in the HVAC system.

The pre-treatment section is represented by two different elements, in which the interaction between outdoor air and exhaust air occurs: energy recovery and mixing. Energy recovery is recommended by technical standards [28], with the aim of reducing energy consumption. Many technologies have been developed for this purpose, such as sensible or total energy recovery and active or passive techniques [29]. Plate heat exchangers, rotary wheels, heat pipes, and run-around loops can be used.

In the treatment section, a series of transformations occur. These transformations can include pre-heating, cooling and dehumidifying, humidifying, and re-heating, and are usually performed by means of coils and humidifiers.

The generation section, which is a relevant part, can be characterized by different devices, such as, in general, a boiler for heating and a chiller for cooling. However, as stated above, electric HP are gaining attention for their improved performance compared to classical generation systems and their possible integration with RES as PV systems. The transport section consists of all the fans, pumps, ducts, dampers, and loops which allow the movement of the fluids present in the plant. The control section consists of all the devices with the function of equipment regulation, namely sensors, controllers, and actuators.

2.1. Heat Pump Potentialities for HVAC Generation Section and Real Operation

One of the main advances regarding the generation section of HVAC systems proposed in the present paper was represented by a systematic use of HPs. The use of this technology is considered today a very efficient way for increasing energy efficiency, but if, and only if, they are properly sized and operated. The HP system performance can be highly reduced under many circumstances, such as high temperature lift, defrost operation, partial loads, and frequent on-off cycles. These conditions often occur when the HP must match variable heating demands in response to the building thermal dynamics.

Considering the coefficient of performance (COP), four different indexes can be defined to take into account and compare the different operational modes of the same HP systems:

- COP_{nom}, or nominal full-load COP, provided by the HP manufacturer at the reference sources temperature, according with technical standards (e.g., [30]).
- COP_{DC}, the full-load performances provided by the manufacturer at the different external and supply temperatures provided by [30,31]. The data are marked with the subscript DC and are evaluated at maximum compressor speed.
- COP_{op}, or the HP operative part-load COP, experimentally measured or simulated through a validated HP model. This index accounts for both external and supply temperatures, together with the effects of the capacity control.
- COP_{sys}, or the HP overall system performance, calculated as the ratio between the heat provided to the building by the emission system, $\dot{Q}_u$, and the electric input used by the HP, $\dot{W}_{HP}$. This value considers HP performances and all the thermal losses in the other pieces of equipment (e.g., pipework, thermal inertial storages, or puffers).

The just-mentioned indexes refer to the instantaneous thermal or electrical power exchanged by the HP unit or system. In general, it seems to be particularly interesting to analyze their average value over a reference period, τ (e.g., a month, a year, a season). Some of the most common time-integral coefficients of performance are the seasonal coefficient of performance SCOP and the capacity ratio. The seasonal coefficient of performance, SCOP, is defined as the ratio between the thermal energy output and the electrical energy input of the HP device ($SCOP_{HP}$) over the considered operational time, τ.

$$SCOP_{HP} = \frac{\int_\tau \dot{Q}_{HP} d\tau}{\int_\tau \dot{W}_{HP} d\tau} = \frac{\int_\tau \dot{m}_{HP} c_f (T_{w,out,HP} - T_{w,in,HP}) d\tau}{\int_\tau \dot{W}_{HP} d\tau} = \frac{Q_{HP}}{W_{HP}} \quad (1)$$

The capacity ratio, defined as CR, is a parameter that allows the analysis of the real operation of the HP unit or system. It can be evaluated according to the delivered heat, Q_{HP}, and the maximum available energy output at nominal full-load power at the given external and supply temperatures over the considered time (off periods included):

$$CR = \frac{Q_{HP}}{\int_\tau \dot{Q}_{DC}(T_{w,out,HP}; T_{air,ext}) d\tau} \quad (2)$$

According to the above-mentioned technical standards, it is possible to evaluate a penalization factor, f_{CR}, as a function of CR and its capacity control system, that assumes values in the range between 0 and 1. This penalization factor should be applied to the full-load COP_{DC} at the specific external and supply temperatures to obtain the operative part-load COP_{op}.

2.2. HVAC Control System

Due to the high number of elements present in the HVAC system, HVAC control plays a very important role by coordinating the operation of the elements in the different sections, primarily with the aim of satisfying the building demands, but also to do that in an efficient way if a supervisory control is applied [32].

Three main categories of variables are involved in HVAC control: Firstly, the conditioned space variables; Secondly, the weather and occupancy variables; Finally, the operational variables. The monitored conditioned space variables can be the internal temperature, relative humidity, and pollutants' concentration, usually represented by carbon dioxide concentration; the monitored weather and occupancy variables can be the external temperature and relative humidity, solar irradiance, and number of occupants; the monitored operational variables can be flow rates, fluid temperatures and pressures, and components' state. The conditioned space variables represent the objective of control, the operational variables represent the controlled variables, and the weather and occupancy variables represent the disturbances. A detailed consideration of these disturbances can help to determine the operational variables able to satisfy the objective of control at the minimum cost possible.

3. HVAC with HP for Large-Scale Buildings: Modelling of the Real Operation and Optimized Control Strategy

The objective of this analysis was to evaluate the positive effects on the overall energy efficiency of the building that can be obtained by using a HVAC system with HPs and an optimal control of the HVAC system operation through a real-time control of real occupation of the building.

The control practices that were developed represented both the best available techniques and at-hand techniques that have still not been implemented in real systems but that could lead to a significant increase in the building energy efficiency at minimal complexity and cost. The techniques proposed were the demand-controlled ventilation, an energy-aware control of the energy recovery system, and a control of the HP and chiller supply water temperature.

The first measure aimed to introduce the energy perspective when dealing with indoor air quality (IAQ) and works by minimizing outdoor air flow rate when not justified by IAQ requirements. The second measure works by selecting the best configuration for the use of the heat recovery based on an energy-based criterion. It points out the importance of a supervisory control able to select the best working state of some of the components involved based on performance of the overall system. Finally, the third measure works by adapting the operating temperature of the generation section to reduce energy consumption. The above-mentioned measures only represent some of the possibilities that can be obtained with supervisory control. This also highlights the complexity of developing a comprehensive supervisory control. In fact, since the number of components and interactions among them are very high, it is common practice to apply it only to some elements.

All the measures proposed were applied in cascade to a Variable Air Volume (VAV) system, representing the energy savings that can be achieved with an increased complexity of the HVAC control system. Moreover, all the Control Strategies (CS) were compared to a Constant Air Volume (CAV) system to show the improvements that have already been achieved with the implementation of VAV systems. The strategies analyzed are summarized in Table 1. Strategy CS 1 represents an outdated configuration based on a CAV system; CS 2, considered as a benchmark for its wide spreading, involves the use of a VAV system; CS 3 adds DCV for the outdoor flow rate; CS 4 adds an energy-aware controller for the pre-treatment section to CS 3; CS 5 adds an exact supply water temperature control of the generation section to CS 4.

In order to evaluate the real advantages in terms of energy saving obtainable in a specific application, the methodology proposed has been traduced in a mathematical model in which both the building under analysis and the HVAC, with all the components involved, were considered. The simulation was carried out using physically based models [33]. The components' models were complex enough to represent the main phenomena involved and catch the interactions between the different pieces of equipment, but quite simple enough to be able to provide insights on the energy savings [34].

Table 1. Control strategies of the proposed HVAC.

Strategy	Supply Flow Rate	Outdoor Flow Rate	Pre-Treatment Section Control	Generation Section Control
CS 1	Constant (CAV)	Based on peak occupancy	Temperature-based	Weather-based
CS 2	Variable (VAV)	Based on peak occupancy	Temperature-based	Weather-based
CS 3	Variable (VAV)	Based on real occupancy (DCV)	Temperature-based	Weather-based
CS 4	Variable (VAV)	Based on real occupancy (DCV)	Energy-aware	Weather-based
CS 5	Variable (VAV)	Based on real occupancy (DCV)	Energy-aware	Optimized

Concerning the HVAC system layout, the analysis was applied to a single zone VAV system, which represented a step forward compared to the more classical CAV systems. The difference is based on the presence of variable speed drive (VSD) fans able to reduce the airflow, when possible, without degrading their performance. The CAV system nominal flow rate is 35,000 m^3/h, and the VAV system flow rate can modulate up to 25% of this value.

The pre-treatment section was equipped with a sensible plate heat exchanger with a nominal flow rate of 17,000 m^3/h and a mixing chamber, which can both be bypassed. The treatment section was equipped with a preheat, a cooling and dehumidification, and a reheat coil, all using water as heat transfer fluid along with an adiabatic humidifier. The generation section was equipped with a chiller serving the cooling and dehumidifying coil, and a HP serving the reheat and preheat coils. Both the chiller and the HP considered were air-to-water machines, meaning that they exchanged with outdoor air on the external side and with a water loop on the internal side. The heat pump size was 310 kW with a nominal COP of 3.2, while the chiller size varied for the different CSs and ranged from 96.9 kW to 190 kW. The transport section had variable speed drive motors for the supply and return fans, each with a nominal input power of 15.5 kW. The system performance was evaluated by means of the integration of the HVAC model with a building model. The building energy demand was calculated based on the setpoints suggested from technical standards, namely 20 °C for heating, 26 °C for cooling, and 1000 ppm for carbon dioxide concentration [35].

4. Building and HVAC Components Modelling: Specific Details

In this subsection, the models for the building and HVAC components, according to the illustration of Figure 2 are discussed and presented in detail in the next subsections. All the models involved are written in MATLAB code to ensure easy intercommunication between them.

4.1. Building Model

To define the interaction between the HVAC system and the control system, a relevant part was represented by the model of the building. It consisted of a resistance-capacitance thermal network, which models the building through several nodes for the envelope and one node for the indoor air, assumed as a single thermal zone [36]. Opaque capacitive walls were modeled as a series of multiple resistance-capacitance nodes, while glazed elements were modeled as solely resistive. Various heat transfer modes were considered: the conductive heat transfer among the capacity layers of opaque walls; the convective and the radiative heat transfer between the envelope surfaces and the indoor/outdoor environments; radiant heat transfer with the sky; absorption of solar radiation; radiant heat transfer among internal surfaces; and radiant heat transfer with internal gains (e.g., the number of occupants for each hour). The sensible energy demand of the internal air node was calculated, accounting for all the mentioned energy exchanges, ventilation fluxes, and internal gains, using an hourly timestep. The latent heat energy demand was evaluated through the vapor mass balance of the indoor air.

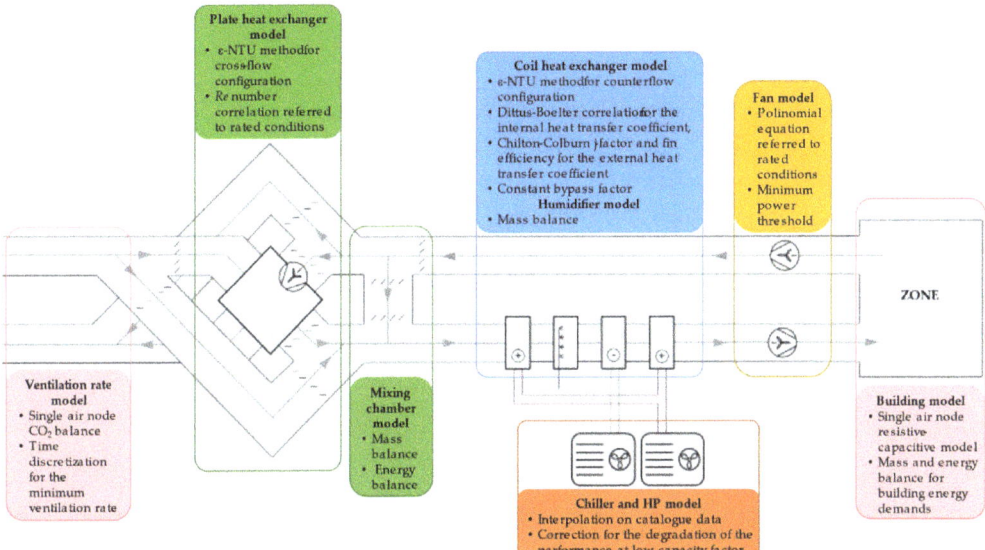

Figure 2. HVAC system components and details of the model.

4.2. Model of the HVAC and of the Various HVAC Components

The model of the HVAC was represented by the equations that describe the behavior of the various components: in particular, plate heat exchanger, mixing chamber, coils, humidifier, HP, chiller, and fan, according to the scheme provided in Figure 2.

The model consists in all the balance equations and in the definition of the various heat transfer coefficient. The analysis of the heat exchangers was carried out using the ε-NTU method, available in all the textbooks about heat exchangers, such as [37]. The heat transfer coefficients on the two sides were defined using the typical correlations for the Nusselt (Nu) number as a function of the Reynolds (Re) number, using the typical correlations available in [37], or developed by one of the authors of the present paper with reference to a similar kind of heat exchangers [38].

In case of finned surfaces, the typical analysis based on the definition of the efficiency of the finned surfaces was used, as largely discussed in the textbook [39]. Concerning the HP and chiller, their performances were based on the data obtained from a technical catalogue from the manufacturer. The modulating capacity for both the components was obtained by means of an inverter, which allows the performance not to be degraded until a capacity ratio of 0.3 is reached. Under this threshold a penalization factor was evaluated for the normalized capacity ratio, as described in [31].

$$f_{CR} = \begin{cases} \frac{CR/0.3}{0.1+0.9CR/0.3} & if\ CR < 0.3 \\ 1 & if\ CR \geq 0.3 \end{cases} \quad (3)$$

This penalization factor must be applied to the declared COP of the HP or efficiency energy ratio of the chiller to represent the degradation of the performance at part load and obtain the operative coefficient of performance or energy efficiency ratio. In the non-optimized situation, the chiller supply water temperature has been considered 5 °C, while HP supply water temperature (in °C) was defined as:

$$T_{w,HP} = \begin{cases} 45 \text{ if } T_{air,ext} \leq 45 \\ 45 - 3(T_{air,ext} - 15) \text{ if } 15 < T_{air,ext} < 20 \\ 30 \text{ if } T_{air,ext} \geq 20 \end{cases} \quad (4)$$

Finally, the fan was described by means of a cubic equation evaluated on rated conditions considering a minimum required power, equal to that corresponding to 20% of nominal flow rate:

$$\dot{W}_{fan} = \dot{W}_{fan,RC} \left(\frac{\max(\dot{m}_{air}, 0.2\dot{m}_{air,RC})}{\dot{m}_{air,RC}} \right)^3 \quad (5)$$

$\dot{W}_{fan,RC}$ and $\dot{W}_{fan}$ are the power at rated conditions and at reduced flow rate, and $\dot{m}_{air,RC}$ and $\dot{m}_{air}$ are the flow rate at rated conditions and reduced flow rate, respectively. The determination of the minimum outside air ventilation rate was based on the balance of carbon dioxide concentration within the control volume and considered human breathing, which acts as a source with a production rate of 0.3 L/min per person, and natural and mechanical ventilation, which acts by means of the renewal of ambient air with external air, assumed at a concentration of 400 ppm. The analysis was based on the methodology discussed in [40] by one of the authors of the present paper. Considering the building volume, the carbon dioxide concentration, the human production rate of CO_2 for each occupant, and the real number of occupants, the model operated the discretization of the time derivative, and allowed us to obtain the ventilation rate for maintaining the carbon dioxide concentration under the value of 1000 ppm [40].

5. Application to Specific Test Cases and Evaluation of Energy Saving Potentialities

In this section, the application of the HVAC control strategies described in Table 1 and defined in the previous sections are evaluated with reference to a specific test case, an educational building. After a short description of the building, the results achieved are shown by analyzing the estimated values of the overall energy consumptions.

5.1. Test Building: Detailed Description and Main Geometrical Data

The building used for testing the proposed methodology and to evaluate the possible energy demand reduction was an academic building located in Pisa (1694 Heating Degree Days), used by the University of Pisa. The considered building had five levels, a total floor surface of about 1560 m², and a gross volume of approximately 6200 m³, of which about 4540 m³ were occupied by classrooms. A total number of 12 classrooms were present in the building; the sizes ranged from 18–25 seats (3 classrooms), 36–38 seats (3 classrooms), 65–72 places (3 classrooms), 160 places (2 classrooms) to 180 places (1 classroom).

Figure 3 provides the typical layout of one on the five levels, in which two different classrooms can be well identified: one of quite high size and one of the smallest. The external building surface was approximately 2000 m² (both opaque and glazed), corresponding to an aspect ratio (surface/volume) of about 0.33 m^{-1}. The glazed surface represented about 16% of the total surface. The envelope was characterized by stone walls with face bricks and a reinforced concrete structure, which resulted in a total transmittance of about 1.1 m²K/W, while the windows were double-glazed. At a full occupancy level, the number of students present inside the building was estimated in the number of 878. Concerning the climatic conditions typical of Pisa, which can be derived by common databases available online; for example, in the proposed case, the reference data are presented in Table 2 and were all derived from a specific database [41], while solar radiation was obtained from [42]. Concerning the test case, it has been connected to the typical occupancy level observed before the COVID-19 pandemic experience.

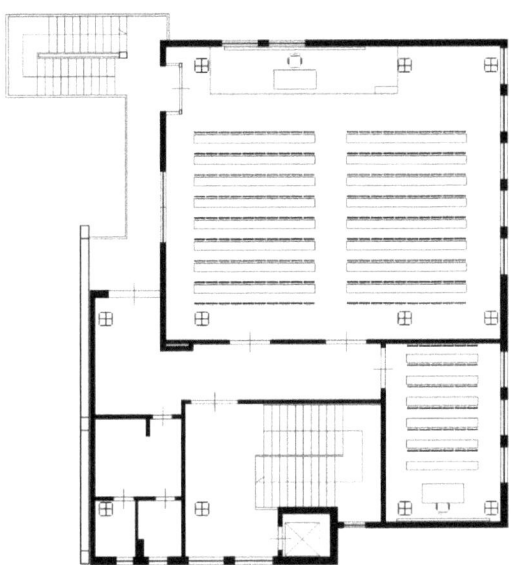

Figure 3. The test building with details about one of the five levels.

Table 2. Reference outdoor climatic data used for the analysis of the test case.

	January	February	March	April	May	June	July	August	September	October	November	December	Year
G [kWh/(m^{2}d)]	1.6	2.3	3.3	4.3	5.9	6.4	6.8	5.9	4.4	2.6	1.9	1.4	3.9
T_{avg} [°C]	5.9	8.4	10.5	12.7	14.1	22.8	25.2	25.1	21.3	17.5	12.5	9.5	15.5
RH_{avg} [%]	77.7	77.3	75.3	75.5	78.5	65.1	64.3	65.6	68.5	77.4	87.1	86.6	74.9

The building was open from 8 a.m. to 8 p.m. during weekdays (Monday to Friday), from 8 a.m. to 1 p.m. on Saturdays, and closed on Sundays. The occupancy of the twelve classrooms was organized in four periods, two trimesters of lessons (March–May and October–December) and three periods of examination (January–February, June–July, and September); these were characterized by different occupancies, with the major ones occurring during the period of lessons. The predicted hourly occupancy was determined, based on the real building schedule, by multiplying classroom capacities, using a reduction coefficient, different for considering period of lessons (1 October to 20 December and 1 March to 30 May) and exams (7 January to 28 February, 1 June to 30 July, and September). The reduction coefficients considered were 0.8 for the period of lessons and 0.5 for the period of examination, respectively. The effective hourly occupancy has been obtained by multiplying the predicted hourly occupancy with a randomization coefficient, obtained by a uniform distribution between 0.5 and 1.

5.2. Overall Energy Saving and Comparison with Fossil Fuel-Based HVAC System

In this subsection, the overall energy saving is presented comparing all the control strategies of the HVAC system obtained considering the five different possibilities described in Table 2. The comparison with the use of conventional boiler, based on fossil fuel (as natural gas), common in Italy, is evaluated later. Figure 4 provides the correspondent electric energy intensity referred to in the application of the five different control strategies, moving from the less structured one (CS 1) to the more complex one (CS 5). Considering the first case represented by the control strategy CS 2, it is shown how the energy consumption was significantly reduced by 50%, from 118 to 69 kWh/(m^{2}y), with respect to the basic one (CS 1), mainly due to the fan energy consumption but also to the heating and cooling energy consumption to a lesser extent. This was due to the reduction of the supply air flow rate when the heating and cooling demand were not the nominal ones. This comparison represents the energy savings achievable thanks to the use of a VAV system instead of a CAV system, a measure already implemented today.

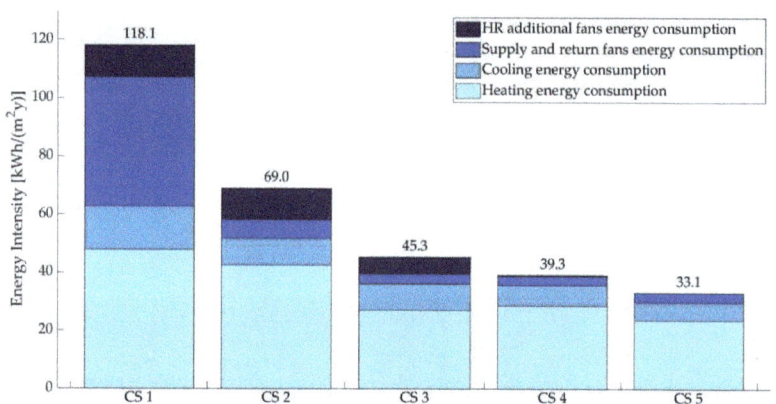

Figure 4. Overall energy consumption intensity obtained with the different control strategies.

Comparing the effect determined by control strategy CS 3 to CS 2, it is shown how the overall energy consumption can be reduced by up to 35% due to a lower heating energy consumption and fan energy consumption, especially in the HR one. This is due to the reduction of the outside air flow rate when the occupancy is low and thus carbon dioxide concentration can be maintained under the threshold of 1000 ppm with a reduced air flow. This comparison represents the energy savings arising from the implementation of DCV in a VAV system. Comparing CS 4 to CS 3, the overall energy consumption can be reduced by an additional 8%, thanks to a reduction in the HR additional fan and cooling energy consumption, which face a slight increase in the heating energy consumption.

Comparing the two strategies CS 5 and CS 4 involving heat recovery, it is shown how the overall energy consumption can be reduced by another additional 9%, summing up to a total of 52% of energy savings, with respect to option CS 2. This is due to the reduction in the heating and cooling energy consumption for the improved performance of both HP and chiller. The energy saving is clearly evident in the graph of Figure 5, which shows, for each day of the year, the different electricity consumption for the basic control strategy (CS 1) and for the optimal control strategy (CS 5). The X-axis illustrates the working days, representing the daily total energy consumption for the system using the basic control strategy CS 1. The corresponding daily total energy consumption in the case of the more advanced control strategy, CS 5, resulted in an oscillating trend, because the energy saving is dependent on the meteorological and occupancy-related conditions, and therefore on the variable building energy demand. This graph shows how the improvement is guaranteed throughout all of the HVAC system operational days, highlighting the relevance of the proposed measures.

The implementation of the real occupancy profile of the building can lead to relevant energy savings for the academic building analyzed. In addition, it seems appropriate to underline the importance of shifting from a conventional boiler, using natural gas, to a HP as a generation unit in HVAC, which is highlighted by its use of energy, as reported in Table 3. For making the comparison with the HP performance, a steam boiler seasonal efficiency of about 0.85 has been obtained by considering the characteristics of the building and the degree days of Pisa: The advantages in terms of energy saving coming from the use of HP, which were in the range of 60% for every CS, are clearly apparent. Moreover, the use renewable energy (e.g., PV power), permits a reduction of the carbon footprint of the building.

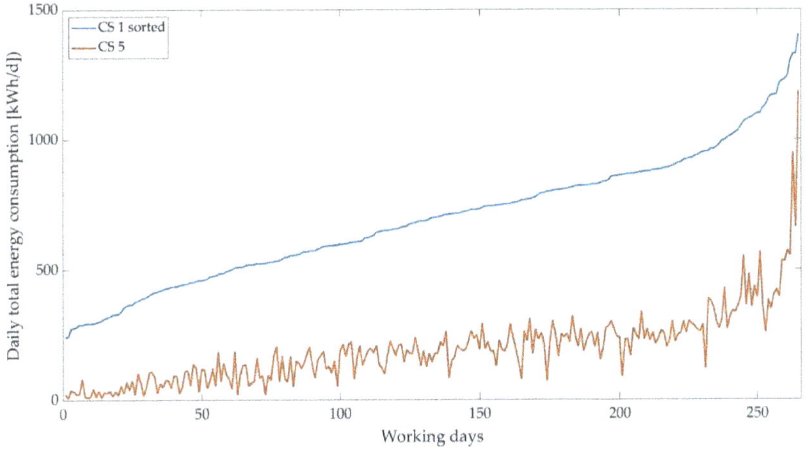

Figure 5. Comparison between daily electricity consumption for control strategies CS 1 (basic) and CS 5 (optimized).

Table 3. Comparison between energy intensity in case of using conventional gas-fired boiler and HP with the different control strategies.

	CS 1	CS 2	CS 3	CS 4
HP [kWh/m^2]	118.1	69.0	45.3	39.3
Conventional boiler [kWh/m^2]	272.3	197.5	118.6	114.2
Energy saving with HP [%]	56	65	62	66

All the remarks about energy savings should be further integrated with economic considerations of the operational costs by comparing the use of fossil fuels and of electricity and the investment costs for the different control strategies.

6. Consideration about the Measure of Energy Saving with Reference to the Specific Building and General Perspectives

In this section, the proposed measures are discussed individually, using as reference data those derived from the simulation, to gain insights about the various energy saving strategies and discuss their extensibility to other types of buildings and climates.

6.1. The Beneficial Effect of Demand Controlled Ventilation

One of the most relevant elements of the control strategy that permits relevant reduction of energy consumption is the application of DCV based on the effective occupation of the building. Let us start by analyzing in detail the positive effect of application of DCV, thus comparing, in terms of annual total energy intensity, the two strategies evidenced as CS 3 and CS 2. This measure can be available if the effective number of occupants is controlled, or by means of specific counters based on smart systems like smartphones or other devices, as described in [16], or by direct measurement of CO_2 concentration and the inclusion of carbon dioxide concentration in the conditioned space monitored variables. This measurement was used to decrease the amount of outside air processed by the HVAC system when indoor air quality can be guaranteed with limited air renewal because of the small number of occupants inside the building. The energy saving achievable with DCV is shown in Figure 6 on a monthly and daily horizon. Considering the various months, it is shown how during periods of exams, when the level of occupation was reduced, the energy saving was significant, reaching a maximum value of 48% and a mean value of 40%, while when the building was used for lessons, the energy savings were quite lower, with a maximum value of 32% and a mean value of less than 30%. Higher values of energy savings were obtained in months in which the real occupancy is much smaller than the peak one, because of greater reductions in the air flow rates. Considering the various days of the week, it is evident how energy saving could be almost constant during the weekdays, with a slightly lower (<7%) value on Mondays, and are relevant on Saturdays due to the reduced number of occupants.

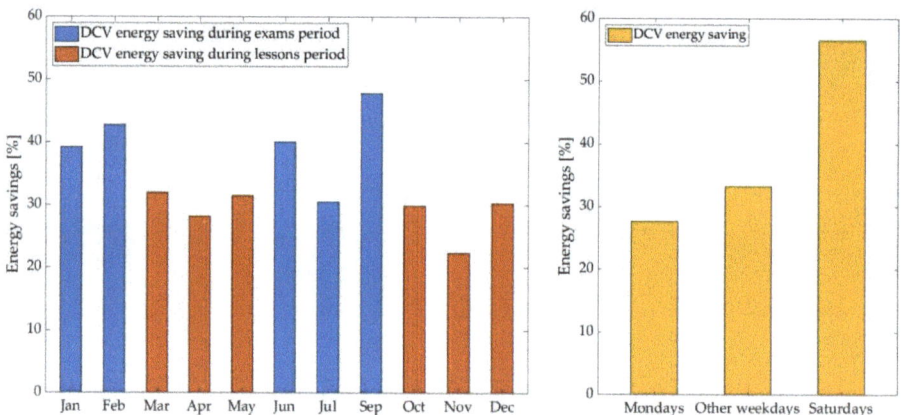

Figure 6. Monthly and daily cumulative energy savings with DCV (strategy CS 3).

In general, according to the relevance of the present analysis, it is possible to affirm that DCV could be more helpful in buildings with long opening hours—academic buildings which are open in the afternoon but also until midnight surely represent a better choice than scholastic ones which are open only in the morning. Event-related buildings such as theaters or sports halls do not represent a good choice, as their use is limited to the hours in which the event takes place. Following from the considerations about Figure 6, the ideal building should be characterized by a large discrepancy between peak occupancy and real occupation; this must be considered on different time horizons, as for the test case. Based on this consideration, buildings such as supermarkets and shopping centers, besides academic ones, could heavily benefit from the implementation of demand-controlled ventilation, for they have long opening hours, and their occupancy assumes very different values, based on the day of the week but also on periods. Since the simulation has been carried out on a single zone, the balance of carbon dioxide represents an average value on the total volume, which does not consider the different occupancy in each classroom. The energy savings obtained in this study suggest a great convenience of implementing these type of measures, even if they should be adapted and tested for each specific application.

6.2. Considerations about Heat Pump and Chiller Operation

The analysis of this measure regards the comparison between the two control strategies named CS 4 and CS 5. There was an observed decrease in the cooling and heating consumption amount to 13% and 18%, respectively. HP and chiller performances are shown in Figure 7, sorted in ascending order for CS 4 coefficient of performance and presenting the oscillating trend for CS 5 due to the variability of the energy achievable in the different working hours. For both the components, when the working conditions were more favorable than those provided in the catalogue, e.g., request of reheat temperature lower than external air temperature, they have been adjusted to the ones in which the performance data are provided. In Figure 7, it is shown how a more dynamic control of both HP (Figure 7a), referred to the operation in heating mode, and chiller (Figure 7b), referred to the operation in cooling mode, can lead to improved performances during all the operating hours, including about 3000 h for heating period and about 1400 h for cooling period.

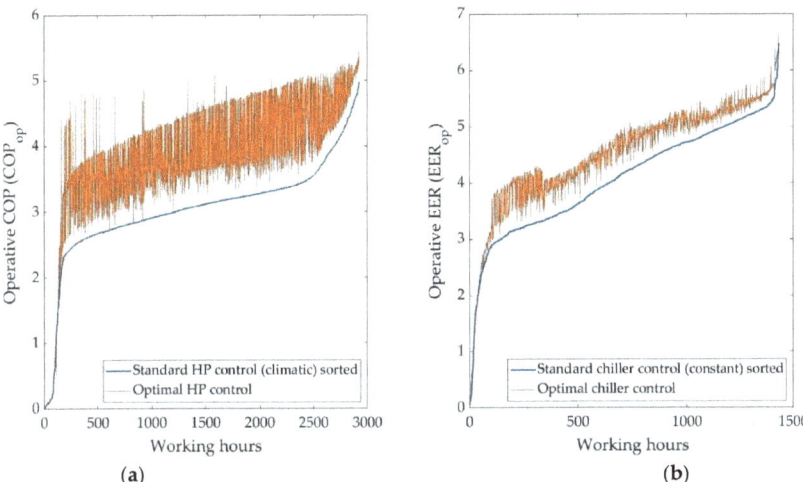

Figure 7. Comparison between typical operation and optimal control for HP and chiller in terms of COP and EER duration curves.

For the chiller, this is because the optimal supply water temperature is always higher than the standard operation one, leading to a smaller temperature difference between the source and the sink, and therefore to a better energy efficiency ratio (EER). Instead, for the HP, the optimal supply water temperature is always lower than classic one, leading to a better COP. The improvements achieved with this solution, but also the overall chiller and HP performances, are strongly dependent on the climatic conditions, so that an extension to other case studies should consider the reference climate of the site.

The proposed optimal supply water temperature control strategy relies upon the exact knowledge of the building energy demand and the possibility of varying the machine's working condition according to these needs. This has been proposed to overcome the actual standard operation which is blind to this context, and which results in continuous on-offs and performance degradation. However, even with accurate predictive models and advanced components, the working conditions required for following the demand could lead to performance degradation and the measure should therefore be experimentally tested to evaluate the actual performance on the field.

To overcome this problem, a feasible solution could be the integration of thermal storages to adjust the performance at part load, and to eliminate the continuous on-offs which strongly degrade the performance of these machines. The presence of a thermal storage would decouple the demand and the generation, allowing for a more continuous operation, limited to the most favorable hours. Regarding HP operation, the presence of a thermal storage could in fact shift the energy generation during daylight hours, in which the external temperature is closer to the supply water one due to the daily oscillation, leading to better COP. The same applies for the chiller during night hours. However, this would bring the need to produce water at a higher temperature to face the additional heat losses of this system, and requires a specific evaluation.

7. Conclusions

In this study, a set of measures to improve the overall energy performance of HVAC systems used for climatic control of large buildings subjected to variable occupation was proposed and examined. The main idea has been that of analyzing solutions which could be easily implemented without expecting technological improvements in future HVAC system controls, without considering overly-advanced controllers based on predictive and adaptive strategies. The paper has analyzed the energy consumption obtainable when operating HVAC systems with HPs under five different control strategies and different levels of complexity. The application to a specific building located in Pisa with HPs as thermal generators has shown that the electric energy intensity of the building can be reduced by over 60%, from 118 to 45 kWh/(m^{2}y), simply by changing the basic control logic that is based on the peak occupancy (CAV) with a logic that considers the real occupation of the building (DCV). Operating a more structured control, the energy intensity can be reduced even further, reaching a promising level 33 kWh/(m^{2}y), which is about 30% of the energy consumption required without specific attention to the real occupation profile of the building.

Among the various elements considered in the analysis, we can state that the installation of HP has been proven beneficial compared to the use of conventional boilers using fossil fuel. The energy savings achievable with the implementation of heat pumps reached about 60%, showing a great importance of these machines in increasing the energy efficiency. Considering that a reference in terms of primary energy consumption for the building in object is about 270 kWh/m^2 per year, calculated for the building energy label prior to the implementation of heat pumps as generation system, even with the simpler control strategy (CS 1), it is evident that the installation of HPs can be highly beneficial for these systems, especially with the increasing share of electric energy from renewables.

Considering the use of HVAC with HPs and the specific effects of the different control strategies, since the achieved energy savings are mainly due to the implementation of DCV, which represents the less advanced of the measures proposed, a certain hierarchy of these approaches is emphasized. In fact, the lack of the energy perspective in dealing with IAQ has been proven to be more critical than considering optimal operation at a supervisory control level. This simple insight should be put at the center of both engineering design and policy actions, because it represents the major weakness in the HVAC control. Since many technologies for the occupants' monitoring are currently available on the market, ranging from indirect measures by means of carbon dioxide concentration to direct techniques as infrared cameras or personal device counting systems, its large-scale deployment for HVAC control and the consequent significant energy saving are within reach in the next years.

Author Contributions: Conceptualization, A.F., D.T.; methodology, A.F., L.M., D.T.; formal analysis, A.F., L.M., D.T.; data curation, A.F., L.M.; writing—original draft preparation, A.F., writing—review and editing, A.F., L.M., D.T. All authors have read and agreed to the published version of the manuscript.

Funding: The authors gratefully acknowledge the financial support of the Italian Ministry of Education, University and Research (MIUR), in the framework of the Research Project of Relevant National Interest (PRIN) "The energy FLEXibility of enhanced HEAT pumps for the next generation of sustainable buildings (FLEXHEAT)" (PRIN 2017, Sector PE8, Line A, Grant n. 33).

Institutional Review Board Statement: Not applicable.

Informed Consent Statement: Not applicable.

Data Availability Statement: Not applicable.

Acknowledgments: The authors would like to thank the Ph.D. student of University of Pisa, Roberto Rugani for his help in the definition of Figure 3.

Conflicts of Interest: The authors declare no conflict of interest.

Nomenclature

c_f	Water specific heat [kJ/(kg K)]
COP	Coefficient of Performance [-]
CR	Capacity Ratio [-]
EER	Energy Efficiency Ratio [-]
f_{CR}	Capacity ratio penalization factor [-]
G	Daily solar radiation on horizontal surface [kWh/m²]
$\dot{m}$	Flow rate [kg/s]
Q	Thermal energy [J]
$\dot{Q}$	Thermal Power [W]
RH	Relative humidity [%]
T	Temperature [°C]
τ	Time period [s]
v_{wind}	Wind velocity [m/s]
W	Electric energy [J]
$\dot{W}$	Electric power [W]
Subscripts	
air	Air
avg	Average
DC	Full-load condition different temperatures
ext	External
fan	Fan
HP	Heat pump
in	Inlet
int	Internal
nom	Nominal full-load condition
op	Operative part-load condition
out	Outlet
RC	Rated conditions
sys	System
w	Water
Abbreviations	
BEMS	Building Energy Management System
CAV	Constant Air Volume
COP	Coefficient of Performance
DCV	Demand-Controlled Ventilation
CS	Control Strategy
EER	Energy Efficiency Ratio
EI	Energy Intensity
GHG	Green-House Gas
HP	Heat Pump
HR	Heat Recovery
HVAC	Heating, Ventilation, and Air Conditioning
IAQ	Indoor Air Quality
MPC	Model Predictive Control
nZEB	nearly-Zero Energy Building
PV	Photo-Voltaic
RES	Renewable Energy Source
SCOP	Seasonal Coefficient of Performance
VAV	Variable Air Volume
VSD	Variable Speed Drive

References

1. Hollberg, A.; Lützkendorf, T.; Habert, G. Top-down or bottom-up?—How environmental benchmarks can support the design process. *Build. Environ.* **2019**, *153*, 148–157. [CrossRef]
2. Wu, J.; Lian, Z.; Zheng, Z.; Zhang, H. A method to evaluate building energy consumption based on energy use index of different functional sectors. *Sustain. Cities Soc.* **2020**, *53*, 101893. [CrossRef]

3. Zou, P.X.; Xu, X.; Sanjayan, J.; Wang, J. Review of 10 years research on building energy performance gap: Life-cycle and stakeholder perspectives. *Energy Build.* **2018**, *178*, 165–181. [CrossRef]
4. Geraldi, M.S.; Ghisi, E. Building-level and stock-level in contrast: A literature review of the energy performance of buildings during the operational stage. *Energy Build.* **2020**, *211*, 109810. [CrossRef]
5. D'Agostino, D.; Zangheri, P.; Castellazzi, L. Towards Nearly Zero Energy Buildings in Europe: A Focus on Retrofit in Non-Residential Buildings. *Energies* **2017**, *10*, 117. [CrossRef]
6. Groissböck, M.; Gusmão, A. Impact of renewable resource quality on security of supply with high shares of renewable energies. *Appl. Energy* **2020**, *277*, 115567. [CrossRef]
7. Franco, A.; Fantozzi, F. Experimental analysis of a self consumption strategy for residential building: The integration of PV system and geothermal heat pump. *Renew. Energy* **2016**, *86*, 1075–1085. [CrossRef]
8. Fischer, D.; Madani, H. On heat pumps in smart grids: A review. *Renew. Sustain. Energy Rev.* **2017**, *70*, 342–357. [CrossRef]
9. Testi, D.; Urbanucci, L.; Giola, C.; Schito, E.; Conti, P. Stochastic optimal integration of decentralized heat pumps in a smart thermal and electric micro-grid. *Energy Convers. Manage.* **2020**, *210*, 112734. [CrossRef]
10. Conti, P.; Lutzemberger, G.; Schito, E.; Poli, D.; Testi, D. Multi-Objective Optimization of Off-Grid Hybrid Renewable Energy Systems in Buildings with Prior Design-Variable Screening. *Energies* **2019**, *12*, 3026. [CrossRef]
11. Péan, T.Q.; Salom, J.; Costa-Castelló, R. Review of control strategies for improving the energy flexibility provided by heat pump systems in buildings. *J. Process. Control.* **2019**, *74*, 35–49. [CrossRef]
12. Naylor, S.; Gillott, M.; Lau, T. A review of occupant-centric building control strategies to reduce building energy use. *Renew. Sustain. Energy Rev.* **2018**, *96*, 1–10. [CrossRef]
13. KMařík, K.; Rojíček, J.; Stluka, P.; Vass, J. Advanced HVAC Control: Theory vs. Reality. *IFAC Proc. Vol.* **2011**, *44*, 3108–3113. [CrossRef]
14. Franco, A.; Bartoli, C.; Conti, P.; Miserocchi, L.; Testi, D. Multi-Objective Optimization of HVAC Operation for Balancing Energy Use and Occupant Comfort in Educational Buildings. *Energies* **2021**, *14*, 2847. [CrossRef]
15. Al Dakheel, J.; Del Pero, C.; Aste, N.; Leonforte, F. Smart buildings features and key performance indicators: A review. *Sustain. Cities Soc.* **2020**, *61*, 102328. [CrossRef]
16. Anastasi, G.; Bartoli, C.; Conti, P.; Crisostomi, E.; Franco, A.; Saponara, S.; Testi, D.; Thomopulos, D.; Vallati, C. Optimized Energy and Air Quality Management of Shared Smart Buildings in the COVID-19 Scenario. *Energies* **2021**, *14*, 2124. [CrossRef]
17. Mayne, D.; Rawlings, J.; Rao, C.; Scokaert, P. Constrained model predictive control: Stability and optimality. *Automatica* **2000**, *36*, 789–814. [CrossRef]
18. Killian, M.; Kozek, M. Ten questions concerning model predictive control for energy efficient buildings. *Build. Environ.* **2016**, *105*, 403–412. [CrossRef]
19. Hilliard, T.; Kavgic, M.; Swan, L. Model predictive control for commercial buildings: Trends and opportunities. *Adv. Build. Energy Res.* **2016**, *10*, 172–190. [CrossRef]
20. Serale, G.; Fiorentini, M.; Capozzoli, A.; Bernardini, D.; Bemporad, A. Model Predictive Control (MPC) for Enhancing Building and HVAC System Energy Efficiency: Problem Formulation, Applications and Opportunities. *Energies* **2018**, *11*, 631. [CrossRef]
21. Noris, F.; Musall, E.; Salom, J.; Berggren, B.; Jensen, S.Ø.; Lindberg, K.; Sartori, I. Implications of weighting factors on technology preference in net zero energy buildings. *Energy Build.* **2014**, *82*, 250–262. [CrossRef]
22. Franco, A.; Fantozzi, F. Optimal Sizing of Solar-Assisted Heat Pump Systems for Residential Buildings. *Buildings* **2020**, *10*, 175. [CrossRef]
23. Wang, X.; Xia, L.; Bales, C.; Zhang, X.; Copertaro, B.; Pan, S.; Wu, J. A systematic review of recent air source heat pump (ASHP) systems assisted by solar thermal, photovoltaic and photovoltaic/thermal sources. *Renew. Energy* **2020**, *146*, 2472–2487. [CrossRef]
24. Franco, A.; Schito, E. Definition of Optimal Ventilation Rates for Balancing Comfort and Energy Use in Indoor Spaces Using CO_2 Concentration Data. *Buildings* **2020**, *10*, 135. [CrossRef]
25. Cho, J.; Kim, Y.; Koo, J.; Park, W. Energy-cost analysis of HVAC system for office buildings: Development of a multiple prediction methodology for HVAC system cost estimation. *Energy Build.* **2018**, *173*, 562–576. [CrossRef]
26. Perez-Lombard, L.; Ortiz, J.; Maestre, I.R. The map of energy flow in HVAC systems. *Appl. Energy* **2011**, *88*, 5020–5031. [CrossRef]
27. McDowall, R. Introduction to HVAC, Chapter 1. In *Fundamentals of HVAC Systems*; McDowall, R., Ed.; Elsevier: Amsterdam, The Netherlands, 2007; pp. 1–10. ISBN 9780123739988. [CrossRef]
28. Goel, S.; Rosenberg, M.; Eley, C. *ANSI/ASHRAE/IES Standard 90.1-2016 Performance Rating Method Reference Manual*; U.S. Department of Energy: Washington, DC, USA, 2017. Available online: https://www.pnnl.gov/main/publications/external/technical_reports/PNNL-26917.pdf (accessed on 27 June 2021).
29. Liu, Z.; Li, W.; Chen, Y.; Luo, Y.; Zhang, L. Review of energy conservation technologies for fresh air supply in zero energy buildings. *Appl. Therm. Eng.* **2019**, *148*, 544–556. [CrossRef]
30. *EN 14511-2: Air Conditioners, Liquid Chilling Packages and Heat Pumps for Space Heating and Cooling and Process Chillers, with Electrically Driven Compressors—Part 2: Test Conditions*; European Committee for Standardization (CEN): Brussels, Belgium, 2018.
31. *EN 14825: Air Conditioners, Liquid Chilling Packages and Heat Pumps, with Electrically Driven Compressors, for Space Heating and Cooling. Testing and Rating at Part Load Conditions and Calculation of Seasonal Performance*; European Committee for Standardization (CEN): Brussels, Belgium, 2018.

32. Aste, N.; Manfren, M.; Marenzi, G. Building Automation and Control Systems and performance optimization: A framework for analysis. *Renew. Sustain. Energy Rev.* **2017**, *75*, 313–330. [CrossRef]
33. Afram, A.; Janabi-Sharifi, F. Review of modeling methods for HVAC systems. *Appl. Therm. Eng.* **2014**, *67*, 507–519. [CrossRef]
34. Satyavada, H.; Baldi, S. An integrated control-oriented modelling for HVAC performance benchmarking. *J. Build. Eng.* **2016**, *6*, 262–273. [CrossRef]
35. Gholamzadehmir, M.; Del Pero, C.; Buffa, S.; Fedrizzi, R.; Aste, N. Adaptive-predictive control strategy for HVAC systems in smart buildings – A review. *Sustain. Cities Soc.* **2020**, *63*, 102480. [CrossRef]
36. EN 16798-3: *Energy Performance of Buildings. Ventilation for Buildings. For Non-Residential Buildings. Performance Requirements for Ventilation and Room-Conditioning Systems (Modules M5-1, M5-4)*; European Committee for Standardization: Bruxelles, Belgium, 2017.
37. Shah, R.K.; Sekulic, D.P. *Fundamentals of Heat Exchanger Design*; John Wiley & Sons, Inc: Rochester, NY, USA, 2003.
38. Franco, A.; Giannini, N. Optimum thermal design of modular compact heat exchangers structure for heat recovery steam generators. *Appl. Therm. Eng.* **2005**, *25*, 1293–1313. [CrossRef]
39. Kern, D.Q.; Kraus, A. *Extended Surface Heat Transfer*; McGraw-Hill Chemical Engineering Series; McGraw-Hill: Chicago, MI, USA, 1972.
40. Franco, A.; Leccese, F. Measurement of CO2 concentration for occupancy estimation in educational buildings with energy efficiency purposes. *J. Build. Eng.* **2020**, *32*, 101714. [CrossRef]
41. Global Modeling and Assimilation Office (GMAO) (2015), MERRA-2 tavg1_2d_slv_Nx: 2d,1-Hourly, Time-Averaged, Single-Level, Assimilation, Single-Level Diagnostics V5.12.4, Greenbelt, MD, USA, Goddard Earth Sciences Data and Information Services Center (GES DISC). Available online: https://doi.org/10.5067/vjafpli1csiv (accessed on 9 June 2021).
42. CTI (Italian Thermotechnical Committee). Italian Typical Meteorological Years. 2016. Available online: https://www.cti2000.it/index.php?controller=news&action=show&newsid=34985 (accessed on 9 June 2021).

MDPI
St. Alban-Anlage 66
4052 Basel
Switzerland
Tel. +41 61 683 77 34
Fax +41 61 302 89 18
www.mdpi.com

MDPI Books Editorial Office
E-mail: books@mdpi.com
www.mdpi.com/books

www.ingramcontent.com/pod-product-compliance
Lightning Source LLC
LaVergne TN
LVHW070502100526
838202LV00014B/1771